Safety and Reliability in the 90s

Will Past Experience or Prediction Meet Our Needs?

T0330953

ESRA
ICE
IChemE
IEE
IMechE
INucE
IQA

SARSS '90

Proceedings of the Safety and Reliability Society Symposium 1990, held at Altrincham, UK, 19–20 September 1990.

Organised by
The Safety and Reliability Society, Clayton House, 59 Piccadilly, Manchester M1 2AQ, UK.

Co-sponsors
The European Safety and Reliability Association
The Institution of Civil Engineers
The Institution of Chemical Engineers
The Institution of Electrical Engineers
The Institution of Mechanical Engineers
The Institution of Nuclear Engineers
The Institute of Quality Assurance

Organising Committee
Dr M.H. Walter (*Chairman*)
Mr D.W. Heckle (*Secretary*)
Mr R.F. Cox
Mr N.J. Locke
Ms B.A. Sayers
Dr G.B. Guy
Ms A. Enderby
Mr M.A. Alderson (*NW Branch Representative*)

Safety and Reliability in the 90s

Will Past Experience or Prediction Meet Our Needs?

Edited by

M.H. WALTER

Electrowatt Consulting Engineers and Scientists, Warrington, UK
and
Safety and Reliability Society, Manchester, UK

and

R.F. COX

AEA Technology, Safety and Reliability, Warrington, UK
and
Safety and Reliability Society, Manchester, UK

CRC Press
Taylor & Francis Group
Boca Raton London New York

CRC Press is an imprint of the
Taylor & Francis Group, an **informa** business

First published 1990 by CRC Press
Taylor & Francis Group
6000 Broken Sound Parkway NW, Suite 300
Boca Raton, FL 33487-2742

Reissued 2018 by CRC Press

First issued in paperback 2020

British Library Cataloguing in Publication Data

Safety and reliability in the 90s.
  1. Reliability engineering
  I. Walter, M. H.   II. Cox, R. F.   III. Safety and
  Reliability Society
  620.004529

A Library of Congress record exists under LC control number: 90044368

Publisher's Note
The publisher has gone to great lengths to ensure the quality of this reprint but points out that some imperfections in the original copies may be apparent.

Disclaimer
The publisher has made every effort to trace copyright holders and welcomes correspondence from those they have been unable to contact.

ISBN 13: 978-1-138-56178-6 (pbk)
ISBN 13: 978-1-138-10547-8 (hbk)
ISBN 13: 978-0-203-71042-5 (ebk)

Visit the Taylor & Francis Web site at http://www.taylorandfrancis.com and the
CRC Press Web site at http://www.crcpress.com

PREFACE

During the last decade a number of significant accidents have occurred which have given rise to grave public concern. Piper Alpha, Chernobyl, Challenger, Kings Cross and the M1 air crash are recent memories. Other less severe accidents in terms of loss of human life have caused major financial loss. These accidents have occurred despite advances in the "tools" available to the safety practitioner in the analysis of complex systems. There is still much to learn and over the last few years organisational and management factors and in particular the role and response of the human element are seen to provide a major contribution to both accident prevention and post-accident response.

This book represents the proceedings of the 1990 Safety and Reliability Society Symposium held in Altrincham, UK on the 19th and 20th September 1990. This is the tenth annual symposium of the Society and in previous years the topic has been based on specific areas of interest. Examples, include Statutory Safety, Transportation, Human Factors and Computers in Safety Related Systems. This year the topic is wider ranging than in previous years and as we enter a decade of more advanced technology and the prospect of further legislation and regulation it is appropriate to ask the question "will past experience or prediction meet our needs?".

The structure of the book represents the structure of the symposium itself and the papers selected provide current thinking on improved methods for identification, quantification and management of risks based on the safety culture developed across a range of industries during the last decade. In particular organisational and management factors feature in a large number of the papers.

The response to the Call for Papers produced more good papers than could be included in the Symposium. I must thank all the authors who submitted their work, the review panel, the presenters of the papers, the organising committee and the co-sponsors for their support. Through the effort and dedication of all these people an interesting and instructive Symposium has been achieved and this book compiled.

M.H. Walter
Chairman of the National
Conference Organising Committee

CONTENTS

LIST OF CONTRIBUTORS

J.A. Astley
Four Elements Ltd, 25 Victoria Street, London SW1H 0EX, UK

M.J. Baker
Department of Civil Engineering, Imperial College of Science, Technology and Medicine, London SW7 2BU, UK

P. Ball
British Nuclear Fuels plc, Sellafield, Seascale, Cumbria CA20 1PG, UK

L.J. Bellamy
Four Elements Ltd, 25 Victoria Street, London SW1H 0EX, UK

D.I. Blockley
Department of Civil Engineering, University of Bristol, Queen's Building, University Walk, Bristol BS8 1TR, UK

M.H.J. Bollen
Eindhoven University of Technology, Faculty of Electrical Engineering, P O Box 513, 5600 MB Eindhoven, The Netherlands

C.J. Bullock
Jenbul Consultancy Services Ltd, 4a Old Market Place, Ripon, North Yorkshire HG4 1EQ, UK

J.P. Byrne
Hazard Assessment Department, AEA Technology, Wigshaw Lane, Culcheth, Warrington, Cheshire WA3 4NE, UK

M.S. Carey
R M Consultants Ltd, Genesis Centre, Garrett Field, Birchwood Science Park, Warrington, Cheshire WA3 7BH, UK

B.A. Coxson
Nuclear Electric plc, Booths Hall, Chelford Road, Knutsford, Cheshire WA16 8QG, UK

C. Dale
Cranfield IT Institute, Fairways, Pitfield, Kiln Farm, Milton
Keynes MK11 3LG, UK

L.P. Davies
NNC Ltd, Booths Hall, Chelford Road, Knutsford, Cheshire WA16
8QZ, UK

N. Davies
Department of Mathematics, Statistics and Operational
Research, Nottingham Polytechnic, Burton Street, Nottingham
NG1 4BU, UK

P. Dixon
Department of Mathematics, Statistics and Operational
Research, Nottingham Polytechnic, Burton Street, Nottingham
NG1 4BU, UK

L.C. Dunbar
R M Consultants Ltd, Genesis Centre, Garrett Field, Birchwood
Science Park, Warrington, Cheshire WA3 7BH, UK

B.W. Finnie
The Centre for Software Engineering Ltd, Bellwin Drive,
Flixborough, Scunthorpe, South Humberside DN15 8SN, UK

T.A.W. Geyer
Four Elements Ltd, 25 Victoria Street, London SW1H 0EX, UK

Z.A. Gralewski
R M Consultants Ltd, Suite 7, Hitching Court, Abingdon
Business Park, Abingdon, Oxon OX14 1RA, UK

D.B. Harries
Blaen Bedw, Halls Road, Trecelyn/Newbridge, Gwent NP1 4FY, UK

M.J. Harris
Department of Engineering, Manchester University, Simon
Building, Oxford Road, Manchester M13 9PL, UK

N.J. Holloway
Safety and Reliability Directorate, Wigshaw Lane, Culcheth,
Warrington, Cheshire WA3 4NE, UK

N.W. Hurst
Health and Safety Executive, Broad Lane, Sheffield S3 7HQ, UK

I.H.A. Johnston
The Centre for Software Engineering Ltd, Bellwin Drive,
Flixborough, Scunthorpe, South Humberside DN15 8SN, UK

J. Jowett
Hazard Assessment Department, AEA Technology, Wigshaw Lane,
Culcheth, Warrington, Cheshire WA3 4NE, UK

B. Kirwan
Safety Department, British Nuclear Fuels plc, Risley,
Warrington, Cheshire, UK

C. Leighton
British Nuclear Fuels plc, Sellafield, Seascale, Cumbria CA20
1PG, UK

D.A. Lucas
Human Reliability Associates Ltd, 1 School House, Higher Lane,
Dalton, Wigan, Lancashire WN8 7RP, UK

C. McCollin
Department of Mathematics, Statistics and Operational
Research, Nottingham Polytechnic, Burton Street, Nottingham
NG1 4BU, UK

J.M. Marriott
Department of Mathematics, Statistics and Operational
Research, Nottingham Polytechnic, Burton Street, Nottingham
NG1 4BU, UK

P. Massee
Eindhoven University of Technology, Faculty of
Electrical Engineering, P O Box 513, 5600 MB Eindhoven,
The Netherlands

F.R. Mitchell
Jenbul Consultancy Services Ltd, 4a Old Market Place, Ripon,
North Yorkshire HG4 1EQ, UK

J.C. Naylor
Department of Mathematics, Statistics and Operational
Research, Nottingham Polytechnic, Burton Street, Nottingham
NG1 4BU, UK

R.P. Pape
HSE Technology Division, St Anne's House, Stanley Precinct,
Bootle, Liverpool L20 3RA, UK

N.F. Pidgeon
Department of Psychology, Birkbeck College (University of
London), Malet Street, London WC1E 7HX, UK

G.A. Richards
R M Consultants Ltd, Suite 7, Hitching Court, Abingdon
Business Park, Abingdon, Oxon OX14 1RA, UK

R.L. Skelton
Jenbul Consultancy Services Ltd, 4a Old Market Place, Ripon,
North Yorkshire HG4 1EQ, UK

E.J. Smith
Technica Consulting Scientists and Engineers, Lynton House,
7-12 Tavistock Square, London WC1H 9LT, UK

J.P. Stead
R M Consultants Ltd, Genesis Centre, Garrett Field, Birchwood
Science Park, Warrington, Cheshire WA3 7BH, UK

J.R. Stone
Department of Civil Engineering, University of Bristol,
Queen's Building, University Walk, Bristol BS8 1TR, UK

M. Tabaie
Electrowatt, Consulting Engineers and Scientists, Stanford
House, Birchwood, Cheshire WA3 7BH, UK

B.A. Turner
Department of Sociology, University of Exeter, Amory Building,
Rennes Drive, Exeter EX4 4RJ, UK

T. Waters
Safety Department, British Nuclear Fuels plc, Risley,
Warrington, Cheshire, UK

S.P. Whalley
R M Consultants Ltd, Genesis Centre, Garrett Field, Birchwood
Science Park, Warrington, Cheshire WA3 7BH, UK

R.B. Whittingham
Electrowatt, Consulting Engineers and Scientists,
Grandford House, 16 Carfax, Horsham, West Sussex, UK

D.W. Wightman
Department of Mathematics, Statistics and Operational
Research, Nottingham Polytechnic, Burton Street, Nottingham
NG1 4BU, UK

J.C. Williams
Technica Ltd, Highbank House, Exchange Street, Stockport,
Cheshire SK3 0ET, UK

W. Zhao
Department of Civil Engineering, Imperial College of Science,
Technology and Medicine, London SW7 2BU, UK

OPERABILITY STUDIES AND HAZARD ANALYSIS IN THE MANAGEMENT OF SAFETY

D. B. HARRIES
Blaen Bedw, Halls Road, Trecelyn/Newbridge, Gwent, NP1 4FY.

ABSTRACT

The paper presents a safety management approach based upon the potential of the Operability and Hazard Analysis method. It relates the method to the general principles of interdisciplinary team approaches and distributed control. It draws analogies with the concepts and methods of the Total Quality Assurance management system.

The paper addresses the philosophy of the approach. The principles and characteristics are presented and exemplified by reference to real life applications. The applications are taken from the author's experience in operations and project safety engineering and in safety engineering education.

INTRODUCTION

In response to the scope presented by the organizers of the symposium the paper shall:

Let readers know of methodologies which the writer has demonstrated as being successful in improving safety management.

Concentrate on the pragmatic aspects of applying the methodologies.

The title of the paper refers to Hazard Operability Studies (Hazop) but it treats this particular method within the genre of interdisciplinary methods. The intent is to emphasise the flexibility of the approach in suiting the nature of the problem and the timing of the application rather than limit the treatment to a specialist technique for a particular process problem. The latter application has been treated lucidly and comprehensively in the classic papers of Lawley [1, 2]. The paper is intended to contribute to the management of safety in applying the principles of the method in different systems and organisations.

The paper addresses the basic philosophy of the approach. The Hazop is shown to be a key element of safety management by reference to principles and practice. The principles and characteristics are presented and exemplified by reference to examples of applications. The applications chosen were aimed at improving the safety performance of the following systems:

1.   A Company Wide Relief Valve System.
2.   A Rapid Mass Transit System.
3.   A Gas Plant Start Up System.
4.   An Undergraduate Process Engineering Design Education System.

PHILOSOPHY

Safety cannot be treated in isolation from the function of the system and must be linked to cost and time elements. Complex systems require complex control systems with balanced responses to internal and external changes to maintain safe operation [3]. The task of safety management is to provide distributed management control systems of this kind which will demand continuous and adaptive improvement of safety standards. The operability study addresses the problem directly by involving participants in applying changes, analysing for cause and effect, and deciding actions in design or control to prevent or mitigate any hazardous effects. It allows management to coordinate interdisciplinary resources to provide optimum improvements in authoritative, logical and imaginative ways.

PRINCIPLES

Operability studies provide a means of distributed control and the authority required to produce timely as well as balance safety responses to change. The method is an element of the systems approach. It is suited to the providing of improvements to complex problems because of its emphasis on wholeness and interdisciplinary interactions. It is based on control theory particularly on the adaptive control diagram of continuous improvement [3].

The Hazop approach is exactly analogous to the Quality Assurance task force team attacks on quality problems. The characteristics of Japanese applications of total quality assurance are pertinent in refining the approach. This is particularly so in its stress on early application of the interdisciplinary effort, involvement at all levels, the principle of prevention rather than mitigation, the importance of training and the decisive need of a lead from senior management.

The essence of the Hazop is the focussing of concentration of the interdisciplinary group on solving the problems. In process problems the P & ID provides the focus of the Hazop and the deviations give the prompts. The crossword diagram and clues gives a common but vivid example of how a focus allows a group to become far more effective than the sum of the individuals in solving the problem. The art of achieving a successful Hazop depends significantly on the skill of the organizer in providing the focus (usually a diagram) and a clear definition of the system. Enthusiasm and ingenuity are qualities of the participants that can produce dramatic improvements but the enabling function is the focus.

The medium is the message is a philosophy worth remembering in designing the Hazop. The spoken word is a good vehicle usually in promoting creative interaction of the participants. The focus diagram allows thoughts to range freely but cohere on the subject. The slavish writing of details can stultify progress. In safety work however an audit trial has become of increasing importance in assuring the quality of the safety input. The use of short notes by a secretary reinforced by taped recordings have proved to be a good compromise.

CONSEQUENCES

An important consequence of the philosophy and principles is that flexibility is a key characteristic in successful applications. Means of focus, prompts and procedures must be developed to suit the particular problem while satisfying the principles. The following case histories show how this has been achieved.

CASE STUDIES

Improving the Safety of a Relief Valve System

The relief valve system under enquiry was the pressure relief system of a major oil company. The Company management were concerned about the quality assurance of the system and called upon its technical consulting division to address the matter. The total inventory of relief valves throughout the company numbered 24535. The design criteria was provided by the engineering standards of the company and the maintenance of the relief valve system was established according to Company General Instructions.

The scheduling of relief valve testing and repair was controlled by a main frame computer. Details of each valve in the system was established in a database holding repair records over a period of 16 years.

Regulation of test intervals was achieved via the computer using a so called 3:1 ratio linked to the repair record of the valve. The test interval would be extended by 6 months if the valve was tested and found to be within test specifications after each of three consecutive test intervals. The interval would be reduced by 6 months after any one unsatisfactory test. The General Instructions allowed for a modifying role by unit inspection supervision with authorisation by operations engineers in making sensible adjustments according to local knowledge but in practice this was seldom exercised.

In our work of monitoring the relief valve system during 1985/86 we had found disturbing evidence of inadequate testing and repair facilities, bad practice in field maintenance and cases of wrong relief valve settings. There had been no equipment failures but we considered that to be a tribute to the efficacy of our control system rather than to that of the relief valve system.

Statistical analysis of the database showed great variation in test intervals at different plants for identical service whereas the variation of failure rates of different types of valves were not reflected in their test intervals.

Application of Principles

We held what we called Discussion Group Meetings (DGM) to which we invited maintenance supervisors from our eleven test shops and operations and instrument engineering specialists. We carried out coarse operability studies on our relief valve system using the control diagram as a focus [Fig 1]. The diagram illustrates the relationships between fundamental elements of a robust continuously improving system. Figure 2 is an application of the diagram to the relief valve system. The deviations used in the Operability Studies referred to information flows - NO FLOW - MORE FLOW - LESS FLOW - WRONG COMPOSITION. We only recorded faults and remedies.

The faults were:

a)   Inadequate demand for improvement from management. It was not difficult gaining the support of maintenance staff who were well aware of the inadequacies of the relief valve system. It was much more difficult to get real support from senior management who believed that our record of no accidents from failure of relief

4

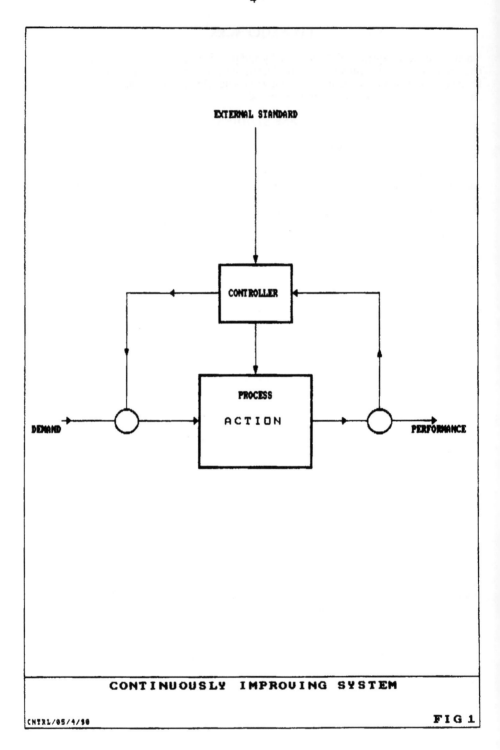

EXTERNAL STANDARD

CONTROLLER

PROCESS

ACTION

DEMAND

PERFORMANCE

CONTINUOUSLY IMPROVING SYSTEM

CNTRL/05/4/90

FIG 1

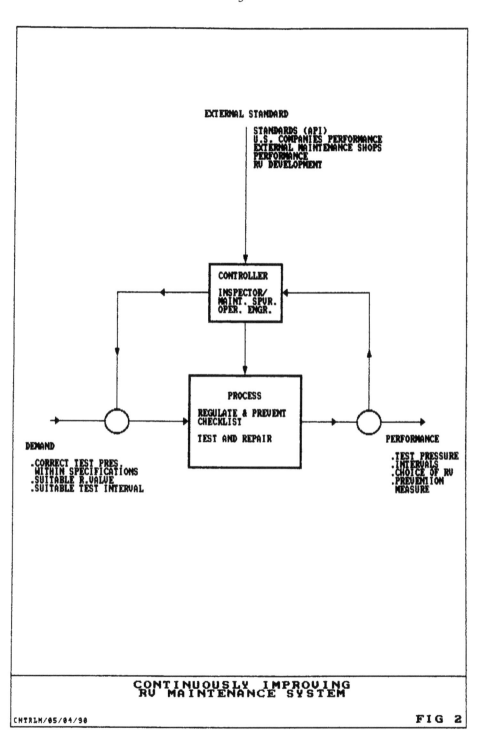

CONTINUOUSLY IMPROVING
RV MAINTENANCE SYSTEM

FIG 2

valves was proof that everything was satisfactory.

b)   Inadequate information from outside of the Company. The Company test standards were related to API standards but we were not receiving information about valve developments, laboratory equipment, test intervals on comparable service, and test and repair procedures. Even within the Company there were no channels of communication between maintenance test shops.

c)   The demand on the system was for test and repair when it should be for TEST and PREVENT DAMAGE as well as REPAIR.)

d)   The performance was adequate in terms of scheduling testing and repair but no improvements were occurring. In many cases equipment was inadequate for testing and repair purposes.

e)   The control loop of repair information and computer programme was adequate for maintaining the scheduling of tests but stifled any initiative for improvements and consolidated faults.

f)   There was a lack of knowledge in many maintenance shops and among inspectors of the relationships between valve damage and cause. There was little or no communication between the maintenance and inspection staff and the operations engineers.

In short the system was static and this was largely because of the basic systematic faults, a), b) and c). The remaining faults were almost inevitable given the existence of the basic faults.

We decided to make improvements by:

a)   organizing a major international conference on relief valves. The conference was aimed at:

raising awareness of the company's relief valve problems particularly among senior management;

informing company staff involved with the relief valve systems of knowledge of preventive measures as well as repair techniques.

b)   Preparing studies of ways of improving:

-       the General Instructions
-       information flows from outside and within the company
-       the interval regulating system
-       the test shop equipment
-       the in-place testing procedures.

Results

The relief valve conference was held in 1986.

The contents of the programme were:

-       Opening Remarks by the Chief Engineer
-       Pressure Relief Valve Selection
-       Metallurgical Aspects of Relieving Devices
-       In-place Testing

- Field Improvements to a Pressure Relief System in a Producing Area
- Acoustically Induced Vibration
- Shop Testing and the Training of Relief Valve Technicians
- Summary of the Proceedings

The shop testing and training presentation was given by a specialist from a leading U.S. valve company. The in-place test presentation was made by a company maintenance engineer. The remaining presentations were made by specialists of the company's consulting services. The conference summary was given by the safety/inspection specialist.

The conference was attended by representatives of manufacturers from Japan and the U.K. Over 90 members of management and staff engaged in relief valve work in the company attended the conference.

In the summary to the conference, proposals for four DGM, and safety studies for the following year were proposed. The year was to be declared "The Year of the Relief Valve". All these proposals were accepted and subsequently implemented.

The operability studies had developed the safety strategy and the conference had enlisted the support and commitment of senior management and staff in an enthusiastic and effective way. The studies were completed using task force teams and the involvement of the person on the job backed by statistical analysis techniques and specialist expertise. The achievements are listed below:

Preparation of the Relief Valve General Instructions with two major Quality Assurance and Regulating Supplements.

The Quality Assurance supplement included requirements and guidelines for:

- a Relief valve technician certification programme.
- a Relief Valve Test Station QA document.
- an Engineering Survey.

The Relief Valve Regulating supplement included requirements and guidelines for:

- Regulation and Prevention Checklists
- Individual Maintenance Unit Test Record

    Note that relief valves are generally damaged when they open as required, when there is leakage, when the valve chatters, and during removal, transport and replacement. The provisions of the regulating supplement directed action toward identifying the cause and informing operations management of the required preventive action.

Amendments to the Company Engineering Standards

- Inclusion and Application of Cold Differential Test Pressure.
- Clarification of Definitions.
- Removal of Pressure Constraint from Pilot Operated Relief Valves.
- Relief Valve System Design Checklist (including piping).

The amendments have significant effects on improving safety by correcting pressure settings and stressing piping constraints and supports. The developing of the amendments occurred following alerts from different units brought to notice in the DGMs.

Publication of Histograms and Comparisons of Company Test Intervals by Service and Test Shop. External comparisons were made against results derived from 5 Major U.S. Companies.

Role play simulations of the regulating system were conducted at two gas plants, a machine shop, and producing area. The members of the teams were maintenance lead technician, inspection supervisor, operations engineering and the safety/inspection specialist. The existing system was shown to be grossly inadequate in sensibly treating real life examples taken from different test shop experiences. The lessons learned from these exercises were applied in preparing the regulating and prevention guidelines of the General Instructions.

Task force teams made up of the maintenance foreman, the maintenance shop supervisor, the inspection field supervisor, the operations engineer and the safety/inspection specialist competed engineering surveys on test shops. Following the task force surveys significant recommendations were accepted for improvements to the Refinery, a Producing Area and Terminal maintenance units:

- Transfer of Terminal inspector to test shop to "get it right first time".

- Refinery test shop to control removal - transport - replacement steps to attain improved quality assurance.

- Appointment of lead technician at a producing area test shop.

- Training improvements:

  - Shop supervisors to spend time in different units.
  - Riggers to be trained by shop staff in valve removal, transport and replacement care.
  - Terminal management to arrange pilot operated valve training course to be presented by a manufacturer.
  - Terminal management to arrange for training videos to be made within their department for in - place procedures on tankage and for removal - transport - installation steps.

- Facility and equipment improvements included a new Terminal test shop and:

  - Lapping rings and Lapping Machine
  - Optical Flats
  - Solvent Cleaner Equipment
  - Blast machine for effective and speedy cleaning of parts
  - Quick Clamping Stand
  - Nozzle Opening Chuck Stand
  - Panel Board of Test Gauges
  - Special Truck to transport relief valves between the test shop and the plant valve location in an upright position
  - Spare Part Storage
  - Relief Valve repair software (manufacturer's charts etc.)

  Further new test stands in the central workshops, a producing area and a gas plant were scheduled for completion in the following year.

An in-place testing procedure was developed in one of our gas plants by a plant engineer. A coarse operability study was administered in a DGM following a presentation by the engineer.

The deviation applied was the presence of flammable gas during the test (caused by leakage or human error). A formal operability study and hazard analysis was completed by a team consisting of a specialist instrument engineer, a specialist safety/inspection engineer, and a plant engineer. Only faults and remedies were recorded.

The validity of the tests were challenged on the grounds of insufficient lift and it was required that the tests were to be simulated in the test shop and compared with the conventional tests. The plant engineers developed original tests to compare lift and found the lift to be higher on the in-place test (the capacity between the stop valve and the disc was greater than the test station capacity). However lift measurements for all operating conditions were considered to be too difficult to simulate for all conditions (varying set pressures, piping configurations, types of valves). Plant management therefore decided to limit their procedure to pilot operated relief valves.

As a result of the operability studies and hazard analysis and the engineering research the following constraints were included in the procedures.

- The procedure was limited to one valve manufacturer.

- The procedure was limited to a particular plant area because of service constraints.

- All in-place testing was to be witnessed by an inspector.

- Valves designated for in-place testing were to be removed at every second test interval for disassembly and shop testing.

- A relief valve that did not pass the in-place test or gave any indication of mechanical problems was to be removed to the shop for disassembly.

In the experiments it was noted that a dynamic technique of testing opening pressures in the test station was less damaging and gave better lift than the traditional tentative approach to set pressure opening followed by immediate shut off. The method required a maximum pressure recorder.

Improving the Safety of a Rapid Mass Transit System

The case study is restricted to the development of an operability study approach and procedure in an non process industry project. It is the first formal study of this kind to have been completed in this industry as far as I know. Its impact on safety management has not yet matured to the commanding role which was described in the previous study.

At the time of writing the operability study has been accepted as an important element in the safety strategy. A procedure for its comprehensive role has been required and put forward.

The rapid mass transport system project is complex technically and organisationally. It involves refurbishment of an existing line on a costly and fundamental scale. Innovations and improvements included new rolling stock, and improved signalling/control system, new switchgear units, recabling etc.

Hazard assessments of the overall system had indicated the prime importance of the traditional collision and fire hazards. Disasters in the U.K. at Kings Cross, and at Clapham on a rail system, had raised public and staff awareness of the potential for major hazards of this kind.

The organization of the management of the project was integrated between the client project team and the contractor project support team. The project management safety demands were established according to the philosophy of the Health and Safety at Work Act. The client enjoyed a fine tradition of engineering excellence and a comparable safety record. The engineering organisation was based on a vertical structure with few or no crosslinks. It was considered necessary to raise safety awareness and commitment to meet the challenge of the increased public demand for higher safety standards, the demands of the new technology and to avoid complacency. With the philosophy and principles expounded earlier in this paper in mind the need and opportunity for the application of an operability study was clear.

Application of Principles

We decided on a policy of full involvement and training. The client project team and the client engineering specialist support teams had little or no experience of the Hazop approach. Their engineering expertise was unquestionable and they had the proper caution of professional men to a new approach. They had too a proper pride in their present methods and safeguards which had a proven good record. The problems arising from such a situation have been anticipated in the Guide produced by the Chemical Industry Safety and Health Executive Council (CISHEC) safety committee [4] and confirmed in this study.

Time was an important factor. The project was moving swiftly toward the design stage when changes become costly and there was a temptation to look upon the operability study as an audit method rather than an improving exercise. The client manager had however recognized the need for interdisciplinary awareness of interactive (side effect) hazards and had initiated seminars attended by representative engineers of all disciplines. The seminars of the conceptual plans were presented by the engineering specialists concerned. This was a firm foundation for progress along the path advocated in the "training" section of the CISHEC guide [4]. The project manager for the same reason supported the idea of a trial run of a crude operability study.

First we distributed the Lawley paper [5] to the project manager, the project engineer, and to all the discipline specialist engineers for background purposes. We proposed systems and sub-systems for demonstration of the approach. The project safety specialist organized a preliminary discussion and the demonstration study inviting representative engineers of the disciplines involved. He prepared "focus" diagrams, led the study, completed the secretarial work and published the results for review by the project team. Urgency in making progress was vital and the training by trial was effective and fast. The response to the trial varied from enthusiastic to sceptical and the project manager was left with the decision to proceed further or not - he chose to go ahead.

He acted decisively to accelerate progress. Consultant safety specialists were contracted to organize and administer a Hazop.

The brief for the contracted study included the objectives which were to:

a) review the design intent as outlined in the specifications and identify events which could lead to major hazards and/or cause significant delays;

b) assess system interfaces from both an operational and safety viewpoint;

c) record the first stage of a safety review and thus contribute to the audit trail;

d) provide a detailed basis for the production of detailed fault trees and event trees.

Six morning sessions of 4 hours were scheduled. Engineers of operations, safety, project, signalling, control, rolling stock, permanent way, and communications attended. The chairman and the secretary were provided by the contracted safety consultants.

Word flow sheets (focus diagrams) illustrated in Figs. 3 and 4 were prepared by the consultants e.g.

- Train movement in tunnel
- Train movement in station
- Train movement in depot.

Situations were sub-divided to take account of different operating modes, and signal conditions.

The reporting form developed was a straightforward adaptation of the type referenced [1, 2, 4, & 5]. In addition an action reporting list was introduced. The Forms are illustrated below:

TABLE 1
Illustrative Hazop Reporting Form
Train Leaves Station

GUIDE WORD	DEV.	CAUSE	CONSEQUENCE	RECOMMENDING	COMMENTS	ACTN
NOT	Train does not leave.	Psgrs on track.	Injuries and delay	Reduce train speed thro stations.	Causes delay with no sure benefit	P.M. S.C
				Consider platform barriers making use of new stop accuracy.	Protects hazardous platform/ car gap. This is a serious hazard.	

12

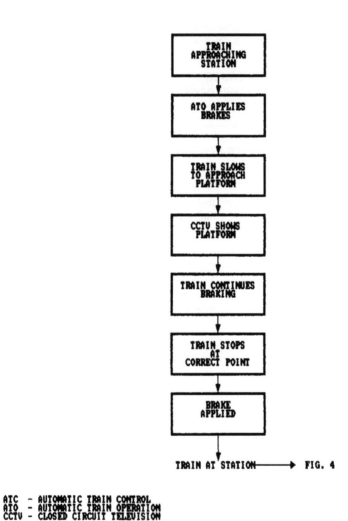

TRAIN
APPROACHING
STATION

ATO APPLIES
BRAKES

TRAIN SLOWS
TO APPROACH
PLATFORM

CCTV SHOWS
PLATFORM

TRAIN CONTINUES
BRAKING

TRAIN STOPS
AT
CORRECT POINT

BRAKE
APPLIED

TRAIN AT STATION ⟶ FIG. 4

ATC - AUTOMATIC TRAIN CONTROL
ATO - AUTOMATIC TRAIN OPERATION
CCTV - CLOSED CIRCUIT TELEVISION

TRAIN ENTERS TUBE STATION − ATC

RMT/11/04/90

FIG 3

13

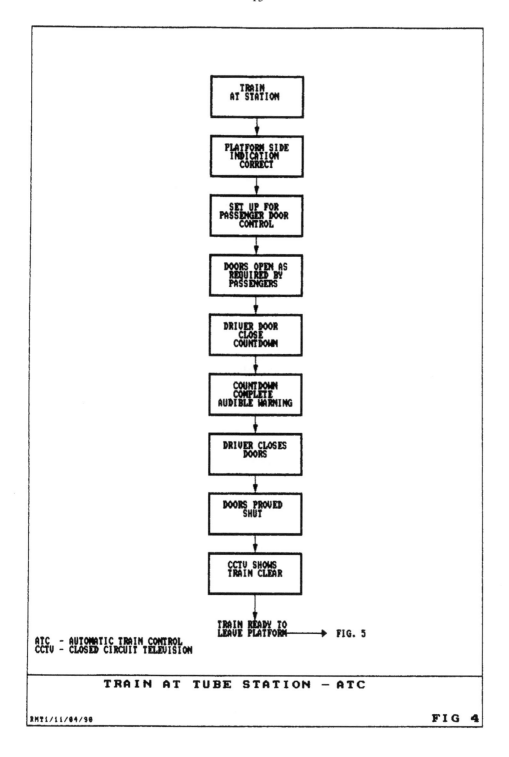

TRAIN
AT STATION

PLATFORM SIDE
INDICATION
CORRECT

SET UP FOR
PASSENGER DOOR
CONTROL

DOORS OPEN AS
REQUIRED BY
PASSENGERS

DRIVER DOOR
CLOSE
COUNTDOWN

COUNTDOWN
COMPLETE
AUDIBLE WARNING

DRIVER CLOSES
DOORS

DOORS PROVED
SHUT

CCTV SHOWS
TRAIN CLEAR

TRAIN READY TO
LEAVE PLATFORM ──────▶ FIG. 5

ATC - AUTOMATIC TRAIN CONTROL
CCTV - CLOSED CIRCUIT TELEVISION

TRAIN AT TUBE STATION - ATC

RMT1/11/04/90   FIG 4

TABLE 2
Illustrative Action List Reporting Sheet

REF	RECMNDD ACTN	NAME	TGT DATE	RESPONSE	END DATE
XXA	Research records for fire loss of track code	A.P.	Not specified	Not new or significant hazard	xx/xx/xx
XXB	Study way of controlled deflation of airbag after over inflation	A.B. C.D.	Not specified		
XXC	Decide whether or not to provide radio - portable or demountable handset	E.F. G.H.	Not specified	They shall be provided by passenger services	xy/xx/xx

Guide words which were to be applied to design conditions e.g. speed, working condition of door sub-system, were listed as MORE, REVERSE, NOT, ALSO, OTHER.

The ideas were speculative and the brief encouraged further development and refinement of diagrams and guidewords.

Results

In general the operability study led to minor improvements in specifications which taken in total contributed significantly to the quality assurance of the project. It confirmed the key role of reliability specifications and it fulfilled its function in the audit trail. In one particular instance it laid the foundation for a brainstorm of a significant advance in the safety of the system and it is interesting to explore this development.

In the illustration of Table 1 a recommendation is shown as 'Consider platform barriers making use of new stop accuracy.' It was decided during the operability study to explore this recommendation in a brainstorm session during one of the morning periods set aside for the operability study.

The brainstorm was introduced by the project safety specialist. He stressed that:

- its role was especially suited to the interdisciplinary composition of the team and was a complementary activity to the operability study. Its role was to help improve safety once a hazard had been defined.

- its stress was on creative aspects

- the engineers would need to be enthusiastic in nurturing the drive toward creative improvement

- the value of curbing the critical faculty was vital in making progress, and that we could bring our critical faculties to bear later. In short don't say 'that's a good idea but' say rather 'that's a good idea AND' - then add to it.

Following his own advice he gave reasons why the idea of the sponsors was an outstanding one, namely:

1. It tackles a root problem of prevention rather than mitigation.

2. It is a simple idea of proven worth.

3. It saves lives and prevents injuries on a scale that was significant given the high safety standards attained by the system.

4. It takes advantage of the technological advances introduced into the project.

The problem was defined as one of development of the idea into hardware.

The 'brainstorm' decided that the engineering of a functional barrier was practicable and several competing structures were proposed. The tolerances on stopping the train at a point were considered to be close enough to warrant a feasibility study of matching train door and platform door openings. Maintenance problems were considered to be capable of solution.

At the time of writing studies were being carried out to consider the movement of passengers on the platforms in the new system and in preparing a specification.

Following completion of the study a procedure for operability studies on rapid mass transit systems was prepared.

Improving the Safety of Gas Plant Start Up System

Gas Plant A (GPA) and Gas Plant B (GPB) were identical plants located in different parts of the country. The section of plant of interest is shown in a schematic flowchart, Figure 5 which is referenced in location aids in the text which follows. Off-specification C3 is transferred to the depropanizer for reprocessing. The start up procedure requires the smoothing of any surge problems by introducing a minimum flow facility. Thus on start up the minimum flow valve (F) is open while the flow to the modules are shut off. When flow has been established the flow to the modules is commenced and brought up to operating rates and the minimum flow valve is closed.

Three identical hazardous incidents had occurred in these plants over a period of four years. The incidents had caused relief valve piping to fail at the junction of the relief valve inlet pipe and the exit pipe from a preheater (point P). The incidents occurred on start up of the plant.

The first two incidents, one at GPA and the other at GPB has caused concern but had been detected early after breakage of containment. Emergency shut down procedures had been initiated and ignition had not occurred. A third incident at GPA however had resulted in ignition followed by a major fire which led to fatalities and $Ms. loss.

The problem facing the Company was to derive solutions quickly and implement them immediately at GPB which was still operational.

16

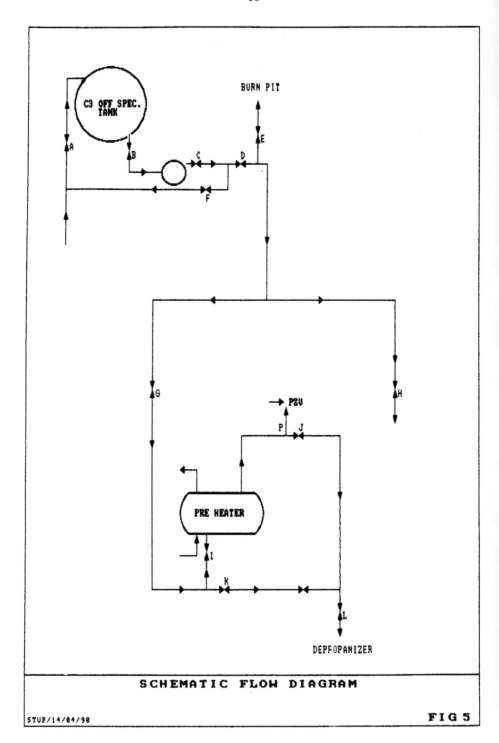

SCHEMATIC FLOW DIAGRAM

STUP/14/04/90

FIG 5

Application of Principles

Involvement of all Company expertise was sought in defining the root cause of the disaster. A seminar was prepared and presented at all gas plants and to the consulting services department. The seminar was aimed at presenting the facts and inviting explanations. The analysis of the cause became interesting. Some evidence seemed to indicate that the flange connection at P had not been connected adequately after maintenance. A theory of acoustically induced vibration following relief valve opening seemed to be popular as an explanation of failure by fatigue. However in the work achieved over the year in researching relief valve related failures we had learned of many similar incidents in different parts of the world. The incidents could be explained in terms of the release of large forces when a relief valve opened and high pressure gas or vapour was released. We did an analysis and showed that such forces could be multiplied by inadequately supported pipework to give maximum bending stresses at point P which were at two thirds of the yield stress. The relief valve maintenance team at GPA had records which showed that chatter was a common occurrence on the relief valves in question. We confirmed that this was to be expected on theoretical grounds. The chatter meant that the forces were frequent, changeable in direction and intermittent as well as being large. The analysis predicted that fracture would occur at point P and that it would be achieved rapidly and be accompanied by violent movement of the pipework. This satisfied the evidence and it attained consensus acceptance. Following this an interdisciplinary task force was set up to derive optimum solutions for GPB.

The task force was compromised of a process engineer, an instrument engineer, a piping engineer and the safety/inspection specialist. They spent 3 days at GPB, interviewed the operations engineers and operating staff and researched the P & IDs. The operations staff had spent a considerable amount of time in developing solutions and played a lead roll in stimulating ideas and the task force team analysed them for hazards. This was not a formal Hazop, the problem was straightforward and we could focus on the simple P & ID represented by Figure 5.

Results

1. The Depropanizer preheater was to be abandoned in place. This was a major decision which effectively eliminated the problem The requirement for heating was unique to GPA and could be ignored at GPB. In addition blocked discharge relief became unnecessary because the vessel with the limiting pressure had been removed.

2. When recycle through the minimum flow valve is not possible a revised start up procedure shall be used. The revised procedure requires establishment of flow conditions to the burn pit while the modules are isolated. Controlled and gradual introduction of the C3 feed to the depropanizers is obtained by cracking open the manually operated valve to the module and closing the valve to the burn pit. In addition it was recommended that a maintenance bypass for the minimum flow valve was installed. This measure minimized the need to use the burn pit option with a consequent saving of feedstock.

3. The inadequate relief valve piping support noted at GPB supported the findings of the need for a company wide review and remedy programme for this matter. Specifications for the design and location of U bolts were prepared for the conditions at GPB should the pre-heater be recommissioned.

**Improving the Safety Input to an Undergraduate Process Engineering
Design Education System**

In the early nineteen seventies the undergraduate school of chemical engineering at the
college had a basic structure of:

Part 1 -   Mathematics, Chemistry, Process Principles, Thermodynamics, Separation
            Processes, Material Science, Energy Systems, Information Presentation,
            Laboratory

Part 2 -   Mathematics, Chemistry, Process Principles, Thermodynamics, Transport
            Phenomena, Separation Processes, Plant Design, Business Principles,
            Laboratory.

Part 3 -   Transport Phenomena, Separation Processes, Particulate Systems, Energy
            Systems, Process and Plant Design, Laboratory, Project.

Two six month periods of industrial experience were included in the four year course.

The course followed the guidelines outline by the Institution of Chemical Engineers and
was validated by the Centre of National Academic Awards.

The philosophy of the course was to educate and train minds by studying chemical
engineering. In the first two years lectures, tutorials and laboratory methods were
stressed. Industrial training was also introduced to stimulate student interest in the
principles behind real life applications and to develop an appreciation of the means of
engineering solutions. In the final fourth year the emphasis was on self development and
project work was emphasized.

Prior to 1973 the topic of Safety was not treated as an academic discipline. We
considered that the general treatment of safety in design was not suited to the needs of
undergraduate education. Its importance was obvious but we considered that it was a
proper role of the project to cover particular applications of safety. The process and plant
design topics would embrace control and safety factors which were central principles of
safety in the process industry.

It was about this time we became aware of the fine work of the staff of the Atomic
Energy Authority in this field and the development of it by Kletz and Lawley. The
philosophy of the approach was attractive to us in terms of education as well as in terms
of its effectiveness.

Application of Principles

We sought and obtained the support of the Head of Department and the Principal in
organizing post graduate workshops in Hazops and invited H.G. Lawley to lead them.
We sent notices to local process industries and national design houses and administered
well attended courses which were well received by the participants.

In 1977 we introduced the approach to our undergraduates using the tried and tested
methods of the post graduate workshop. Principles of hazard analysis were presented in
lectures and a two day workshop on operability studies and hazard analyses was held in
which real life problems were tackled.

The project work was reappraised. Student presentations in front of staff and colleagues
were required at the conceptual, design and final report stages. At each stage process and
safety matters were questioned and defended. In the final stage a Hazop of a single line
was required.

We introduced safety discussion periods into the degree scheme. We used methods developed by the Institution of Chemical Engineers. Common operating hazards were shown to our students on slides. Members of staff played the roles of Plant Manager, Operator, Instrument Engineer, and Safety Engineer and students were able to cross examine them to determine the root cause of the hazard.

Results

We found the Hazops to be an excellent vehicle of learning as well as an effective safety approach. It stimulated improvements in design, detected errors and was a powerful learning forum for undergraduates and staff.

REFERENCES

1. Lawley, H.G., Operability Studies and Hazard Analysis, Chem. Eng. Progress, Vol. 70, No. 4, April 1974.

2. Lawley, H.G., The Application of Hazard and Operability Studies and Hazard Analysis to the Offshore Industry, Paper for Presentation at the U.K. Offshore Safety Conference, Feb., 10/11, 1982.

3. Young, J.Z., Philosophy and the Brain, OUP, 1987.

4. A Guide to Hazard and Operability Studies, CISHEC Saftey Committee.

5. Lawley, H.G., Size up Plant Hazards This Way, Hydrocarbon Processing, Vol. 55, No. 4, April 1976.

Strengthening the Link Between Management Factors and Quantified Risk Assessment

E J Smith
Technica Consulting Scientists and Engineers

M J Harris
Department of Engineering, Manchester University

Abstract

The findings of a study of maintenance management at hazardous plant are used to illustrate the inter-relationships between management elements. The significance of these complicated interactions for incorporating managerial and organisational factors into Quantified Risk Assessment (QRA) is considered and potential problems highlighted. The potential for risk analysis to play a more active role in ongoing safety management is also explored; it is suggested that the adoption of a hierarchy of robust, safety related, Performance Indicators matched to the hierarchy of managerial responsibility and identified via risk analysis might allow plant managers to exercise more pro-active control of plant safety.

1. Introduction

Recently there has been a growing awareness that organisational and management (O&M) factors can have a very significant influence on the level of risk presented by an industrial plant or transport system. In part this awareness has arisen from the conclusions drawn in the wake of major incidents. Turner and Toft [1], in a study of 19 public inquiries into major accidents, found that over 80% of their recommendations were concerned with organisational and procedural rather than solely technical matters. The public inquiry reports into two of the recent UK disasters, Kings Cross Underground Fire [2], and Clapham Junction Railway Accident [3], focussed strongly on the management inadequacies which extended up the organisational hierarchies and which allowed risky conditions to go unchecked and to multiply.

This desire to better understand the role of management and organisational factors in controlling plant risk has led to the World Bank sponsorship of a series of workshops looking at this issue. In the opening workshop Batstone stated that

"Traditionally, major accident investigations have focussed on identifying the operator and technical failures....., the analytical methods of risk assessment which are currently being used are based entirely on an analysis of operator error and equipment failure."[4]

The use of generic equipment failure rates in risk assessment means that the actual range of risks that is likely to exist between well managed and poorly managed industrial plants will not be truly reflected. Watson [5] estimated that the range from very good to very poor management can be reflected in an order of magnitude increase in the risk of an accident. As yet, however, quantitative techniques to assess the influence of O&M factors are not widely used.

This paper presents the findings of a three-year project which studied one branch of plant management, maintenance management; the role of maintenance management in the safe operation of hazardous systems is considered in a qualitative manner and this work is then used as a basis for considering how the link between QRA and O&M factors could be strengthened.

2. The Role of Maintenance Management Deficiencies in Some Major Accidents

A good illustration of how inadequate maintenance management can dramatically increase the risk of a major accident is provided in the report of the Tribunal of Inquiry into the fire and explosion which occurred at the Bantry Bay oil terminal on 8th January 1979 [6].

An oil tanker, the MV Betelgeuse, operated by Total was unloading her cargo at a jetty of The Gulf-owned terminal. The vessel was incorrectly ballasted, broke her back and spilled a large quantity of flammable liquid. A fire occurred, engulfing a large portion of the ship and jetty and killing over 50 people. The accident tribunal identified the poor maintenance of the vessel as a contributory cause of the accident. At its last drydock survey a decision had been taken by Total managers not to renew corroded longitudinals nor to replace sacrificial anodes in the ship's ballast tanks. The reason given for these decisions, which led to the tanker being in a significantly weakened state on the day of the fire, was that at the time of the drydock survey Total intended to sell the vessel.

The tribunal also criticised the maintenance of the terminal's safety systems. An inspection of the terminal in December 1978 had been carried out by a Gulf Inspector. He commented that,

"Because of the economic situation of the last few years present manpower is frequently stretched to the limit and preventive maintenance of equipment, so necessary in an exposed environment.....appears to be falling behind."

As a result of this inadequate maintenance a diesel fire pump had been out of service for 9 months, a fire monitor on the jetty was out of service, emergency lighting was faulty, a foam tender did not work, violations had occurred of the conditions under which electrical wiring had originally been certificated as explosion proof, and life raft preventive maintenance checks were over one year overdue on the day of the fire.

This accident illustrates how commercial and operational considerations can influence plant safety. In the case of the ship these considerations caused specific decisions to be taken which had an immediate effect on the risk of structural break-up. In the case of the terminal the pressures on maintenance resources that were present throughout the seventies caused a longer-term increase in risk. These observations lead to the conclusion that mechanisms should exist at hazardous plant for controlling plant risk in the short term and in the long term.

Table 1 shows examples of other accidents where maintenance problems have contributed and highlights the managerial influences on the specific maintenance errors. These analyses of accidents provided useful background to our on-site case studies at major hazard plants.

3. A Methodology for Analysing the Maintenance Function

In order to better understand how management and organisational factors can affect the reliability of both plant and personnel Kelly's structured systems approach [10] for analysing maintenance management was adopted. One of the basic ideas in this approach is the structure of relationships, between the key elements in maintenance management, which is shown in Figure 1. The ideas encapsulated in this figure have proved useful in assessing the contribution of maintenance to achieving economic goals and we believe it also aids appreciation of the role of maintenance in safety management.

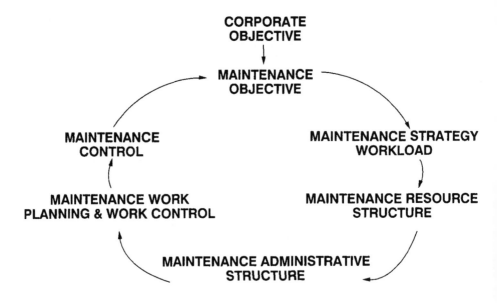

FIGURE 1. ELEMENTS OF MAINTENANCE MANAGEMENT (adapted from Kelly [10])

The functions of the maintenance system and the ways in which these functions might be affected by the dynamic relationship between maintenance and production need to be clearly understood. From this, a definition of the **maintenance objectives** which is compatible with the

TABLE 1

THE ROLE OF MAINTENANCE IN THE CAUSATION OF SOME ACCIDENTS

Accident	Maintenance Deficiencies	Management Factors
Grangemouth 1987 (Three incidents) [7]		
(i) Fire during removal of valve from flare system.	-Failure to identify hazards of maintenance task. -Failure to check that valve had been correctly isolated. -Poor condition of some gate valves.	-Planning of task by supervisors, not senior engineers. -Inadequate job supervision.
(ii) Explosion in hydrocracker caused by over-pressure due to removal of low-level trips.	-Ill conceived plant modification. -Failure to repair control room indicators promptly.	-Deficient plant modification procedures. -Failure to learn from previous near miss.
(iii) Fire in crude oil storage tank.	-Contractor employee smoked while carrying out cleaning task in tank.	-Inadequate control of contract work.
King's Cross 1987 [2] Fire, starting under escalator, spread to ticket hall, killing 31.	-Lack of escalator cleaning. -Lack of inspections underneath escalators. -Lack of testing of water sprays.	-Failure to set quantitative safety objectives comparable with operational and cost objectives. -Failure to learn from previous fires and from hazard surveys. -Mismatch between maintenance workload and resources.
Chicago DC-10 1979 [8] Engine detached from wing during take-off. Pilots lost control and plane crashed, killing 271.	-Ill conceived change to maintenance procedure for dismantling engine and pylon led to cracks in engine pylon.	-Failure to adequately consider safety consequences of changed procedure. -Failure to learn from earlier cracks.
Three Mile Island 1979 [9] Poor condition of plant plus operator errors led to a plant trip developing into a core meltdown.	-Repair of leaking relief valve deferred to summer shutdown. -Auxiliary feedwater pumps mistakenly left valved off after preventive maintenance. -Condensate filters caused repeated problems during maintenance, but these were never reviewed.	-Failure to enforce operational limits. -No checks of preventive tasks. -No control of maintenance backlogs. -Failure to identify safety critical items. -No analysis of failure data.

24

corporate and production objectives can be identified. These objectives, together with the production strategy and plant type determine the **maintenance strategy**. The maintenance strategy has a strong influence on the level and nature of the **maintenance workload** which will, in turn, affect the **maintenance resource structure**. An **administrative structure** is needed to guide the efforts of the resources towards achieving the objectives. The **work planning and work control** system should be built around the resource structure and determines the details of how work is scheduled, allocated and controlled. Finally, a **maintenance control** system is needed to ensure that the maintenance system is working towards the set objective and to provide corrective action if it is not, eg. by changing the strategy or resource structure. Clearly, for a major hazard plant a maintenance control system must include a plant safety and reliability control system.

Figure 2 indicates how this logic was used to structure the analysis of maintenance management inadequacies in NASA's space shuttle programme, as revealed by the Presidential Commission investigation into the Challenger disaster [11]. It highlights the inter-relationships between the maintenance system elements and the importance, for safety related maintenance, of production and resource cost objectives. This same logic was also used to structure the observations from the on-site case studies, as will now be shown.

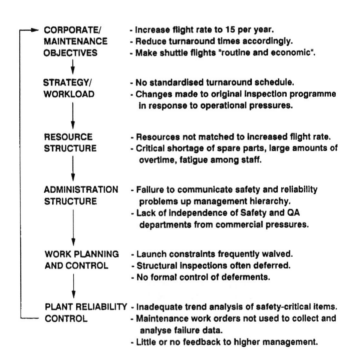

FIGURE 2. MAINTENANCE MANAGEMENT INADEQUACIES
IN SPACE SHUTTLE PROGRAMME.

4. Methodology Applied to On-Site Case Studies
(a) Corporate and Maintenance Objectives and Maintenance Control

The setting of corporate and maintenance objectives is a vital element in maintenance management because it represents the first step in establishing control (see Figure 3). Explicit objectives act as yardsticks against which an organisation can assess its degree of success or failure. Regarding the commercial aspects of a company's operations, managers usually seem to have clear, detailed, objectives eg. quantitative targets for overall production output, plant availability and maintenance resource costs. Such objectives are often translated into more specific sub-objectives concerning production targets for plant units and resource budgets for individual plant areas. Such detailed quantitative sub-objectives allow tight **pro-active** control of the commercial aspects of plant operations, with corrective action being taken **before** production drops to unacceptable levels or before resource costs spiral out of control.

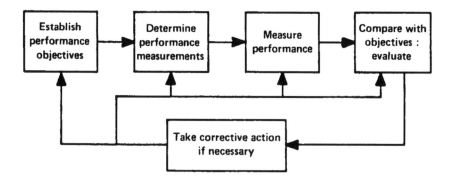

FIGURE 3. THE PROCESS OF CONTROL [12]

It was observed that safety and reliability objectives, by contrast, are often less explicit and of a qualitative nature. For example, one of the five stated maintenance objectives of the Mobil company in the North Sea was "to ensure that appropriate standards of safety, environmental pollution control and good housekeeping continue" [13]. While it was commendable (and rare) that a company had produced a written safety-related objective, such a qualitative objective was very open to subjective interpretation.

An exception to the general lack of detailed safety objectives was found, in an on-site study carried out by the authors, at one of the recently commissioned UK Advanced Gas Cooled Nuclear Reactors. A set of detailed, quantifiable, sub-objectives (see Figure 4), related to the availability of safety systems and the frequency of plant trips, had been developed and were designed to meet the overall plant safety objective that the probability of failing to prevent an unacceptable release of radioactivity should be less than 10^{-6} per reactor year. Planned or forced unavailabilities (outages) of individual safety system components could be entered into computerised fault tree programs from which overall safety system availabilities could be derived. This facilitated pro-active control of plant safety, ie. if trends were detected which indicated that safety system availabilities were deterioriating then corrective actions could be initiated before plant risk rose to unacceptable levels. Clearly such tight control of safety system availabilities should have a very positive effect on plant safety; yet a traditional "hardware only" QRA would not take this into account.

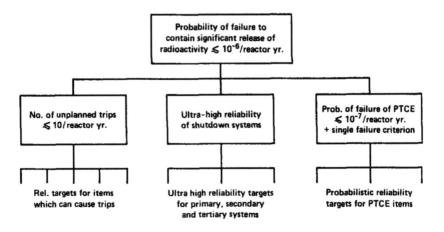

FIGURE 4. HIERARCHY OF AGR SAFETY OBJECTIVES

In most industries, however, such detailed probabilistic measurement of ongoing plant safety is not, at present, appropriate or easily applicable. That there is a need for readily measurable quantitative and robust parameters of plant safety performance has been recognised by the US nuclear industry and by the aviation industry. Both have developed sets of safety-related performance indicators (PIs) which enable trends to be monitored and objectives to be set. Table 2 shows the set of PIs used by the US Nuclear Regulatory Commission to monitor plant safety performance [14]; they are directed primarily at measuring frequencies of plant trips and safety system availabilities.

Maintenance can be regarded as "...the essential means of perpetuating the design intent..." [15]. Thus, explicit safety and reliability plant objectives are needed to assess whether the maintenance function is succeeding in achieving its task and whether safety standards are acceptable. Without such objectives control of plant safety and reliability is likely to be highly subjective.

TABLE 2
PERFORMANCE INDICATORS CHOSEN BY NRC FOR NUCLEAR POWER STATIONS

Performance Indicator	Definition
Automatic Scrams	Unplanned automatic reactor protection system actuations which produce control rod motion.
Safety System Actuations	Emergency core cooling system actuations resulting from spurious or genuine signals, and diesel generator actuations.
Significant Events	Events judged by NRC to have particular safety significance such as degradation of fuel integrity or of primary coolant pressure boundary.
Safety System Failures	Events that could prevent (a) safe reactor shutdown, (b) residual heat removal, (c) control of radioactive release.
Forced Outage Rate	Forced outage hours divided by the sum of the operational hours and forced outage hours.
Enforcement Action Index	Weighted index of safety violations detected by NRC inspectors.
Mean Time Between Forced Outages Induced by Equipment Failure	The number of hours the reactor is critical in a given period divided by the number of forced outages.
Maintenance Backlog	The percentage of total outstanding corrective maintenance work requests not requiring a unit outage, that are greater than three months old.

(b) Maintenance Strategy

The maintenance strategy is critically affected by objectives and maintenance control. Without explicit objectives strategies will be based on precedent. In the case of well established technologies precedent might be based on sound engineering judgement informed by considerable accumulated knowledge. However, with newer technologies, or those which present a major hazard, then clearly

a) extra care needs to be taken in the initial formulation of strategies (with use of structured techniques of hazard identification such as HAZOPS, FMEA, FTA),

b) there is a need for high quality feedback from returned maintenance work cards, incident reports and operational logs to enable a judgement to be made as to the suitability of maintenance tasks and frequencies.

At one of the major hazard plants that was studied the instrument
maintenance department had an extremely large workload, mostly testing the
large number of instrument loops that were critical to plant safety. A
mismatch between this large workload and the available manpower resources
was producing increasing maintenance backlogs. We studied one
particularly critical system in order to analyse the rationale for its
maintenance regime. The system consisted of electronic subsystems and
electro-mechanical and mechanical subsystems (see Figure 5). Analysis of
returned maintenance work orders revealed that faults in the mechanical
and electro-mechanical subsystems were responsible for well over 90% of
the complete system's unavailability and that the test frequencies of the
electronic sub-systems were excessive, employing resources that could have
been more profitably used elsewhere. At this plant there were no clear
availability objectives for safety systems and no effort was made to
analyse failure trends, i.e. safety related reliability control was highly
subjective and unstructured.

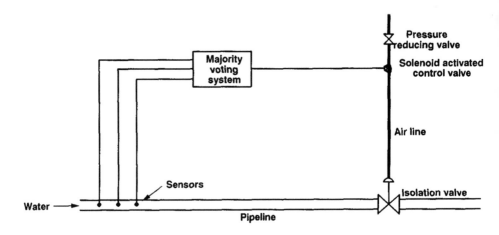

FIGURE 5. SAFETY CRITICAL SYSTEM (SIMPLIFIED)

Our review of major accident documentation revealed examples of
changes being made to maintenance strategies (eg. reductions in escalator
cleaning at King's Cross; reductions in space shuttle inspections) in
response to production or finance pressures rather than as a result of the
analysis of collected data or the application of sound engineering
judgement. Clearly, at major hazard plant such changes should be as
strictly controlled as modifications to plant hardware already mostly are,
i.e. only after thorough review by experienced engineers and plant
managers. Given the importance of such changes for plant risk then there
is a definite need for consideration of such factors to be incorporated
into QRA.

(c) Maintenance Resources

It is often very difficult to assess the contribution that the
maintenance function makes to preserving plant reliability and safety.

Thus there are usually strong pressures to reduce maintenance costs, particularly when the economic climate is unfavourable. The result can be a mismatch between the resources actually available and the resources needed to meet all the scheduled and unscheduled workload. This can lead to ill-considered changes in the maintenance strategy (as mentioned before) or to increasing deferments of planned maintenance.

Some companies are trying to balance the potentially conflicting objectives of minimising resource costs and preserving inherent plant reliability and safety by making greater use of contractors, developing multiskilled tradeforces and removing restrictive demarcations. Such measures, combined with reductions of in-house manning levels, have clear financial and operational advantages allowing a better matching of the resource structure to a dynamic workload. However, these measures may also carry various negative implications for plant safety. Eberlein has stated that some managements view the introduction of contractors as a way of "avoiding some of the more demanding safety policy requirements, on the basis that these become the responsibility of the contractor's management." [16]. Our on-site studies revealed that a high proportion of incidents and accidents involved contractors. Their widespread use demands structured pre-employment selection procedures, the provision of training and tight control and auditing of all their activities.

There are also long term implications of employing contractor labour. Their use enables rapid adaptation to changing economic conditions e.g. manning levels can be quickly reduced if the economic climate becomes depressed. There need to be mechanisms in place for assessing and controlling the possible negative consequences, for long term plant reliability and safety, of such dramatic changes.

(d) Administrative Structure

In many of the companies studied there appeared not to be a clear translation of overall, corporate, safety objectives down the management hierarchy. As a result, there appeared to be no motivation to communicate safety-related information up the hierarchy. In many of the major accidents that we studied it was apparent that senior managers had become out of touch with the state of their plants. There appears to be a need for relating hierarchies of line management to hierarchies of meaningful, measurable, safety objectives (see Figure 6). Senior managers would be primarily concerned with overall plant safety objectives - if they were worried about negative trends in accidents, fires or near misses then they might demand more information from lower in the hierarchy. More junior managers would be concerned with more specific, narrower, objectives relating to the reliability and safety of plant units and individual safety systems. Developing such a visible link between these hierarchies should improve upward communication and improve the motivation for collecting and using data from returned maintenance work cards for long-term control purposes.

It was pointed out in the discussion of the Bantry Bay accident that, in addition to long term control, there needs to be short term control of safety; specific errors such as an ill-conceived plant modification or maintenance deferral could seriously damage plant integrity. For this reason many petrochemical plants carry an internal, but highly autonomous, inspection department. Its inspectors are independent of the production pressures on line management and thus should give advice based purely on the adjudged risks and not influenced by commercial considerations. Thus, they provide valuable short term control and should have ready access to senior managers should they wish to express any concern (see Figure 7).

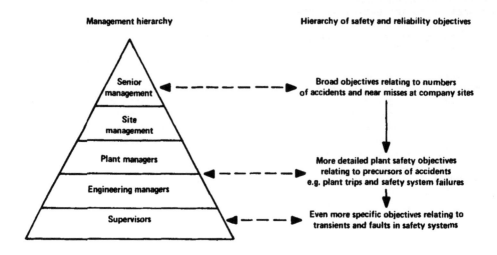

FIGURE 6. MANAGEMENT HIERARCHY AND THE OBJECTIVES HIERARCHY

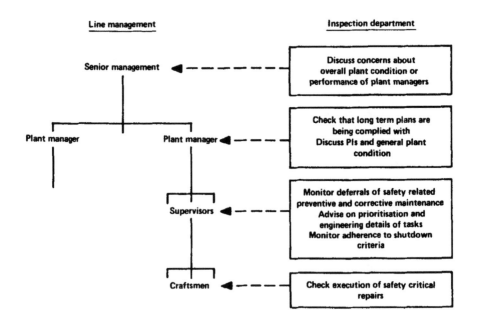

FIGURE 7. LINE MANAGEMENT AND THE INSPECTION FUNCTION

(e) Maintenance Work Planning and Work Control

It is in this element that the higher level managerial problems such as objective setting, strategy selection and maintenance control become evident and visible. Mismatches between the workload and the available resources can lead to planning dilemmas, with conflicts between production priorities and safety-related maintenance. The result can be repeated and widespread deferral of safety-related maintenance and poor quality work arising out of time pressures. Improving safety objective setting, strategy selection and control would have positive effects on maintenance work planning. Explicit feedback regarding the reliability and safety of plant would enable planners to better assess the consequences of their actions and decisions. A maintenance strategy better matched to plant reliability and safety objectives should reduce excessive, needless, maintenance and make work prioritisation easier. In addition, an independent internal inspection function should review safety-critical maintenance and advise on its planning and control.

5. Feeding Management Factors Knowledge into QRA

The previous section has highlighted some O&M factors which can have a very significant influence on plant risk e.g.

- Are there explicit safety objectives?
- Is there thorough feedback of safety and reliability data to compare actual performance to objectives?
- Is the safety-related maintenance strategy based on a structured selection process and formally reviewed in the light of collected failure data?
- Are maintenance resources carefully matched to the necessary workload or have they been changed to meet economic considerations with little feedback from plant safety control?
- Do individual line managers have safety objectives and do their assessments give weight to safety performance?
- Does the administrative structure allow for authoritative assessment of plant safety independent of economic pressures?
- What communications channels exist for the transfer of safety-related information?

It has also been seen that interactions between many of these factors can combine to produce dilemmas for, and pressures on, those at the sharp end of work planning and work control; the result can be poor plant reliability and reduced human operator reliability.

Given that these effects on plant risk can be very significant (e.g. as at Bantry Bay, Three Mile Island, Kings Cross) there is a natural desire to incorporate consideration of them into QRA. Attempts at doing this have been based on the idea of a **management audit**. Canter and Powell [17] described the development of a set of questions about O&M factors for inclusion in the Instantaneous Fractional Annual Loss (IFAL) method of the Insurance Technical Bureau. Their method involved an initial on-site visit by an expert surveyor who would allocate gradings to 46 different features related to safety management and risk control. These gradings would then be weighted (the weightings having been initially agreed by a group of experts in the field of process plant safety) to give an overall management quality coefficient which could be combined with the result of the plant hardware assessment. Such a technique bears similarities to the International Safety Rating Scheme (ISRS) which uses a much larger questionnaire (over 600 questions on 20 topic areas) to grade a company's safety management on a scale of 0 to 5 stars [18]. It is generally agreed that ISRS can have a useful influence in improving the safety culture of a firm and giving safety a higher profile. However, the use of an audit technique such as this as an input to QRA would need to address some important issues, viz.

(a) An effective weighting system would be needed to ensure that those factors which could have great significance for plant safety are highlighted. With a large question set (such as that used in ISRS) weak performance in a crucial area might be disguised by adequate performance in less important areas.

(b) The scope and level of detail of an O&M assessment would need to strike a balance between the desirability for it to be concise and easily handled and the need for it to be thorough and comprehensive. A contributory cause of a recent large tank farm fire was an inadequate schedule for inspecting the floating tank roof; however, an ISRS audit had failed to pick this up because it could not go into that level of detail given the practical time constraints [19].

(c) The subjectivity of an assessor cannot be ignored when considering audit type techniques. The extent to which an O&M assessment is sensitive to such influences would need to be thought about carefully.

All these issues may have been of relevance in the Kinross mine fire; the main cause of the high number of fatalities, 177, was the use of an insulating material which gave off highly toxic fumes in a fire and which was therefore not suitable for use underground. And yet apparently the mine had been awarded 5-stars under ISRS [20].

It seems commonsense that the nature of the operational and maintenance regimes can influence the reliability and availability of equipment. In some situations it may be possible to obtain a quantitative feel for how the choice of preventive maintenance regime, for example, might affect equipment reliability; an understanding of such a "cause and effect" could be built up in terms of comprehensible physical phenomena. However, if the subject area is broadened to include a quantitative understanding of how a whole range of plant management and organisational factors could influence overall plant reliability and risk then the problem clearly becomes much more complicated. No longer are there clear "causes and effects". As has been mentioned, many of the causes (O&M deficiencies such as poor training of operators, failure to set up adequate plant modification procedures, lax control of safety-related maintenance deferrals, lack of explicit safety objectives, poor definitions of authority and accountability etc) may be closely related and may interact in unpredictable ways.

An attempt was made in the previous section to highlight the links between maintenance management elements, but the model of Fig. 1 is clearly simplified; in a specific, real life, organisation it is likely that the influences of the various branches of plant management on each other could be very complicated and difficult to model. Yet given the importance of O&M factors it is vital that progress in improving this aspect of QRA is made. A further consideration is that improving the management at a hazardous plant may prove to be a very cost effective method of reducing plant risk compared to upgrading plant hardware. In order for this area to be effectively explored, however, a credible technique for quantitatively evaluating the reductions in plant risk due to O&M changes needs to be available.

6. Feeding QRA into Management Factors

Much of the debate about management factors and QRA has centred around the issue discussed in the previous section, viz. how to incorporate O&M aspects into QRA. However, our project maintenance management studies have provoked another line of thought which is, "Could QRA be better utilised as a tool to aid more effective safety and risk management?" The difficulty of measuring ongoing plant safety has been highlighted as a fundamental problem in managing hazardous plant - the build up of risky latent conditions can be overlooked if there are no clear yardsticks for safety assessment. It is suggested that the adoption of robust safety-related Performance Indicators (PIs) derived from a detailed safety assessment might prove useful in this respect. This is illustrated in the following case study.

PI Case Study

This study was concerned with aspects of the safety of London Underground (LUL) railway stations. Two accident scenarios were examined, viz.

(i) fire originating underneath an escalator,
(ii) uncontrolled descent of a station passenger lift.

With the help and co-operation of staff at LUL's Lift and Escalator Engineers (L&EE) Division fault trees which showed the various possible events and combinations of events which could lead to an accident were developed. Figure 8 shows a simplified fault tree for an escalator fire. (A formal QRA would provide a more detailed set of fault trees with the relative contribution of each branch quantified). The process of selecting appropriate PIs mirrored the progressive sub-divisions of the fault tree (see Figure 9). Clearly, the incidence rate and severity of escalator fires should be the central PI for this accident scenario. However, if this was the only PI then it would form the basis of only reactive control. By selecting supplementary PIs which relate specifically to the identified contributory causes it should be possible to establish proactive safety control.

The hierarchy of PIs in Figure 9 could be linked to the management hierarchy. Senior LUL managers might study the higher level PIs such as the rate of incidence of escalator fires. However, as well as receiving information from the L & EE Division, senior managers will also be receiving safety-related information from other sections and thus the level of detail cannot be too great or else senior managers will become overwhelmed by the volume of data. Further down the administrative hierarchy, in the L & EE Division itself, managers should not be just monitoring the higher level PIs, but also those lower level PIs shown in Figure 9.

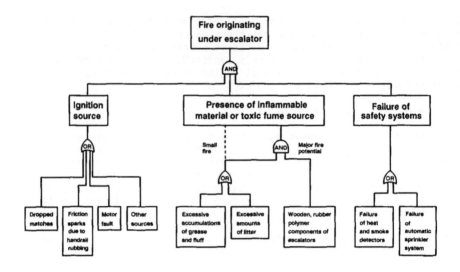

FIGURE 8. FAULT TREE OF ESCALATOR FIRE

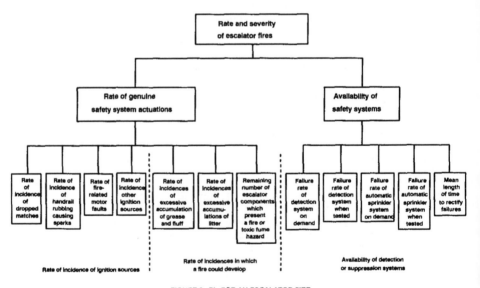

FIGURE 9. PIs FOR AN ESCALATOR FIRE

Relating the management hierarchy to meaningful "measures of plant safety" should provide a structured framework for communication of safety-related information between management levels and should make explicit the managerial safety responsibilities at each of the various levels of control. There are 275 escalators and 70 lifts at LUL stations and thus collected data should quickly enable trends, alert levels and safety objectives to be established.

Such tight plant reliability and safety control might enable the adequacy of the safety-related maintenance strategy (e.g. the frequency of testing of smoke detectors and automatic sprinklers), of the maintenance resource structure (eg. manning levels) and of the work planning and work control systems (eg. deferral procedures) to be assessed. Thus, correlations between the safety-related maintenance "input" and the resulting plant safety "output" could be developed.

Conclusions

(i)   There are strong inter-relationships between the key elements of maintenance management. Failure to set detailed, explicit, reliability and safety objectives and to establish tight control affects the maintenance strategy, the maintenance resource structure and ultimately work planning and control.
(ii) Maintenance management, as with the other branches of plant management, takes place within a dynamic environment and must respond to changes in the production, sales and finance systems of an organisation. Changes in sales targets or resource budgets are likely to affect the key maintenance management elements. This heightens the vital importance of plant reliability and safety control - this element must enable maintenance management to develop iteratively without allowing reliability and safety standards to fall to unacceptable levels.
(iii) Understanding these inter-relationships between O&M elements is vital to understanding how management factors can affect plant reliability and human operator reliability. Any method which attempts to allow for O&M factors in QRA must address this issue and develop a credible technique which does not depend exclusively on subjective judgement.
(iv) There appears to be considerable scope for greater use of QRA in safety and risk management. In some of the major accidents studied, there was very little indication of tight plant reliability and safety control and this can be partly attributed to the difficulty of quantifying ongoing "plant safety" in a meaningful manner. Safety-related PIs derived from QRAs could greatly improve managers' "feel" for intrinsic plant safety and increase their ability to exercise proactive rather than re-active control.
(v)   Such an application of QRA demonstrates that risk analysis need not just be a tool for providing numbers at the design stage, but that it can fit into the wider framework of **quantitative safety management,** helping managers to implement practical improvements.

Acknowledgements

To: RM Consultants for sponsorship and technical advice; SERC for financial support via the Total Technology scheme; LUL, in particular the L and EE Division and Richard Lashwood of the Safety Audit Department, for their co-operation and interest during the PI study; Tony Kelly, of the Department of Engineering at Manchester University, for help and guidance.

References

1. Turner, B.A. and Toft, B. Organisational Learning from Disasters, Emergency Planning for Industrial Hazards, Gow, H.B.F. and Ray, R.W. (Eds.), Elsevier Applied Science, 1988.
2. Fennell, D. Investigation into the King's Cross Underground Fire, HMSO, 1988.
3. Hidden, A. Investigation into the Clapham Junction Railway Accident, Department of Transport, HMSO, 1989.
4. Reason, J. Safety, Human Fallibility, Proof of evidence at Hinckley Point C Inquiry, 1989.
5. Watson, I.A. and Oakes, F. Human Reliability Factors in Technology Management, IEE Proceedings, 1989, Vol. 136, No. 1.
6. Tribunal of Inquiry, Report on the Disaster at Whiddy Island, Bantry, Co. Cork on 8th January 1979, The Stationery Office, Dublin, 1979.
7. HSE (Health and Safety Executive). The Fires and Explosion at BP Oil (Grangemouth) Refinery Ltd, HMSO, 1989.
8. Wallich, P. Keep Them Flying, IEEE Spectrum, November, 1986.
9. Kemeny, J. et al. The Need for Change: the Legacy of TMI. Report of the President's Commission on the Accident at Three Mile Island, Washington, D.C. Government Printing Office, 1979.
10. Kelly, A. Maintenance and its Management, Conference Communication, 1989.
11. Rogers, W.P. et al. Report of the Presidential Commission on the Space Shuttle Challenger Accident U.S. Government Printing Office, 1986.
12. Megginson, L.C. et al. Management: Concepts and Applications, Second Edition, Harper and Row, 1986.
13. Johannessen, B. Offshore Inspection, Maintenance and Repair Noroil, September 1982.
14. NUREG (U.S. Nuclear Regulatory Commission), Inter-office Task Group on Performance Indicators, Performance Indicator Program Plan for Nuclear Power Plants, September 1986.
15. Pilborough, L. Inspection of Chemical Plant, Leonard Hill, London, 1971, p. 372.
16. Eberlein, A.J. Managing Safety in a Multi-disciplinary Organisation. The King's Cross Underground Fire: Fire Dynamics and the Organisation of Safety. I.Mech.E conference, June 1989.
17. Canter, D and Powell, J. Quantifying the Human Contribution to Losses in the Chemical Industry, Journal of Environmental Psychology, 1985, 5, pp. 37-53.
18. Bond, J. The Industrial Safety Rating System, Loss Prevention Bulletin, 1986, pp. 23-39.
19. Private Communication.
20. Byart, L. The Safety Council's 5-star System, Safety Management, May 1988.

"SURVIVE"

A Safety Analysis Method
for a
Survey of Rule Violation Incentives and Effects

NIGEL J HOLLOWAY
Safety & Reliability Consultant
SRD
Wigshaw Lane, Culcheth, Warrington, WA3 4NE, UK

ABSTRACT

The safety of many potentially hazardous plants is normally assessed on the basis that they will be operated in accordance with a set of rules, which are the various instructions given to the human elements in the plant operation.

Violations of these rules may well invalidate important assumptions made in the design and safety analysis of a plant, leading to unanticipated and unanalysed situations which may be much less safe than those which arise when the rules are obeyed.

In recognition of the potentially serious effects of rule violations on plant safety, a methodology has been developed for the qualitative investigation of such violations. The method covers the identification of violations, their effects on safety, and qualitative assessment of the incentives and disincentives for such violations, including the degree to which the violations would be recorded.

In its current state of development, the method is intended to provide an approximate ranking of the importance of violations, but does not offer a numerical quantification of probabilities. Its use should be limited, at this stage of development, to qualitative investigations intended to identify violations worthy of further analysis or anticipatory preventive measures.

INTRODUCTION - THE SAFETY SIGNIFICANCE OF RULE VIOLATIONS

The safety of many potentially hazardous plants is normally assessed on the basis that they will be operated in accordance with a set of rules, which are the various instructions given to the human elements in the plant operation.

Violations of these rules may well invalidate important assumptions made in the design and safety analysis of a plant, leading to unanticipated and unanalysed situations which may be much less safe than those which arise when the rules are obeyed.

Violations of rules have been important contributors in major accidents. Some illustrative examples of major accidents where rules were violated are listed below. Had the rules not been violated, none of these accidents would have happened, and the number of major accidents in recent times would have been drastically reduced.

Accident	Rule Violations	Effects
Bhopal	MIC refrigeration turned off Scrubber out of service	No defence to release after water ingress to MIC tank Thousands of deaths
Chernobyl	Reactor trips turned off Control rod positions beyond permitted limits	Reactor explosion 32 early deaths Significant radioactive contam- ination of Russia and Europe
Zeebrugge	Ship sailed with front access door left open	Ship overturned and sank Over a hundred drowned
Flixborough	Improper modification procedure used on pipework	Pipework failure caused plant to explode 28 deaths
Three Mile Island	Operator turned off important safety system (ECCS)	Partial reactor meltdown Billions of dollars lost

Rule violations are also major contributors to the much larger accumulated consequences of lesser accidents. For example, the many deaths attributable to drunken driving and speeding on roads are directly associated with rule violations.

Therefore, it is desirable to recognise the possibilities and dangers of rule violations in the formal analysis of safety. It will be difficult to quantify the corresponding risks in numerical terms as there are many non-random cognitive human processes involved and probabilistic risk analysis has very limited capabilities in this area. However, an awareness of possible violations, their potential effects, and the influences on their likelihood of occurrence, will be a useful step to anticipating and perhaps preventing some of the more serious cases.

(Note : In the above, and throughout this report, we take the term "rules" to encompass all the instructions given to plant staff on the operation, maintenance, testing, inspection, etc of the plant, and not only those instructions which might have particular titles including the word "rules".)

THE "SURVIVE" METHODOLOGY

An Overview

The "SURVIVE" methodology involves a survey of the various rules which should constrain or direct the human elements in plant safety, and an assessment of the violations of those rules which could seriously degrade safety. The following stages are involved in the overall process.

1   The survey initially identifies those rules which , if violated, will allow a fairly immediate and significant degradation of safety to arise. The possible violations are given against each identified rule.

2   For each violation, the magnitude of the effect on safety is assessed. The effect may be assessed in terms of increased probabilities of accidents, increased conseqeucnes of accidents, or combinations of both.

3   For each violation, the incentives and disincentives for the violation are assessed. These may be judged high if there are considerable perceived advantages and few perceived dangers or other disadvantages in the rule violation. A judgement of low incentives may be made if there are few perceived advantages and considerable perceived dangers and disadvantages.

4   The particular disincentives associated with the "recording" (interpreted widely) of violations are assessed. "Recording" may be only in the memory of one person (the perpetrator of the violation), or may be in the memories of several people, or may be writen down - for example by a computer based monitoring system. Recording is particularly important as a lack of written recording may also result in a false assessment of the extent of historical violations.

5   The overall ratings of the Effects, Incentives and Disincentives are combined in a final assessment. High (bad) ratings on all three will highlight the possibilities of most concern while low (good) ratings may give reassurance of safety even though violations are possible. The level at which "concern" is registered is a matter of judgement for the particular analysis. However, for practicality it is expected that the top few rating combinations would indicate concern and the bottom few would not.

5a   It is optional to use some numerical ranking scheme to add together the assessments, in order to obtain an overall ranking. However, there is as yet no measure of each effect which would make such a scheme fully self-consistent and the validity of such ranking schemes is a matter for judgement by the analysts using the method.

The results of the analysis will normally be a set of rule violations with assessments against the two attributes of effect and incentive (+ve and -ve). A selection of the violations of highest concern will then normally be made and these will be preserved for some further analysis or assessment. The set of violations, with ratings for the two aspects of effect and incentive is the end result of the SURVIVE method as currently defined.

Identification of Rule Violations Important to Safety

This process will normally be applied to a set of documents which describe the rules applicable to the operation of the plant. The situation at most plants is that some rules are written down in close accord with the actual operations and some are not. Thus, the plant documentation may need to be supplemented by further descriptions of the rules normally in effect.

Once the documentation of the rules is assembled, it should be inspected for violations which have an important effect on safety. In one sense, most of the rules will have some effect on safety, but the main concerns are with rule violations where the following criteria apply:

> There is a fairly immediate degradation of safety by removing or degrading some clearly identified defence against accidents

and

> The removal of the defence would cause at least an order of magnitude increase in probability or consequence of an accident.

Thus, for example, violations of rules concerning with periodic inspections or authorizations will not normally feature in those identified. Violations of these rules will eventually produce a reduction of safety, but there is no immediate step or rapid reduction. In contrast, a violation disabling a main safety system which should be ready for action is likely to increase the probability of some accidents by several orders of magnitude, with immediate effect, and should very definitely be in the identified list.

At this stage, violations are postulated only with regard to only two factors:

> The EFFECT on safety, as discussed above, exceeding some defined mini-

mum

> The POSSIBILITY of occurrence, however chimerical it may seem.

and not with regard to any notions of likelihood. Factors influencing likelihood are assessed in the later stages of the methodology.

The completion of the process should result in the identification of a subset of the operating rules with one or more violations associated with each rule in the subset. Rules might be violated in both negative (ignoring a requirement) or positive (doing something forbidden) ways.

The basic application of the methodology does not address the various combinations of independent rule violations which might lead to worse results than single violations. However, mutiple violations emanating from a single plan, or violations requiring essentially the same multiple actions by more than one person (or at least co-operation of more than one person) should be identified. The effects of involving more than one person are assessed later.

Notes: (1) The process of identifying rule violations is most effectively done when rules are being drawn up, as the reasons for the rules provide simultaneous indications of possible violations and their effects on safety.

(2) The methodology is not intended for analysing situations on currently operating plants where it is already known that a rule is frequently violated. analysis of such situations (if not corrected) should be done in the traditional safety assessment of a plant.

Assessment of the Effects of Rule Violations on Safety

The identification process described above assumes that there will be at least a minimum effect on safety associated with each identified rule violation. Above this minimum, the effects need to be further assessed. A simple scheme, which may be extended if there is a particularly large range of magnitudes of effects, involves simply categorizing the effects as high, medium and low, corresponding roughly to the following concepts of magnitude:

HIGH EFFECT (E3)
Removes a main safety system or main defence from its role in preventing the escalation or reducing the consequences of an accident

Corresponds to increasing the probability or consequences of an accident by about a factor of 1000 or more.

Example : Switch off a redundant safety system which should be on

MEDIUM EFFECT (E2)
Removes a safety system of moderate reliability or moderate effect in reducing accident consequences.

Corresponds to increasing the probability or consequences of an accident by about a factor of 100

Example : Leave a reactor containment door open

| LOW EFFECT (E1) | Removes a safety factor of some identified value, but not a main safety system |

Corresponds to an order of magnitude increase in probability
or consequences of an accident

Example : Fail to check on a system status (where the system would
usually be in the correct status anyway)

Where effects corresponding to removal of substantially more than one safety system or line of defence are identified, assessments such as E4, E5, E6 etc can be added to the top of the scale. If, for example, a violation switching off two diverse reactor shutdown systems were identified as a possibility, then a much higher effect category would be justified (see also the discussion on overall assessment and ranking).

(Note 1: The processes of judging the magnitudes of the effects on safety will normally require some expertise in reliability and/or accident consequence assessment, to the level required to judge the approximate reliabilities or accident consequence mitigation capabilities of the systems affected by the violations, and the corresponding increases in probability x consequences if the systems are negated or degraded.)

(Note 2: There is of course the possibility of a "NO EFFECT" category (E0) but it is assumed that violations identified against this category would be dropped from any analysis.)

Assessment of the Incentives to Violate Rules

The assessment of incentives must be done as far as possible taking into account the operator's perception of both the incentives and disincentives. This may involve interviewing the operators with regard to their appreciation of the effects on safety of the violations. If their perceptions of the effects on safety are different from the assessed effects then this may influence the incentives.

Although the rating of incentives is eventually a matter of overall judgement, it is useful to consider a list of aspects which raise and lower the incentives for violations. Some of the aspects typical of many plants are:

Raise Incentives

- Convenience (i.e. inconvenience of following the rules)

- Economics (eg plant remaining on-line, especially for plant with long minimum shutdown times)

- Management pressure to meet performance targets rather than obey rules

- Perception that there is still a reliable safety system left if things go wrong

- Lack of knowledge or awareness of dangers involved in rule violations

- Pride in managing without the rules/safeguards (eg - "we can do it without these beginners' procedures")

Lower Incentives

- Lack of perceived benefits

- Awareness of immediate personal danger (eg danger from materials being handled)

- Difficulty, or lack of opportunity to commit the violation

- Punishments/recriminations (which may depend on recording aspect)

- Management pressure to prioritise obedience to rules

- Pride in obeying the rules

* Recording of the violation (written or in other peoples' memories)

As yet, the SURVIVE methodology has not developed any scoring or ranking scheme for combining the above aspects and this is a matter of judgement by the analyst. It is the intention of further developments to refine the assessment of incentives. The current methodology assumes a multi-level rating for the overall combination of incentive/disincentive aspects (including recording, for which see next section) i.e. :

HIGH	INCENTIVE	(I 2)
MEDIUM	INCENTIVE	(I 1)
LOW/NO	INCENTIVE	(I 0)
MEDIUM	DISINCENTIVE	(I-1)
STRONG	DISINCENTIVE	(I-2)

As with the effects, it is optional to extend the scale to I 3, I-3 etc.

Particular attention is given to the matter of recording as a disincentive, because of its other implications. This is discussed next.

Assessment of the Disincentives in Recording Processes

The existence of records of a rule violation is an important disincentive unless a widespread conspiracy to ignore records of violations is in effect.

The methodology recognises three levels of "recording", as follows:

U1   UNRECORDED, 1 person : The weakest level is an effective lack of record, except in the memory of the perpetrator of the violations.

U2   UNRECORDED, 2 or more persons : An intermediate level is still un-recorded in writing but "recorded" in memories when more than one person is aware of the violation, even if that person concurs with the violation. This creates a situation in which the occurrence of the violation is more likely to be revealed than if only one person were involved.

WR   WRITTEN RECORD : The strongest level of recording is a written record, which is open to eventual scrutiny by a number of persons. However, the situation in which a written record is easily falsified by the violators of the rules must be asssessed as U1 or U2 according to whether one or more than one person is involved.

The assessment of these aspects of the violations is normally fairly straightforward once familiarity with the plant and procedures is obtained directly or from interviews. The methodology does not as yet offer a direct numerical probabilistic interpretation of the effectiveness of these levels of disincentive.

The assessment of recording, although usefully conducted separately, should be fed into the overall assessment of incentives. The measures for incorporating the various incentives and disincentives are not yet quantified, but an initial guide would be to reduce the otherwise assessed incentive level by one for WR recording, and to use U2 to mark incentive downwards in marginal cases. Where recording is seen as a particularly strong disincentive, then U2 could mark down by one point and WR could mark down by two points on the Ix scale.

Options for Ranking and Quantification of the Assessments

The procedure as described above will result in each rule violation being accorded two assessments on the basis of effects and incentives (including the disincentives in "recording" of the violation). There are options to extend the range of ratings on both effects and incentives.

The overall ratings may simply be used directly as "scores" , using the numbers included in the assessments. The numbers in the assessments can simply be added to arrive at an approximate ranking. Thus, for example, assessments of E2/I1 would score 3, while E3/I-2 would score 1, etc.

However, although it is reasonably clear that assessment E3/I2 (score 5) is higher than E2/I2 (score 4) and that assessment E2/I2 (score 4) is higher than E2/I1 (score 3), it cannot be deduced that E3/I1 (score 4) is equal to E2/I2 (score 4) or even that E2/I2 (score 4) is higher than E3/I0 (score 3). In order for the inequalities and equalities implied by the simple scores to be valid, the factors of effect and probability corresponding to one point on the scale must be about equal. Such an equality has not yet been established by any quantification or calibration scheme. Therefore, numerical "scoring" of the assessments must be used with caution and in recognition of the lack of a basis for strictly numerical ranking.

Until a clear basis for quantification is established, it may be sensible to leave assessments as pairs, which can be conveniently plotted on a graph such as that illustrated in Figure 1. In such a 2-dimensional scheme, the areas of the plot corresponding to "assessments of concern" can be delineated without any need for either equality or uniformity in the quantitative equivalents of assessments.

At this early stage of development, the main purpose of the method is to identify violations which give cause for concern and/or merit further analysis, rather than to provide any definitive quantification.

CONCLUSION

In recognition of the potentially serious effects of rule violations on plant safety, a methodology has been developed for the qualitative investigation of such violations. The method covers the identification of violations, their effects on safety, and qualitative assessment of the incentives and disincentives for such violations, including the degree to which the violations would be recorded.

In its current state of development, the method is intended to provide an approximate ranking of the importance of violations, but does not offer a numerical quantification of probabilities. Its use should be limited, at this stage of development, to qualitative investigations intended to identify violations worthy of further analysis or anticipatory preventive measures.

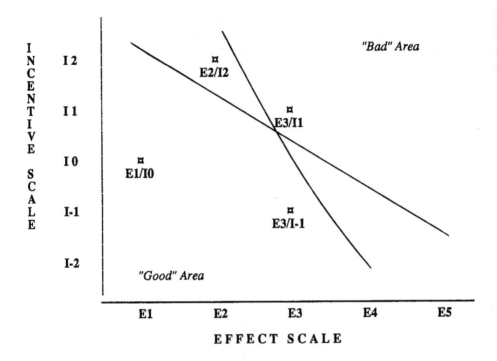

Notes: *The continuous lines illustrate typical lines which may be used to designate boundaries between "good" and "bad" i.e. between those potential violations which are of little
concern and those which are of sufficient concern to merit further assessment and/or action. The positions of the lines will depend on the particular situation.*

Figure 1. A 2-dimensional plotting scheme for "SURVIVE" assessments

EXAMPLES OF "SURVIVE" APPLICATION

Everyday situations - Road Traffic Rule Violations

Although the methodology is designed for application to violations in potnetially hazardous
plants such a nuclear and chemical plants, it can also be used in more familiar situations.
For example, it could be used (admittedly with the benefit of hindsight) to assess two violations which actually contribute a substantial proportion of the accidental mortality of the
UK population, viz:

<u>Drunken driving</u>   and   <u>Speeding</u>

The assessment of these violations would be as follows:

Drunken driving (divided into moderate/gross categories)

Effect : E1 or E2 - for moderately drunk drivers, the risk only rises by somewhat less than an order of magnitude, for seriously drunk drivers, the risk of accidents rises by more than an order of magnitude

Incentive : initially I2 - there is a low perceived risk (when drunk), and a common perception that the driver's skill is still an adequate "safety factor". The inconvenience of not driving when one wishes to do so is high.

Recording : U1 - there is still only a relatively small probability than any record of the violation will be made.

Thus the initial evaluation of I2 is not reduced.

Speeding (divided into moderate/gross categories)

Effect : E1 or E2 - normally, speeding has rather less than an order of magnitude effect on risk. Gross speeding may merit an E2 rating.

Incentive : initially I1 or I2 - in most cases there is a modest incentive (saving time), a very low perception of risk in modest speeding. In some cases, gross speeding is an end in itself, with a high incentive value (I2).

Recording : U1 for modest speeding - the proportion of offenders noticed or booked is very small. However, gross speeding is a clearly visible offence (at least when other road users are nearby) and the chance of being booked (WR) is much higher for gross speeding, so this may merit a reduction in the incentive assessment from I2 to I1.

Using the scoring and categorization introduced earlier, the rankings for the four violations considered above are:

	Drunken driving	Speeding
Moderate	E1/I2 (3)	E1/I1 (2)
Gross	E2/I2 (4)	E2/I1 (3)

These correspond reasonably well with the actual dangers associated with the violations, and with society's views of the seriousness of the threats involved. Modest speeding (2) is largely tolerated, excessive speeding and moderate drunken driving (both 3) are considered to be anti-social, and deserving of significant punishment, while gross drunken driving (4) is considered to be serious criminal behaviour and a notable threat to the health of others.

Plant Situations - Hazardous Substance Handling Rule Violation

An example is given here, which has been generalized from a particular plant application.

The situation is that sealed containers (glass bottles) of a hazardous volatile substance arrive at a storage depot. The containers themselves are not capable of withstanding severe impacts, and are thus transported in heavy outer containers which protect them.

The rules require the heavy container to be off-loaded from its lorry on arrival, and taken into the building before being opened. This is a tedious operation and it would be simpler to leave the heavy container on the lorry, take out the small container, and carry it by hand into the building. This violation is analysed as follows:

EFFECT   The violation would have the effect of exposing a relatively weak inner container to possible damage if handled clumsily. This damage could not reasonably arise when following the rules. The release of the contents to the environment would also no longer be subject to the mitigation of the building. In combination, the effect is rated as E3.

INCENTIVE The process of taking a heavy outer container into the building, unloading it, and taking it back out, is quite onerous compared with simply removing the inner container and carrying it in, so the incentive to do this could be quite high, particularly if there is time pressure. Given that there is no reason why the inner container should not be intact, the operator would perceive that the situation was still safe because of the effectiveness of the inner container (in any case, when following the rules, the operator will handle the container when it gets into the building, so handling it outside may not be perceived as any more dangerous to the operator). Thus I2 is initially assigned if time pressure is anticipated to be common, I1 if not.

RECORDING There would need to be several people co-operating in such a violation, but they may all have a similar incentive, and the action would not be recorded. This leads to a U2 assessment and a reduction in incentive to I1 or I0.

Thus, the rating is normally E3/I0, but would rise to E3/I1 if time pressure were anticipated to be a common feature of the operation. In the latter case, the rating would normally be regarded as sufficient to highlight this violation for further analysis. Figure 2 illustrates the positions of each of the violations discussed above on the 2-D plot.

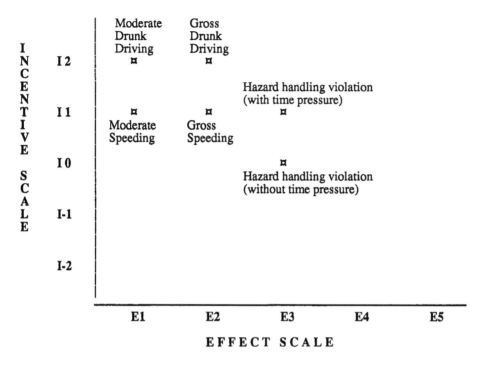

The plot above shows the positions of the assessments of various driving violations and a postulated violation of hazardous substance handling procedures.

Figure 2. "SURVIVE" plot for driving and dangerous substance rule violations

Wise men learn by others harms, fools by their own:
Organisational barriers to learning the lessons from major accidents

Deborah A. Lucas, Ph.D.

Human Reliability Associates Ltd.,

1 School House, Higher Lane, Dalton, Wigan, Lancs. WN8 7RP, UK

ABSTRACT

After virtually every major disaster of the last decade there has been a full scale investigation resulting in a lengthy report and a comprehensive list of recommendations. The philosophy of this investigative process, as it is for those incidents with less serious consequences, is that we must "learn from experience" to prevent future crises from occurring. However, there are barriers to such learning in organisations. Through a discussion of a number of major accidents this paper will outline some recent theoretical studies of organisational safety management. The value of these ideas in summarising the organisational barriers to learning the lessons from past experience will be shown. The accidents which will be referred to include Clapham Junction, King's Cross, and Hillsborough.

INTRODUCTION

Most organisations gather extensive amounts of information. A fairly large proportion of such information relates to safety, reliability and availability matters. Data collection systems covering safety related aspects of an organisation include: accident reports to employees and to non-employees, lost time injuries, significant events, near misses, and dangerous occurrences. Certain industries will also collect safety related data which is unique to their context e.g. air misses, signals passed at danger, reactor trips, etc. Each organisation will expend a substantial amount of effort and considerable resources investigating a subset of such events in detail (usually those with the most serious consequences) and in documenting the findings fully. Time and money will also be spent in collating and publishing accident and event statistics.

Why do organisations collect, analyse and document such cases? There appear to be two main driving forces: one extrinsic to the organisation and the other intrinsic and relating to its safety policy. The extrinsic driving force for data collection comes primarily through legal requirements including demands for recording from regulatory bodies. In the UK, for example, the Department of Transport, the Health and Safety Executive, and the Nuclear Installations Inspectorate are just some of the relevant authorities who may demand statistics from different industrial sectors. The intrinsic driving force for data collection emerges from the underlying philosophy of accident investigation, i.e. it is a vital information feedback loop from operational experience. From this perspective data collection systems have two broad uses (Reason, in press):-

- Learning the right lessons from past accidents in order to implement effective error reduction strategies

- Preventing the occurrence of future incidents through monitoring of unsafe acts, near misses and potential problems.

So the fundamental concept underlying the investigation and data analysis of safety related incidents is to find the cause (or causes), to derive effective remedies and to prevent future accidents. The rather less commonly found near miss and dangerous occurrence reporting schemes share this aim but with the emphasis more firmly on prevention. One problem which appears to be present in some organisations is a conflict of goals between the need to satisfy the legal and regulatory requirements for individual incidents and the latter aim of more widespread prevention of accidents with similar causes.

So does it work? Do organisations learn lessons from operational experience? We have little evidence of accidents which were prevented by the promulgation of insights derived from operational experience since such benefits of incident analysis receive little publicity. However we do know of cases which were essentially a repetition of an earlier and often non-serious incident. For example, after the sinking of the Herald of Free Enterprise, previous incidents came to light where other Ro-Ro ferries had left Zeebrugge with their bow doors open. Numerous examples of fires ("smoulderings") on escalators had occurred on the London Underground before the disastrous King's Cross fire. There had been a previous non fatal crush of football fans at the Hillsborough ground in 1981 at the Cup semi-final. In hindsight it was argued in each case that these earlier incidents should have helped to prevent the later disasters.

COMMON PROBLEMS WITH OPERATIONAL FEEDBACK

So why don't organisations always learn effectively from experience? One way of answering this question is to start by looking both at the nature of data collection schemes and at the organisational climate or culture in which they are enshrined.

Reviews of existing data collection systems (see, for example, Lucas, 1987, 1989) have highlighted five central problems:-

• Technical myopia. Most approaches are orientated towards hardware failures rather than human failures. This is despite the known predominance of human performance problems for which figures of between 20% to 80% are cited.

• Action orientated. There is often a strong tendency to focus on what happened rather than why the problem occurred. Data is often collated into classes based on the nature of the incident and/or the severity of the consequences. Statistical treatment of incident data only rarely considers causal categories and, as an added problem, there are differing opinions as to the level of causal explanation that is deemed appropriate. This level is determined by the underlying model of human error causation held by an organisation or its safety department.

• Event focussed. Systems are usually restricted (in practice if not in theory) to looking at individual accidents rather than looking for more general patterns of causes. Hence accident reporting systems are often anecdotal in nature.

• Consequence driven. Incidents with serious consequences are recorded and investigated, near misses and potential problems are not often perceived as necessary or worthy of analysis. Even if the advantages of near miss reporting are appreciated, adequate resources of time or personnel are not always available in existing safety departments.

● Variable in quality. Reports vary considerably in the quality of description of the facts and in any attempted investigation of underlying causes. Such variations occur both within an organisation and between data collection schemes in different companies. This implies problems with the in-company training of accident investigators and in the lack of systematic methods of incident analysis across all sectors of industry.

It is important to note that these features are not only characteristics of the data systems but that they also appear to relate to features of the organisation. For example, underlying causes may not be investigated because of a reactive management attitude to safety which has a remedial, accident driven approach rather than a preventative, problem solving orientation. Such an attitude may result from a "traditional safety culture" in which human error is perceived as being due to the negligence or "carelessness" of an individual. It follows that there will be no attempt to investigate "causes" of the accident other than who did what wrong. Once the "guilty" individual is identified, standard preventative methods (usually designed to increase the motivation of the individual to act safely) are wheeled in. A typical example would be the use of disciplinary actions such as a period of suspension without pay. In such cases, the traditional safety culture and its prevailing "model" of accident causation as being due to a lack of "safety-mindedness" will influence the search for causes and restrict the range of viable alternative remedies. Three of the problems of data collection systems stem from this type of safety culture and its associated model of error causation: action orientated, event focussed, and consequence driven. The safety cultures and the models of human error causation held in an organisation are therefore a key aspect in explaining some of the barriers to effective learning of lessons from the past.

ORGANISATIONAL SAFETY CULTURES

In this section we consider the differing models of safety ("safety cultures") that could be held by an organisation. This issue is addressed by considering the variations between organisations in how they address the issue of "safety" and the common problems that emerge with the different use of safety models. This is a slightly different perspective to recent papers on organisational theory and crisis management. Organisational theory aims to explain why certain organisations are more prone to disasters than others (e.g. Pauchant and Mitroff, 1988; Miller, 1988) or it attempts to model the influence of organisational and management decisions on the probability of a disaster (e.g. Reason, 1989).

Chamber's 20th century dictionary defines safe as "unharmed, free from danger, secure, sound, certain, sure, reliable, cautious". Safety is defined as the "state or fact of being safe". What this definition lacks is the context in which the concept of safety is applied. Firstly, the concept can be applied to individuals as in the case of personal injury. This context leads to an organisational culture of occupational health and safety which is perhaps best encapsulated by the employer's responsibilities laid down in the 1974 Health and Safety at Work act. This model or safety culture is referred to here as "occupational safety management".

Secondly, the concept of safety can be applied in a systems context. This is particularly the case for large scale socio-technical systems such as nuclear power and chemical processing plants which have witnessed such major incidents as Three Mile Island and Chernobyl. In this context the organisational model of safety is one of risk assessment hence we refer to this as the "risk management" culture.

Finally, the safety concept can be applied at the level of an organisation. This relatively recent model of "crisis management" is strengthened by issues such as corporate liability and environmental protection. Recent crises and disasters such as the benzene found in Perrier water, the charges on the management of the Herald of Free Enterprise ferry and the Exxon Valdese oil spillage illustrate the need for such a "crisis management" culture.

TABLE 1

COMPARISON OF THE PARAMETERS OF THE THREE TYPES OF ORGANISATIONAL SAFETY CULTURES

LABEL	Occupational safety management	Risk management	Crisis management
EMPHASIS	The individual Personal injury "Safety is the responsibility of all"	The system System safety "Safety is built into the system"	The organisation Disaster avoidance "Safety is the responsibility of line managers"
OWNED BY	Safety dept. Personnel dept.	Reliability depts. Engineering depts. (inc. human factors engineers)	Top management Directors (e.g. of Safety)
MAJOR ATTRIBUTES	"carelessness" "safety mindedness" accident proneness "of his own making" negligence unsafe acts & conditions	hazard identification quantification of risk man-machine mismatch violations of procedures operator error	costing of safety organisational pathogens "crisis proneness" management decisions latent failures system induced error
KEY CONCEPTS	motivation safe systems of work selection training	assessment/prediction control engineering safeguards redesign (retrofits) automation	preparation prevention planning environmental protection
CRITICAL EVENTS & EXTERNAL PRESSURES	Health & Safety at Work act Insurance premiums PR implications of accidents	Client led requirements Regulatory bodies Public opinion (e.g. TMI)	Management liability e.g. Zeebrugge Environmental disasters e.g. Exxon Valdese Product liability e.g. Perrier Challenger etc. etc.
TECHNIQUES	Safety campaigns Internal safety reviews Accident logging Training Selection	HAZOPS/HAZANS RAMS Fault/event trees PRA Databases Ergonomic guidelines International safety audits Human Reliability Assessment	Near miss reporting "Corrective action plans" Safety systems (5 star, Du Pont)
LINKS WITH	O&M & work study	Loss prevention	Total Quality Management

Thus, an organisation may adopt a model of safety which has as its focus the safe worker, the safe system, or now the safe organisation. Of course, the three cultures may be held simultaneously by different parts of the organisation each with their own concerns.

Table 1 summarises the essential differences between the three types of organisational safety cultures. The aim of the review is to highlight the distinctions between the models rather than to describe each in detail. The key variables between the three cultures of occupational safety, risk and crisis management are: the emphasis placed by each model on the individual, the system or the organisation; the ownership of the model; its major attributes and key concepts; the techniques or methods which are available; the critical events and external pressures which have influenced the development of the culture; and the links each model of safety may have with other management methods and theories. It is clear that there may be some overlapping of the available techniques between the three safety cultures, for example, hazard identification is a fundamental issue for all three types of management of safety. Similarly, the concern for good public relations is one kind of external pressure relevant for all three models. In all other respects however there emerges a quite distinct pattern of differences for the three organisational models of safety.

MODELS OF HUMAN ERROR CAUSATION AND ORGANISATIONAL SAFETY CULTURES

One other important difference which needs to be stressed is the fundamental model of human error causation which usually underpins each organisational safety cultures. Table 2 summarises the differences.

Table 2
Safety cultures and models of human error

Organisational safety culture	Predominant model of human error
Occupational safety management ----->	Traditional safety model
Risk management -------------------->	Man-machine mismatch
Crisis management ------------------>	System induced error concept

---

As has already been mentioned briefly, the occupational safety management model typically accepts a traditional safety philosophy of error causation where the motivation of a person to carry out the system of work safely is automatically questioned in the event of an accident. This view is closely related to the legal idea of negligence and it is therefore not surprising to find that accident investigations based on this model of human error are often concerned with apportioning blame between the worker and management.

The risk management model is usually based on a more sophisticated but passive view of man and human error as a man-machine mismatch. This view maintains that human errors result from a mismatch between the demands of the task, the physical and mental capabilities of the human, and the characteristics of the machine "interface" provided to do the task. This model of error concentrates on the individual and his/her immediate work situation (typically this will be the operator in a control room). Design changes and the provision of job aids (such as procedural support) are typical solutions which this view would produce. Another alternative is, of course, the automation of tasks although the inherent dangers of this approach are becoming increasingly recognised (e.g. Bainbridge, 1987).

The third model of error causation which appears to underlie the crisis management safety culture is the concept of system induced error. This is the notion that human failures are caused by certain preconditions in the work context. These preconditions can range from poor procedures, unclear allocation of responsibilities, lack of knowledge or training, low morale, poor equipment design, etc. One important aspect of system induced errors is that such preconditions can be traced to management decisions and organisational policies. This is why Reason (1979) has argued that organisational and management decision making failures are "organisational pathogens" and may contribute indirectly to accidents through latent failures.

The man-machine mismatch model and the system induced error model are not incompatible indeed the latter may be seen as merely an extension of the former. However, there is a fundamental difference between the traditional safety model and the other two error models. So much so that safety practitioners holding the "carelessness" model of error may dismiss the system induced model as providing an "excuse" for the individual who is, after all, "responsible for the accident".

Whilst there appears to be a general pattern of matching between organisational safety models and models of human error (as shown in table 2), alternative links may also exist. For example, the occupational safety accident investigator who considers an accident was caused by an inherently unsafe system of work is framing the incident within a system induced errors approach (albeit a simplistic form). The risk manager who attributes an incident to an operator violating a procedure is usually acting within a traditional safety model.

ORGANISATIONAL SAFETY CULTURES AND THE INDUSTRIAL CONTEXT

Every organisation can use all three safety cultures at any time. The factor that determines which models are appropriate in which context is the industrial situation. Low technology industries involving high risks of personal safety for workers (for example, the construction industry) need to place emphasis on occupational safety management whereas high technology industries (such as nuclear and chemical processing) rightly place their prime emphasis on risk management. Railways which carry passengers and have large numbers of rail staff combined with a moderately high technology need an equal mix of occupational safety and risk management. Figure 1 shows where some typical industries would lie in terms of two of the organisational models, occupational safety and risk management.

Crisis management is essentially a stage through which an organisation will pass rather than a permanent function. Crisis management is required following a major incident or when coping with new laws (such as the recent EEC product liability directive or the UK's COSHH regulations). Figure 2 shows how crisis management alters the movement of an organisation within its available "safety space". For reasons of clarity, changes to occupational safety management are not given on the figure although modifications to these may also be required. The arrows in figure 2 represent certain industries or organisations before and immediately after a "crisis". It should be noted that a change to crisis management may be due to a problem in-house or to an industry wide crisis.

FIGURE 1

Safety and risk management in different contexts

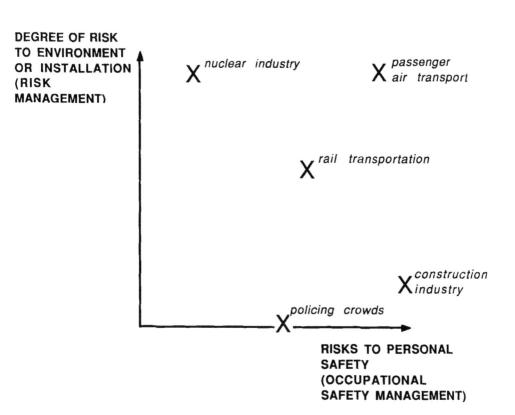

FIGURE 2: THE ORGANISATIONAL "SAFETY SPACE"

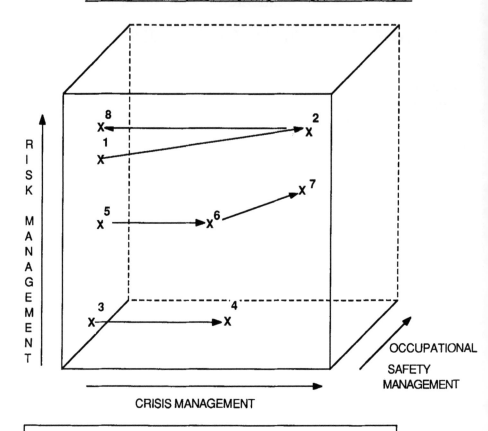

KEY

1. Nuclear industry pre-TMI
2. Nuclear industry post-TMI
3. LUL pre King's Cross
4. LUL post King's Cross
5. BR pre King's Cross
6. BR post King's Cross
7. BR post Clapham Junction
8. Nuclear industry pre Chernobyl

As an illustration of the latter, British Rail moved towards a crisis management model after the King's Cross fire on the London Underground. This move resulted in some positive changes such as the appointment of a Director of Safety.

After a period of time any crisis management state will lapse, primarily through pressure on its "owners" (senior management) to delegate the responsibilities to existing safety and risk departments. Hopefully this crisis management stage will leave a heightened awareness of risk or occupational safety management. This appears to have been true of the nuclear industry some time after the Three Mile Island incident in 1979 but immediately before the Chernobyl fire. The accident at Chernobyl should have triggered a further move to a crisis management stage in the Russian nuclear industry and in the nuclear industries of other countries too.

PROBLEMS

Two types of problems can occur in the response of an organisation to its fundamental safety issues and to its crises. These problems may be specific to a particular organisational safety culture or they may be generic to all such models. Here we look first at the specific problems and then at the generic issues. The list below is not intended to be exhaustive but rather to begin to identify key problems.

Problems with the occupational safety management culture

Two obvious problems with occupational safety management are:-

a.   Situations where there is no formal ownership of safety or where such responsibility is unclear. Perhaps the most recent example of this is the Hillsborough stadium disaster where the responsibility for crowd safety was not clearly allocated to either the police or to the football authorities (Taylor, 1989).

b.   An inappropriate emphasis on safety management where risk management is more appropriate. Consider, for example, the case of a small research laboratory where large quantities of an inflammable liquid are stored and pumped round the building. The major risks are clearly fire and explosion but the emphasis is actually placed on the safety of the individual. This results in a great deal of effort in encouraging the wearing of personal protection equipment. Fire hazard identification such as the risk of subcontractors smoking on site has received rather less attention. The reason for this emphasis is that safety is "owned" by a personnel officer who believes accidents are caused by "carelessness". This limited model of error causation colours his actions dramatically.

Risk management problems

Three problems are specific to the risk management safety culture.

a. Overconfidence in the basic concept that safety is built into the engineering system. The most recent example of this was British Rail's attitude to wrong side signal failures (where a signal fails and shows "green"). It became clear from the inquiry into the Clapham Junction Accident that there was a mindset that signals only ever failed safely to "red". This mindset influenced the organisation to such an extent that drivers who reported signals failing on the wrong side were not always believed (Hidden, 1989).

b. An inappropriate emphasis on the hardware aspects of the system. For example the US nuclear industry spends about 95% of the available funding for probabilistic risk assessments on hardware assessment and only 5% on human reliability assessment. This is despite knowing that over 50% of incidents are "caused" by human performance problems.

c. A lack of emphasis to important hazards. This is evident in the case of the open bow doors on Ro-Ro ferries and in the lack of attention paid to fires on the London Underground. Both of these hazards could quite easily lead to major disasters as indeed Zeebrugge and King's Cross demonstrated (see, for example, Fennell, 1988)

Crisis management problems
==========================

Three problems are identifiable for the crisis management of safety.

a. Failure to adopt a crisis management mode of thinking or "it couldn't
    happen here". One example of this quoted by Reason (1987) was the
    reaction of the UK's then Central Electricity Generating Board to the
    Chernobyl incident.

b. Failure to remain in the crisis management mode for a sufficient
    length of time to effect any significant changes. This failure is
    probably influenced by pressures on senior management to delegate
    such work to existing safety departments.

c. A reductionist approach where it is believed that the problems will
    be solved by applying one "off the shelf" solution. Typically such
    solutions may cover a safety auditing system (e.g. the 5 star system)
    or "total" approaches such as Total Quality Management.

Generic problems
================

In addition to the specific problems evident with the three organisational
safety cultures there are some generic issues common to all three models.
These are outlined below.

a. Complacency ("it's always been like that") and self-satisfaction
    ("we're as good as anyone else"). These attitudes will prove to be
    strong resisting forces to changes in safety management and data
    collection systems.

b. Holding a narrow goal which aims to demonstrate competence in safety
    matters (either to regulatory authorities or to the outside world)
    rather than aiming to solve safety related problems. This
    demonstration of competence can take the form of a stage model of
    accident investigation (Lucas, 1989) as shown in figure 3. In such a
    model any causal analysis is used merely to apportion blame and the
    learning process from the accident analysis is non-existent. Figure
    4 shows an alternative process model of accident investigation where

Figure 3.

Traditional "Stage" Model of Accident Investigation

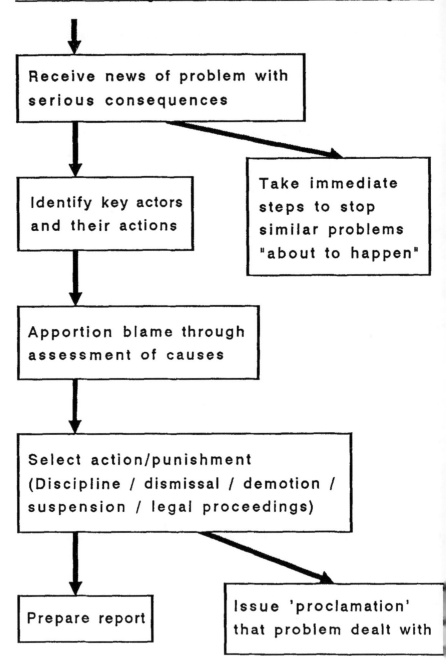

Figure 4.

"Process" model of accident and near—miss reporting.

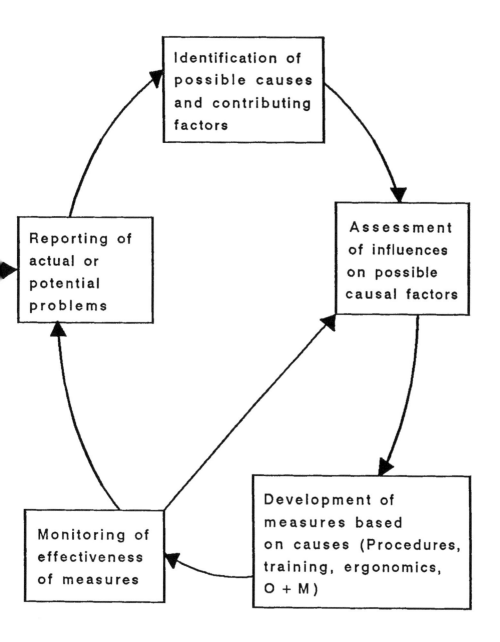

the emphasis is on monitoring of remedial actions and hence on learning from the unfortunate experience of the incident.

c. The rigidity of organisational thought. By this is meant that the same ideas, standard solutions and the same patterns of causes tend to keep recurring.

d. A narrow focus on individual incidents or accidents and the immediate causes of such events. Wider implications particularly organisational and management factors are not usually considered.

e. The failure to seek the underlying causes of incidents or even to ask the question "why"? This obviously relates to points b, c and d above.

The combination of these problems leads to the lack of appropriate feedback and feedforward mechanisms in organisations and hence to the failure to learn from accidents.

These general problems of organisational models of safety are echoed in a tripartite classification of organisations proposed by Westrum (1988). Westrum identified three types of organisations: the pathological, the calculative, and the generative. The pathological organisation is characterised by individuals who tend to deny or suppress information on hazards and may actively circumvent safety regulations. Calculative organisations use "by the book" methods but have few contingencies for unforeseen events or exceptions to the rules. They also tend to have a technical orientation, a rules and regulations culture, a traditional safety view of human error, and a primary emphasis on individual solutions to human failures. Generative organisations accept that a safety problem may be global in character and take appropriate action to reconsider and reform the operational system.

CONCLUSIONS

"To learn from experience one must:

(a) Conceptualise the particular incident so that it becomes part of a general rule rather than an isolated anecdote. A collection of anecdotes is no more than a basis for some stories unless they can be linked by concepts.

(b) Apply the scientific method to life. Hypothesise, test, hypothesise again."

(Handy, 1985, pp385)

The aim of this paper has been to set down some of the reasons why data collection systems and accident analysis are currently not effective tools for learning from experience. It follows that improvements need to be made at the level of the organisational safety culture, and in the underlying human error causation model as well as in aspects of the data collection system. The organisational safety model needs to be appropriate to the industry and responsive to disasters in that industry. They need to be process and not stage oriented (see figures 3 and 4) and to do this there is a need to move away from the anecdotal nature of many current reporting systems. Ideally, the underlying human error model should be that of system-induced errors. Some consciousness raising of the benefits of this approach is probably needed and certainly methods of analysis which are based on this model and which may be used by safety practitioners without extensive psychological training would be desirable.

One practical starting point is to evaluate the existing data collection schemes and to look for their developmental potential. Such audits need to consider two aspects: the technical content of the information recorded and the organisational climate in which the system is conceived and implemented. Five key areas are given below (see also Lucas, 1987):-

1. Nature of the information collected.

   This factor covers such aspects as whether the data is descriptive or causal, whether the system concentrates on reports of significant events, accidents and injuries or if it includes near misses, and whether the data is collected using a written report or through a standardised questionnaire.

2.   Help provided in gathering information.

This includes factors such as the nature of any job aids, training for the analyst, and the ease of use of any methods for determining causal analysis (e.g. algorithms, expert systems).

3.   Use of the information.

Items here include: the nature of feedback from the system, what functions this information serves, and whether the system is used to generate specific error reduction strategies.

4.   Nature of organisation of the scheme.

Factors such as whether the system is plant-based or centralised, whether it contains voluntary or mandatory reporting, and is computerised or paper based are covered under this heading.

5.   Acceptability to personnel.

This final category looks at whether personnel have a "pride of ownership" in the system, whether data is collected by known personnel, management actions on receiving advice of an incident, and the nature of any introductory training provided.

A review of these five critical areas should highlight problems needing attention and positive features of existing schemes which must be retained. The review should also provide insights into the dominant organisational safety culture and the underlying accepted human error causation model. An inappropriate safety culture and/or human error causation model may, as this paper has discussed, lead to organisational barriers to learning the lessons from accidents.

References

Bainbridge, L. (1984) Ironies of Automation. In J. Rasmussen, K. Duncan and J. Leplat (eds.) **New Technology and Human Error**. London: Wiley

Fennell, D. (1988) **Investigation into the King's Cross Underground Fire.** London: HMSO.

Handy, C. B. (1985) **Understanding Organisations**. Harmondsworth: Penguin.

Hidden, A. (1989) **Investigation into the Clapham Junction Railway Accident.** London: HMSO

Lucas, D.A. (1987) Human Performance Data Collection in Industrial Systems. **Human Reliability in Nuclear Power** IBC Technical Services.

Lucas, D.A. (1989) Collecting data on human performance: Going beyond the "what" to get at the "why". Proceedings of workshop on Human Factors Engineering: A Task Oriented Approach. ESTEC, Noordwijk, Netherlands.

Miller, D. (1988) Organisational Pathology and Industrial Crisis. **Industrial Crisis Quarterly, 2,** 65-74.

Pauchant, T.C. and Mitroff, I.T. (1988) Crisis prone versus crisis avoiding organisations: Is your company's culture its own worse enemy in creating crises? **Industrial Crisis Quarterly, 2,** 53-63.

Reason, J.T. (1987) The Chernobyl errors. **Bulletin of the British Psychological Society, 40,** 201-206.

Reason, J.T. (1989) The contribution of latent human failures to the breakdown of complex systems. Paper presented at the Royal Society Discussion Meeting on Human Factors in High-Risk Situations, London, 28-29 June.

Reason, J.T. (in press) **Human Error** Cambridge: CUP.

Taylor, P. (1989) **The Hillsborough Stadium Disaster** Interim report. London: HMSO.

Westrum, R. (1988) Organisational and inter-organisational thought. Paper given to the World Bank Workshop on Safety Control and Risk Management, Washington DC, 18-20 October.

ORGANISATIONAL, MANAGEMENT AND HUMAN FACTORS IN QUANTIFIED RISK ASSESSMENT: A THEORETICAL AND EMPIRICAL BASIS FOR MODIFICATION OF RISK ESTIMATES

N W Hurst   and   L J Bellamy*, T A W Geyer*
Health and Safety Executive   and J A Astley*
Broad Lane,   Four Elements Ltd
Sheffield S3 7HQ   25 Victoria Street
London   SW1H OEX

* The work described in this paper was carried out while these authors
were employed by Technica Ltd, Lynton House, 7/12 Tavistock Square
London   WC1H 9LT.

ABSTRACT

This paper describes work sponsored by the Health and Safety Executive which
is designed to give a better understanding of the effects of organisational
and managerial factors on the levels of risk at industrial and major hazard
plants. Failures of pipework and vessels are analysed and classified, using
descriptions of both the underlying causes of accidents and their immediate
causes. The implications of these classifications on the values of generic
failure rates are discussed with a view to modifying risk calculations in
the light of different standards at nominally identical plants.

INTRODUCTION

The Health and Safety Executive (HSE) makes use of Quantified Risk
Assessment (QRA) methods to give advice to local planning authorities
concerning the use of land around Major Hazard plants (1, 2). In particular
the Risk Assessment Tool, RISKAT, is used to calculate the individual risk
of receiving a specified criterion dose of toxic gas, thermal radiation or
overpressure at various distances from the plant.

The RISKAT code uses various consequence models to calculate these
doses. For example, a 'fireball' model calculates the thermal radiation
dose $(kWm^{-2})^{4/3}S$ from an LPG BLEVE. The risk of an individual receiving
such a dose is calculated by the use of so called generic failure rates,
which give the probability that specific events will occur in a given time.
Such events include catastrophic failures of vessels and pipes.

Generic failure rates are based on historical reports of failures of
plant items among known populations. Various sources of information are

used to provide a best estimate of, say, the frequency of guillotine failure of a six-inch pipe. This figure is of a general, or generic nature and is not specific to a particular plant or service type although some types of service, e.g. LPG, may be distinguished.

It is important to realise that the use of the individual risk of receiving a criterion dose (i.e. a risk criterion) has arisen from the need to make acceptable and defendable decisions about land use around major hazards. Consequence analysis is of limited use for this purpose because of the difficulty of dealing with events of low probability but high consequence. The 'consequence' approach introduces the concept of a 'worst credible accident' which is arbitrary by its nature. Thus the use of risk criteria provides a method of dealing with these decisions in a consistent way. HSE is committed to this approach and improving the methodology where possible.

MODIFICATION OF RISK

The values of risk calculated by RISKAT are clearly determined by the nature of the substances involved and their inventories, as well as details of plant equipment. The use of generic failure rates means that the risk figures calculated for a plant are 'averages' in the sense that for a given plant the risk figures relate to a whole range of failures which have occurred historically. These failures will cover a wide range of causes which will include hardware problems such as corrosion, overpressure and impact and operating errors such as opening the wrong valve. In this sense the risk figures include an 'average' amount of human error and any other causes of the reported failures.

It is possible to make the risk analysis more sophisticated than the approach outlined above. One approach is to carry out sensitivity testing of the original risk assessment assumptions and to model in more detail features of the assessment which prove to be important. In one study (3) it was found that the loading of tankers at a toxic gas producing plant was critical to the overall level of risk and this aspect of the study was enhanced by the use of fault tree analysis and estimates of operator reliability for specific operations. In this way the role of human error at the plant operating level was included in the risk assessment. Indeed, a range of methods are available to aid the analyst in this process (4). Similarly plant specific failure rates might be used, and more detailed fault tree analysis methods can be employed if time allows. These approaches then allow cost effective risk reduction strategies to be identified and implemented.

These approaches do not, however, address the problem of accounting for organisational and management influences on the levels of risk at a particular plant. Recent accident investigations such as those into the disasters at Kings Cross Underground Station, the sinking of the Herald of Free Enterprise and the chemical accident at Bhopal have shown conclusively that management and organisational factors as well as the industrial and regulatory climate will have a strong, if not a dominant influence on the risk from a major hazard site. This is now widely recognised by both regulators and industry.

Because of these considerations the concept of 'risk modification' has developed over the past few years (5, 6). Modification of risk methods start with a risk assessment based on generic values of failure frequencies, leak durations etc. and apply a factor, based on a numerical assessment of management and organisational issues to modify the initial risk figure. One approach would be to modify generic failure rate data to account for differences in the way hazardous installations are managed, maintained and operated. This assessment would probably need to be based on an audit method to access the management safety systems at an installation. Of course planning decisions have consequences that are measured over years if not decades and cannot be dependent upon changes in management except within narrow limits. Nevertheless, the extent to which management, organisational and human factors contribute to the risk is not made explicit.

Recently, the HSE has published a booklet (7) which begins to look at some of these important issues "Human Factors in Industrial Safety". It provides an examination of the roles of organisations, jobs and individuals in industrial safety and a practical guide to control. Furthermore, the HSE has started to investigate these issues by commissioning research (8). It is implicit that the research will look, in due course, at the validity of using audit schemes, which are designed to make a measure of the quality of management systems, for the purposes of modification of risk calculations. However, in the first instant it is important to understand how generic failure rates are made-up of components of different contributory factors. As a first step the factors involved in a limited data set have been investigated. References 8 and 9 report an analysis of the contribution made by various factors to pipework failures at major hazard sites. Recently (10) this has been extended to include vessel failures and this paper presents the first, if necessarily limited, account of that work.

THE SOCIOTECHNICAL SYSTEM

In order to make a connection between failures which occur at the plant level (to hardware items of equipment) and organisational and management factors which in a sense are remote from the actual incident requires a framework which describes both the immediate causes of the hardware failure and the root causes or origins of that failure.

The concept of the sociotechnical system (the entire human and technical system and its environment) has proved very useful in extending the analysis of failures from hardware 'forensic' investigations to more broad-based descriptions of systems failures. Such schemes as the systematic cause analysis technique (11) embody this thinking. Sociotechnical factors, such as economic pressures, have been particularly highlighted in large scale accidents and most discussions of accidents such as Challenger, Chernobyl, TMI and Flixborough make some kind of reference to these factors. Bignell and Fortune (12), for example, review a number of accidents which derive from sociotechnical systems failures. The results of our work (8,9,10) confirm that sociotechnical factors have a causal influence in many accidents, not just large scale disasters. Sociotechnical systems failures emphasise the idea that accidents arise from problems "deeper" in the system than the direct or immediate case would suggest. The International Loss Control Institute (ILCI) loss causation model is based on similar thinking (13) as is the MORT approach (14).

A HIERARCHICAL SCHEME OF ACCIDENT CAUSATION

The concept clearly illustrated by our work and a study of the relevant literature is one of an immediate cause acting as the carrier or symptom of underlying problems in the sociotechnical system (see for example Refs 12-16). This suggests that there exists a hierarchical scale of accident causation from the most immediate causes to increasingly remote causes. This does not imply a sequence of events in time but illustrates the

73

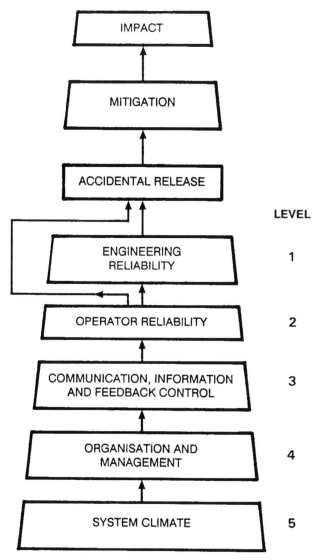

Figure 1. A hierarchical scheme of accident causation. The figure is used
to illustrate the potential effect of actions or inactions at different
levels within the sociotechnical system on the safety of a plant.

potential effects of actions or inactions at various levels within the sociotechnical system on the safety of a plant. This concept of different levels of causes is represented in Fig 1. Thus the sociotechnical "pyramid" (Fig 1) represents levels of increasingly remote cause from an accidental event. This is not remoteness in time, rather that the connection between an event and its cause becomes more remote as the number of intervening variables increases. When an operator incorrectly opens a valve and this causes a release, the opening of the valve and the release are directly connected. However, the event may have occurred because the operator was not provided with an appropriate procedure, or because there had been a failure in communication, or the operator was not adequately trained etc. These causes are more remote. Even more remote, for example, is where management may not have allocated sufficient resources to training, and this may have been due to inadequacies in prioritising brought about by severe production pressures. It is in this sense that there is a hierarchy of causes.

CLASSIFYING PIPEWORK AND VESSEL FAILURES

Although the sociotechnical pyramid is a very useful concept, it is not sufficient to enable the factors which contribute to generic failure rates to be expressed in a mathematical form. To do this requires a second classification scheme which we have developed in analysing over 900 reported accidents involving pipework failures (8,9). This is a 3-dimensional classification describing each incident in terms of a direct (or immediate) cause such as overpressure, a basic or underlying cause such as a design error and thirdly in terms of a preventive mechanism which failed to recover the error such as a HAZOP study not carried out. This classification scheme is shown in Fig 2. A similar scheme is applicable to failures of vessels (10).

The scheme consists of a number of layers of immediate cause. Each immediate cause is overlaid with a two-way matrix of underlying cause of failure and preventive mechanism. Essentially, this gives a 3D scheme whereby every incident is classified in 3 different ways, locating it at some point within the 3D space shown in Fig 2 (e.g. corrosion due to design error, not recovered by routine inspection). The scheme allows contribution counts to be made in a number of different ways. For example, the vertical column indicated in Fig 2 includes all immediate causes whose origin was domino effects with unknown preventive mechanism. These results have been described in detail elsewhere (8,9,10).

GENERIC FAILURE RATES

As explained in the introduction, generic failure rates are used within RISKAT to quantify the risk from Major Hazard plant. However, if nominally identical plants are managed, maintained and operated to significantly different standards but both are above the minimum required by the Health and Safety at Work Act then the use of generic failure rates will fail to reflect these differences. The aim of the programme of research which HSE has commissioned is to find systematic ways to describe this variation.

The numerical value of generic failure rates must reflect the immediate causes which have given rise to the actual incidents. In other words it is the immediate cause such as corrosion which 'causes' the failure to occur. However the immediate cause is the symptom of the sociotechnical system failure which may be the underlying cause of failure or the failure of a

75

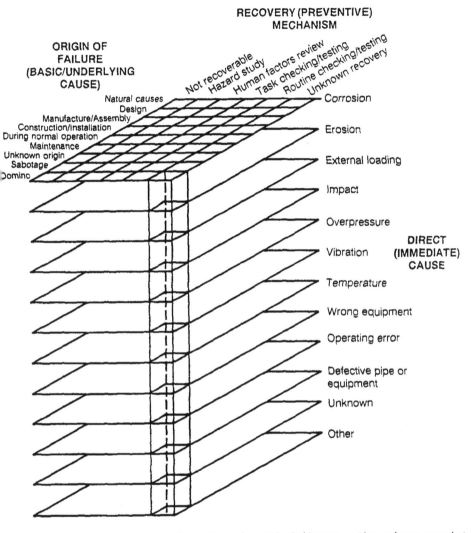

Figure 2. Classification scheme for pipework failures. The scheme consists
of a number of layers of immediate causes (e.g. corrosion). Each immediate
cause is overlaid with a two-way matrix of underlying cause of failure
and preventive mechanism.

potential preventive mechanism. Thus the immediate cause is not a fundamental way to categorise the various contributions to a generic failure rate for the purposes of comparing different plants.

Our work confirms that if improvements are made at a plant to, for example, maintenance procedures then the incidence of failures would be expected to reduce (17). Operators would make fewer mistakes because the procedures would be better designed. This suggests that the more fundamental way of categorising failure rates for our purposes is in the two dimensional way described by the underlying cause of failure and preventive mechanism matrix which is used here to overlay each immediate cause. Thus, for example, we would categorise failure rates in terms of the percentage contribution from design errors which were not recovered by HAZOP studies rather than by the percentage contribution from corrosion.

RESULTS

Taking all the immediate causes shown in Fig 2 together, provides the percentage contributions for pipework and vessel failures from the different underlying causes and failures of preventive mechanisms. This is shown in Table 1. and 2.

TABLE 1

Pipework failures: % Contribution of failures according to underlying cause of failure and failure of preventive mechanism from Ref 8.

UNDERLYING CAUSE OF FAILURE	PREVENTIVE MECHANISM						TOTAL
	NOT RECOVER-ABLE	HAZARD STUDY	HUMAN FACTORS REVIEW	TASK CHECKING	ROUTINE CHECKING	UNKNOWN RECOVERY	
NATURAL CAUSES	1.8			0.2			2.0
DESIGN		24.5	2.0		0.2		26.7
MANUFACTURE				2.4			2.4
CONSTRUCTION	0.1	0.2	1.9	7.5	0.2	0.4	10.3
OPERATIONS		0.1	11.0	1.6	0.2	0.8	13.7
MAINTENANCE		0.4	14.5	12.7	10.3	0.8	38.7
SABOTAGE	1.2						1.2
DOMINO	4.5	0.2			0.3		5.0
TOTAL	7.6	25.4	29.5	24.4	11.1	2.0	100

Thus Table 1, for pipework failures, shows that design errors account for 26.7% of the incidents and that about 92% of these ((24.5/26.7) x 100%) could potentially have been prevented by a hazard study (HAZOP). In Table 2 for vessel failures we see that 32% of the incidents occurred during normal operations and that about 77% of these ((24.5/32.0) x 100%) could potentially have been recovered by a human factors review of operational procedures. This clearly demonstrates the link between management control and actual failures at the plant level.

TABLE 2

Vessel Failures: % Contribution of failures according to underlying cause of failure and failure of preventive mechanism

UNDERLYING CAUSE OF FAILURE	PREVENTIVE MECHANISM						TOTAL
	NOT RECOVER-ABLE	HAZARD STUDY	HUMAN FACTORS REVIEW	TASK CHECKING	ROUTINE CHECKING	UNKNOWN RECOVERY	
NATURAL CAUSES	0.5						0.5
DESIGN		29.0			0.5		29.5
MANUFACTURE							
CONSTRUCTION		0.3		1.8			2.1
OPERATIONS		5.4	24.5	2.1			32.0
MAINTENANCE		2.1	5.7	3.6	10.8		22.2
SABOTAGE	1.0						1.0
DOMINO	11.9	0.3			0.5		12.7
TOTAL	13.4	37.1	30.2	7.5	11.8		100

EXPRESSING GENERIC FAILURE RATES AS COMPONENT PARTS

If the more fundamental way of expressing failure rates is in terms of the matrix of underlying causes of failure and failures of preventive mechanisms shown in Tables 1 and 2, for pipes and vessels respectively, then the numerical values in the tables represent percentage contributions to the generic failure rates. Thus from Table 1 we would expect that a generic failure rate for pipework failures would include a 12.7% contribution from maintenance errors which were not recovered by task checking.

If the entries in the tables are expressed, not as percentages, but as fractions of unity i.e. 24.5% becomes 0.245 then the sum of the entries will be one. In that case if a generic failure rate is x (yr^{-1}) then x can be expressed as the sum of the components in the table.

i.e.   Generic Failure Rate = $x \sum_{ij} N_{ij}$
                 (x)

where $\sum_{ij} N_{ij} = 1$ is the sum of all the table entries.

This analysis presupposes, of course, that the overall statistical distribution obtained for the failure matrix represents the same data that generic failure rates are based on. In this respect the value of $\sum_{ij} N_{ij} = 1$ is termed "average" for the purpose of assessment of a plant. A plant which we would call "average" should therefore show a breakdown of known underlying failure causes according to the matrix and have a "generic" failure rate for pipework failures.

If for a particular plant we would expect a 50% reduction in maintenance failures compared to average, this would reduce the maintenance contribution to pipework failures to 19.4%. In this instance $\sum_{ij} N_{ij}$ would equal 0.806, and the failure rate for pipework for the plant would be 81% of the generic value. Similarly the maintenance contribution to vessel failures would reduce to 11.1% and $\sum_{ij} N_{ij}$ would equal 0.889. The failure rate for vessels for the plant would be 89% of the generic value.

A risk assessment calculation could then be carried out using the modified generic failure rates to provide a modified risk value which would indeed reflect the different standards of maintenance of the two plants. It is the intention of HSE to consider if this approach can be linked to audit methods to provide a sound method to modify risk values.

REFERENCES

1. Hurst, N.W., Nussey, C. and Pape, R.P. Development and application of a Risk Assessment Tool (RISKAT) in the Health and Safety Executive. Chem. Eng. Res. Des. 67 1989.

2. HSE 1989. Risk criteria for land-use planning in the vicinity of major hazards. (HMSO).

3. Purdy, G., 1988. A practical application of quantified risk analysis. Human factors and decision making - their influence on safety and reliability. Ed. B. A. Sayers. Elsevier Applied Science.

4. SRD RTS 88/95/Q. Human reliability assessor's guide. Ed. P. Humphreys. UKAEA.

5. Bellamy L. J. and Geyer T. A. W (1988). Techniques for assessing the effectiveness of management. European Safety and Reliability Research Development Association (ESRRDA) Seminar on Human Factors. Bournemouth, March 1988.

6. Powell, J. and Canter, D. (1985). Quantifying the human contribution
    to losses in the chemical industry. Journal of Environmental
    Psychology, 5, 37-53.

7. Human Factors in Industrial Safety. HSE (HMSO) HS(G)48 1989.
    ISBN 011 885486 0.

8. Bellamy L. J., Geyer T. A. W and Astley J. A. Evaluation of the human
    contribution to pipework and in-line equipment failure frequencies.
    Contract Research Report No 89/15. Published by HSE. ISBN 071
    7603245.

9. Hurst, N. W., Bellamy, L. J., Geyer, T. A. W. and Astley, J. A. (1990).
    A classification scheme for pipework failures to include human and
    sociotechnical errors and their contribution to pipework failure
    frequencies (draft). Paper prepared for the Journal of Hazardous
    Materials.

10. HSE Contract Research Report. 1990. To be published.

11. The International Safety Rating System. Institute publishing 1988.
    ISBN 0.88061.002.7.

12. Beignell, V. and Fortune, J. (1984). Understanding system failures.
    Manchester University Press.

13. Bird F. E. and Germain G. L. Practical loss control leadership.
    Institute publishing 1988. ISBN 0.88061.054.9.

14. Johnson W. G. MORT The Management and Oversight Risk Tree. J. Safety
    Research $\underline{7}$, 4

15. Reason, J. (1988). Published 1989. Human factors in nuclear power
    operations. pp 238-242 in House of Lords Select Committee on Science
    and Technology (Subcommittee II), Research and development in nuclear
    power, Vol. 2 - Evidence. House of Lords Paper 14-II, London: HMSO,
    1989.

16. Bowonder, B. and Miyake, T. (1988). Managing hazardous facilities.
    Lessons from the Bhopal accident. Journal of Hazardous Materials,
    Vol. 19. No. 3, pp 237-269, November 1988.

17. Dangerous Maintenance. HSE 1987. HMSO ISBN 011 883957 8.

HUMAN FACTORS TECHNOLOGIES
FOR SAFETY AND RELIABILITY IN THE '90s

JEREMY C. WILLIAMS
Technica Ltd.
Highbank House, Exchange Street,
Stockport, Cheshire, SK3 0ET

ABSTRACT

It is believed that the new decade is likely to see a maturation of human factors techniques in many areas of safety and reliability engineering. The 'demographic time-bomb' and society's continued expectations for still greater safety and reliability are thought likely to stimulate demand for rapid corporate training in the appropriate human factors technologies, the development of rapid, cost-effective, assessment techniques and the distillation of a currently complex data-base into one from which easily assimilated messages can be extracted. Five themes seem probable in the field of human reliability assessment; cross-validation of error mechanism predictions, progress in human dependency modelling, collation of human reliability databases, the development of 'super methods', and more open exchange of information concerning human failure. It is thought likely that the greatest improvements in human factors technology will develop from a more astute use of industrial resources rather than major innovation in this area.

INTRODUCTION

Over the last three decades the formal safety and reliability assessment industry has grown from being a small group of visionaries to a sizeable group of professionals with a highly developed, specialist technology and culture. The human factors profession which dates back more than forty years has progressed in similar fashion, focusing its attention and technology on product and systems development as appropriate to the needs of industrial investment in manufactured goods. As a result the human factors engineering community has tended to develop techniques and technologies which find ready application to small and medium size systems, but which do not always translate easily to the large-scale system operation and design issues, that have tended to become the concern of safety and reliability engineers.

tendency has been to regard large problems as being an amalgam of smaller problems, which are then seen to be amenable to solution via the battery of human factors engineering, assessment and analysis techniques, which have been developed over a long period.

The assumption that large scale problems can be dealt with by analysis and breakdown into smaller more tractable problems is not a view confined to the human factors profession. The concept is endemic in most large scale systems design and development work, and is used to considerable effect to ensure the success of these systems. The principle has also been used to make progress in research into the understanding of fundamental phenomena, and can make the difference between successful and unsuccessful implementations of new products and systems.

The scientific reductionist approach described above, however, although necessary, is not sufficient. The application of such an approach, it will be argued, can waste resources and divert attention away from the larger strategic concerns of organisations, which are to maximise overall performance and achieve the highest possible integration, synergy, and profit.

Whilst considering these issues the author has been struck by the fact that there are a large number of system and product design/development failures, which co-exist with the reported and visible successes. From a safety and reliability perspective, when one looks hard at the evidence provided by system and product development programmes, it is clear that the actual (but currently hidden) cost of failure may be considerable. (The author estimates this cost as being of the order of at least 40 million pounds per annum in the U.K.). It is rare, for example, for a major system to be put into full-scale production at anything less than an Issue 4 level of design documentation, and this characterises only those systems which have actually passed from concept, through feasibility, to a pre-production status (generally less than 40% of those started). Although the high mortality rate associated with new systems design could be interpreted as being a good error-trapping and removal device, it is usually the case that project cancellation occurred, not because inherent design, production or operational weaknesses were identified early on in the project's life, but because investment decision-making criteria were changed.

In meeting the criteria for progression through various stages of project documentation it is assumed that the design, commissioning, operational, maintenance and testing requirements have been fully satisfied at each stage. Although the systematic design approach argues that this should be the case, it is sometimes difficult to find persons within development teams, who will agree, under oath, that a project is achieving the intended level of quality, even though they may have endorsed the documentation. Apparent adherence to quality management procedures, therefore, for some, may have become a substitute for the actual achievement of project quality, which ironically was their initial purpose.

Examples of this type of quality management failure may be seen in the inappropriate processing of information associated with the performance of the Challenger space shuttle 'O' rings, the reassessment of the resistance of highways to earthquake damage in the San Francisco Bay Area, and most recently, in the apparently incorrect grinding of the Hubble Space telescope's mirror.

Despite advances in the design of complex systems, products and processes, it is very rare that any of them will work, as intended, during the commissioning and/or optimisation phases. Often they will fail (sometimes catastrophically) within a depressingly short period of initial use and they will be subject to frequent refurbishment, recall or up-grade. Such potential for failure has become an expected, and formalised, part of system deployment.

The engineering community has even gone to the trouble of giving many of these failed mode phenomena names, such as "de-bugging", "shake-down", "teething troubles", "burn-in", "system development", "prototyping" and so on.

It is obvious that if some reduction could be made in the amount of time, effort and resources that are consumed in rectifying these operational problems, the savings that might flow, should be enormous. Industry recognises that these problems exist and takes upon itself the need to institute "zero defect", and quality assurance programmes. It sets up "competence centres", "centres of excellence", "quality circles", "Tiger Teams" and so on. Thus it is clear that many of the problems are recognised and are under sustained attack from a variety of angles.

There are two issues, however, that many of these remedial measures do not seem able to identify, and tackle.

There is abundant evidence all around us that we are fallible, and prone to human error, whether we be managing director, design engineer, marketing executive, maintenance technician, foreman, installation engineer, or safety and reliability engineer. Secondly the methods used to control, minimise or reduce our potential for error do not always seem well-matched to the types of error that we might make, nor are they necessarily particularly well-suited to the severity of the consequences of failure. For example, quality assurance procedures, although often appearing quite comprehensive, rarely give detailed insights into the types of error that can occur, their severity, or the best ways of detecting them.

As safety and reliability engineers, whilst we are supposed to accept certain types of failure, as being inevitable, or certain likelihoods of failure as being "reasonable", it seems strange that the major improvements in design, development, testing, manufacturing, installation, commissioning, operation and maintenance that our models would perhaps suggest we should seek, are often regarded as being beyond the scope of our task. Even "total quality" and "corrective action" programmes somehow seem to miss the really important basic issues, such as:-

o   "has the systems analysis been conducted properly?"

o   "has the system been specified correctly?"

o   "has the system really been designed properly?"

o   "have appropriate materials been specified?"

o   "has the system been constructed properly?"

o   "is it going to be installed correctly?"

o   "will the system be commissioned as intended?"

o   "will the system be operated correctly, within the anticipated limits?"

o   "is the system really going to be maintained as required?"

.....etc.

Quality Assurance procedures attempt to specify the detailed requirements associated with answering such all-embracing questions, and clearly there is nothing unusual or unreasonable about them. Although the scope of such questions is obviously quite general, it is most noticeable that virtually every single system failure or accident that will be observed will trace, upon close examination to seemingly quite trivial human failures in one or more of these system design/development/operation areas. Once such failures have occurred the question that is most commonly asked is, "how could this have happened, bearing in mind the rigorous nature of our procedures?".

Apart from the obvious conclusions and exhortations to improve, what does this indicate? First, it is apparent that, despite the existence of "Zero Defect" and "Total Quality" programmes the desired levels of reliability in design, manufacture, installation, commissioning, operation and maintenance are still not achieved to a high, consistent standard (for example, it is said that some 80% of Total Quality initiatives will fail). Secondly, it seems that the error-trapping and error-reduction mechanisms currently in place may not be complete, nor may they be wholly effective.

The reason for this preamble, therefore, is to question whether we really do achieve our safety and reliability expectations with respect to human error in general, and to suggest that there is still considerable scope for improvement beyond the immediate confines of the assessment process, even though some would perhaps regard the remedies as being outside the remit of the S&R engineer.

AWARENESS AND INTEGRATION

If the assertions and indications given in the introduction prove to be valid, it is suggested that the S&R community is likely to become much more aware over the next decade that it is impossible to design, operate or assess a plant successfully without considering the full implications of human factors issues in relation to these processes. This awareness is expected to be sharpened by the continuing (and regrettable) observation that major accidents are beginning to seem less amenable than hitherto to straightforward technical prediction and solution, and seem to possess fewer of the "usual" ingredients that engineers can tackle.

The argument is that the '90s will see the occurrence of the Human Factors Accident. These accidents, although having a technical basis, will either be "Beyond the Design Basis", or beyond the ability of the S&R engineer to have predicted, based upon current technical understanding. And yet the most important ingredient, from the public perspective, will be the relative "inevitability" of such incidents, and the concern that the S&R community is currently unable to offer any real assurances regarding their ability to model such accident sequences, assess them accurately, or indicate plausible steps leading to accident potential reduction, mitigation or the modification of probable consequences.

Once it has been demonstrated that existing S&R methods and models will not predict the "real" likelihood of human error in complex systems design, commissioning, maintenance and operation, the question will be, "what does it take, and when will it be possible, to make plausible predictions regarding the likelihood of all forms of human error?"

As the '90s might not see the full development of a set of techniques that could answer this question, we have to think about what a partial answer might be.

In addition we have to consider the fact that there is a "demographic time-bomb" waiting to go off, in the sense that, at the very time when we most need a large number of expert individuals to try to answer this question, we shall find that it is difficult to attract even the few that we needed to keep the basic assessment and analytical processes going, let alone innovate.

The author argues that this situation is not one to be avoided, but one to be welcomed as a challenge to our ingenuity, not only to solve for the absence of key personnel, but also to develop the technologies that the public will expect to see in place by the turn of the century.

From the public's point-of-view the integration of Human Factors Technologies within the framework of the S&R community's remit will, by then, seem self-evident. They will seek assurance that the S&R community has paid full and proper attention to the existence of Human Factors issues within safety and reliability assessments and analyses, and that it has developed sufficient technology that it can argue from a position of being the "Informed Client", rather than a "Technological Dinosaur".

In this climate we will have to have thought through the implications of how we are going to integrate human factors in our S&R work. This is not as difficult as it would appear.

Part of the awareness generating process will create the conditions for integration. It seems likely to the author, at least, that the '90s will require large enterprises to commit to the integration of human factors issues in their S&R work. This will either be brought about by regulatory pressure or the expectations of society. A logical consequence of the requirement to integrate will be the desire on the part of the organisations affected to acquire and spread human factors technology rapidly and effectively within their work-force. Thus the processes of awareness and integration will have an interaction and synergy.

Three of the features of corporate training programmes that are likely to be sought are, the creation of effective frameworks for including human factors considerations, the development of rapid, effective assessment techniques, and the distillation of a currently complex data-base into one from which easily-assimilated messages can be extracted.

It is almost certain that effective rules will be developed which will indicate which areas of systems design and development S&R work will benefit most from the application of human engineering principles and practices. From a global perspective it is already apparent that organisational management and communications processes will benefit most from attention to error recovery techniques, whereas at the design end of systems development, error trapping and error reduction techniques are likely to offer the most cost-effective use of resources.

The reader may wish to contrast the view expressed above with the practical implications of many Quality Assurance processes, which tend to focus attention on error recovery at design review stages, without significant amounts of error trapping and reduction during the design process. Naturally, it would be preferable if all error potential could be minimised at all stages of a project's life cycle. Unfortunately the structure of most current major projects is not always conducive to error trapping and reduction at a macro level because basic structural decisions tend to be taken in the context of a political/economic climate that often has more to do with resource utilisation than immediate correctness of judgement, and therefore it seems likely that this particular application of human factors technology will have to await the full awareness development programme that has been the concern of this section.

HUMAN FACTORS RESEARCH AND DEVELOPMENT

The HF technologies relevant to S&R engineers that are likely to be developed in the '90s will, of course, depend very much on the R&D that is already underway or projected in the next few years. Much of this work is focused on the organisational, management, communications, stress, task analysis and cognitive error issues. Although there is, as always, considerable R&D effort being applied elsewhere (such as man-machine interactions in virtual environments), the applicability of this work as a basis for S&R HF technology is not immediately apparent. Therefore we need to consider where the current thrust is likely to result in technological spin-off.

The first prediction is that it is expected that the principal causes of management failure as they apply to S&R assessments will be identified, characterised, quantified (to a first approximation), and result in accurate audit tools that will find ready application in a variety of business areas.

A second prediction is that the factors influencing communication success and failure and the ways in which they interact will become well-enough understood that it will be possible to create models that will give insights into the potential for communication failure within a system. These models will find ready application to highly-defended/redundant/diverse systems, in particular with regard to the seeking and use of uncorroborated information, and will be relevant to dependency modelling. The reliability of communication will also be assessed in broad terms so that it will become possible for S&R engineers to make plausible predictions about the ability of designers, managers, operators and maintainers to furnish each other with satisfactory accounts of their respective concerns.

Another prediction is that this endeavour will produce sufficient insight into the processes of high level information handling that it will be possible to connect this research to fault diagnostic research, so that the total mission from symptom detection through to a chosen course of action will start to become traceable, and ultimately, predictable. It is possible to speculate also that such a convergence of research thrusts might result in the means to validate some decision support concepts, and even spawn some new, more economical, ways to undertake man-machine interaction testing and evaluation.

On the cognitive and conceptual error side it is already apparent that the means to identify such error potential, and guard against its most deleterious effects, is almost within the grasp of the HF community, certainly at a working level. The theoretical work that would be needed to connect such undertakings into the communications and fault diagnostic frameworks mentioned above does not seem to have commenced yet, but it is possible that papers of this sort will stimulate the necessary expenditure of effort, so that the connections will be perceived and acted upon, perhaps before the decade is out. It is certainly very frustrating that cognitive error and conceptual error of the sort which could be caused by conflicting goals would appear to dominate most fault and event trees that we would care to develop. Yet, at this stage, we do not seem to have any very well-developed means for modelling, let alone quantifying, the likelihood of such errors, even though they quite obviously can, and usually will, dominate other errors, both in terms of influence and frequency. (The author expects this despondent note to stir the HF community into the necessary activity, as it would not be difficult to undertake; just time-consuming).

One other candidate area for major HF Technology growth related to S&R concerns appears to be in the field of Task Analysis. Not only will it be possible to be quite explicit about the nature and application of task analytical techniques, but it will be possible to perform task analyses that can be used without further modification as inputs to human reliability assessments, human engineering verification and discrepancy reviews, procedure structure development tools, and training curricula and training aids development mechanisms. The most exciting prospect in the task analytical area is likely to be the direct, and unambiguous linking of Hazard and Operability Studies to Task Analysis Techniques. It is clear that the opportunities presented by HAZOP studies are ideally suited to the simultaneous gathering of relevant task data, which could find use (as indicated) in a number of areas. The process of HAZOP data gathering is similar to those associated with Task Analysis and the information is of considerable value to HF engineers. It raises similar issues to those raised by the task analytical process, and it needs to be acted upon in a timely fashion. In the author's opinion, there is no question but that the necessary linkages between these data gathering techniques should be established at the earliest possible opportunity.

HUMAN RELIABILITY ASSESSMENT

In the latter part of the '80s considerable progress has been made in the modelling of human error mechanisms, and this impetus is likely to be sustained into the early '90s. The consequence of this work is that attempts will be made to determine whether the predictions of various error mechanism generators are plausible, and consistent. Some of this work has already commenced, and much more work may be anticipated in the '90s

This cross-validation, it is anticipated, will lead to the development of a whole new set of technologies for the use of S&R engineers, especially in the area of fault diagnosis and maintenance error prediction. The degree of concordance reached is expected to be relatively high, and therefore, when completed, this development process should have high value to both the HF and S&R communities. Also, building on this outcome and the success thought likely in the area of communication research, it is fairly certain that substantial progress will be made towards much more accurate models of within and between human dependency.

The outstanding problems of collection and collation of human reliability data are likely to receive some attention during the '90s, but (in the author's view) by no means as much as the subject merits. The problem has been aired on many previous occasions, promises have been made, and data collation systems developed. In almost every case the development thrusts of previous decades have not resulted in any major progress. The author believes these failures cannot be tolerated in the '90s, and the necessary actions to collect and collate only those data that will have immediate application to the field of human reliability assessment will have to be taken. Researchers in the area will have to formulate practical data-base development strategies, confine their attention to the real industrial problems that their sponsors are trying to solve, and only use data that have a sound pedigree.

The outlook for progress in this area is dependent upon the seriousness with which the S&R and HF communities view the problem, and the extent to which they are prepared to recognise the limitations. Major technological innovation in the area of human reliability data collection is thought somewhat unlikely. However, the most profound change will be the recognition that progress is going to be slow, and hard-won.

The major HF technology development in this area will therefore be a step change improvement in the quality of information supplied to the S&R community, rather than a change in scope.

It is anticipated that the direction of human reliability assessment modelling will change in important ways, but that this will not inhibit the further computerisation of existing methods, and their continued refinement. The important changes that are likely to occur will involve a total re-appraisal of the purpose of human reliability assessment, its component parts, and a consideration of the extent to which it will be possible to draw upon the validated parts of each method in order to create "super methods". The re-appraisal of the purpose of human reliability assessment methods is thought likely to result in a shift from the current reasons for undertaking such work, such as those associated with justifying safety arguments and finding areas of potential weakness and improvement, through to the introduction of focused human engineering techniques to ensure that assessment assumptions and principles remain valid, and facilitate justifiable changes to design, operational, procedures and training principles.

These processes will be subject to many iterations, but it is anticipated that by the turn of the century, most of the principal methodological feature extractions will have been accomplished, and the bases laid for the creation of much higher validity methods. In anticipation of this prediction it is also likely that there will be many false starts, and movements towards highly refined methods which will not always enjoy the status of being recognised methods.

To facilitate the types of progress described above it is clear that there will have to be a much more open exchange of information concerning human failure in process operation. As this information improvement is already underway, it seems plausible to assume that it will continue for some time yet. A crucial feature of this predicted more open exchange will be the better interpretation of the significance of such failures with respect to management, safety assessment and modelling.

RECOMMENDATIONS

Because of its speculative nature, this paper has, of necessity, tended to focus on factors, concepts and trends that are difficult to substantiate in a formal scientific sense. In addition it should not be interpreted as distracting attention away from other approaches to achieving safety, such as intrinsically-safe design or passive protection. However, there are some messages that appear to be worthy of statement as stand-alone recommendations. These are:-

o   The HF community should provide assistance to ensure that Quality Management approaches provide better error trapping, reduction and recovery methods within complex systems development.

o   S&R engineers should be alert to the possibility that there are ways in which they can improve safety and reliability, over and above those currently judged to be within their job specification.

o   S&R technologists should be urging the development of better, more insightful human reliability assessment modelling, possibly by reference to, and amalgamation of, existing methods, with a view to creating 'super methods'.

o   Particular effort should be applied by the HF community in concert with S&R engineers to facilitate the development of means for modelling and quantifying dependent, cognitive and conceptual error.

o   Means should be explored to link the opportunities presented by HAZOP studies with the needs of task analytical studies so that common, more penetrating methods of data collection can be created which will satisfy the needs of both risk analysts and human factors engineers, to minimise resource requirements for both groups.

o   S&R engineers should press the HF community to produce practical human reliability data-bases that have a sound pedigree and address real industrial issues.

SUMMARY

It has been argued that there are some areas of S&R technology which would benefit greatly from developments in Human Factors Technology, and it has been suggested that these areas of development are not confined to those more normally associated with S&R technology. In particular it has been suggested that there is scope for very considerable improvement in design, commissioning, management and maintenance. The human error defences available for application in these areas are, it is believed, under-utilised, despite the existence of current industrial preventive measures.

Furthermore it is argued that not only are Human Factors Technologies required to deal with these outstanding issues, but that they must be developed at a time when the pressures to adhere to well-established S&R routines will be considerable. Whilst it will be necessary to keep modestly-defended simple man-machine systems under continuous review, it is suggested that the '90s represent a period when highly-defended complex systems will be seen to fail principally because Human Factors failures can invalidate reliability assumptions and predictions.

In order to make progress the S&R community will have to become more discriminating in their use of HF techniques, and this prospect is seen as a challenge to the HF community to create more economic, more-specific and valid methods. From a range of technologies currently under development, and projected, a pattern can be seen to emerge, indicating that there is unlikely to be major innovation, but rather a synergistic use of existing knowledge that will make the Human Factors Technologies for the '90s more applicable to the S&R community than hitherto.

A METHODOLOGY FOR APPLYING TRADITIONAL SAFETY ANALYSIS TECHNIQUES TO SYSTEMS HANDLING CROWD FLOWS

L C Dunbar, M S Carey, S P Whalley
R M Consultants Limited, Genesis Centre, Garrett Field,
Birchwood Science Park, Warrington, Cheshire, WA3 7BH

ABSTRACT

This paper shows how traditional safety analysis methods can be applied to study the safety of crowd flow in a confined space. Particular attention is devoted to the problem of discriminating between aspects best studied by qualitative and quantitative techniques and to creating models that are representative of the complexity of the system without being too complex for practical use. The process of analysis moves through a series of connected phases. It includes the systematic application of hazard analysis techniques, the creation of models for quantification and the production of a prioritised list of actions in order to increase system safety. Examples of the application of the methodology are provided throughout the paper.

INTRODUCTION

The ultimate aim of safety analysis is to produce a better understanding of the potential safety problems for a given system and to suggest actions which may improve system safety.

To achieve this aim, the process of safety analysis often calls for the use of different techniques, for example, hazard identification, the quantification of the frequency of specific events and the assessment of consequences. Both qualitative and quantitative assessment techniques may be employed.

This composite approach is recommended when the different techniques available are complementary and their parallel application allows the appreciation of the many aspects connected with system safety [1]. This approach can be successfully applied to systems which do not belong to the traditional engineering field. Consider, for example, the analysis of safety aspects associated with crowd flow in a confined space; this could apply to shopping facilities, sport and leisure centres, transport systems or cinema and concert halls. In systems such as these, the flow of people through halls and passageways resembles in some ways the flow of a liquid through vessels and pipes in an engineering system.

Table 1 summarises some of the main differences between engineering systems handling inanimate media and systems handling crowd flows. The characteristics summarised in column 2 of Table 1 show that when considering crowd flows :

a)   Flow dynamics are likely to be very complex because of the possibility of cross and inverse flows, the higher dependency on external factors, which may determine periodical changes of flow regime, and the dependency on psychological as well as physical factors.

b)   The definition of 'near-to-danger' level is difficult because of the complexity of the dynamics, the necessity to rely on judgement for control and the nature of the hazards.

c)   Deviations from design purposes are more difficult to define due to a lack of a unique definition of the purpose of system operation. Several aims may co-exist (eg. the customer uses a transport system to go from A to B whilst the operator expects the operation to produce a profit).

d)   Control of system configurations and recovery from potentially dangerous situations rely on person to person communication. Misunderstandings can arise, depending on the ability of individuals and their expectations. Actual operating conditions may turn out to be different from those intended at the design stage, due to unpredictable factors (eg. changes in government policies or demographic factors could affect both safety standards and the pattern of 'customer' usage).

TABLE 1 Summary of the principal differences between a typical
engineering system and a system with crowd flow

Typical Engineering System	Crowd Flow System
Hazards are mainly to items or people external to the system (e.g. from explosion, fire, etc.).	Hazards are to the flow media, i.e. the crowd itself (e.g. discomfort due to congestion).
The direction of flow is determined by physical constraints (e.g. pressure differentials) and is predictable once these are known.	Cross flows are always a possibility. Flow direction may depend on conditions internal to the crowd which may escape prediction (e.g. panic created by a perceived danger).
Fluid Dynamics are based on pre-determined physical laws and can be re-adjusted by acting on well known physical parameters (e.g. temperature and pressure).	Reaction to abnormal conditions are less predictable as both physical and psychological factors come into play. Granular rather than fluid flow.

The success of control methods is less predictable since it depends on complex interactions of these factors. |
The 'product' of system operation is usually a change in the media (e.g. chemical change) and is well defined at design stage.	The 'product' of system operation may vary according to each individual customer. The behaviour of the crowd can be affected by the expectation of individuals. This may contrast in some cases and may vary with time.
Media are affected mainly by local conditions, i.e. local pressure or temperature.	Crowds may be affected significantly by conditions external to the system. The influence of external factors may last a long time and vary from individual to individual.
The approach to danger point can be recognised by monitoring the trend of selected parameters which can be measured quantitatively.	Conditions of impending danger are difficult to recognise. Their detection relies on judgement rather than observation of measurable parameters. Judgement is also required to assess the speed with which conditions may escalate.
Operators rely on equipment to affect system parameters and conditions. Controls can be uniquely defined (e.g. stop pushbuttons) and descriptions of control tasks and interfaces are determined at the design stage.	Interaction of staff and customers is on a person to person basis and relies on verbal and visual communication. Requirements may vary across the life of the system.

METHODOLOGY

The Need for a New Methodology

Due to the differences described between traditional engineering systems and crowd flow systems, the need exists for a new methodology which may:

1   Provide guidance on the specific technique(s) to apply to each aspect of the system.

2   Suggest a systematic approach capable of dealing with system complexity and with the dual physical and psychological nature of causes and consequences. The approach should ensure that all the most important aspects are covered without unnecessary overlaps.

3   Suggest a format for the presentation of results (qualitative and quantitative). Cross referencing should be possible and the emphasis should be on suggesting practical ways to improve safety management of the system.

Description of the Proposed Methodology

Figure 1 shows the suggested sequence of steps which make up the methodology. This sequence can be grouped into four main phases:

    i) Information Collection
   ii) Hazard Identification and Modelling
  iii) Quantification
   iv) Tabulation and Action Checklists.

Phase I - Information Collection. This consists of a physical survey and of interviews with the staff which together provide the basis for the construction of a preliminary description of the system on which to base the subsequent hazard analysis.

The survey is an essential first step in the analysis of the system, since the relevance of certain details may be difficult to appreciate on the sole basis of system diagrams. For example, if considering the features of a public circulation area, the physical survey would check (among other features):

93

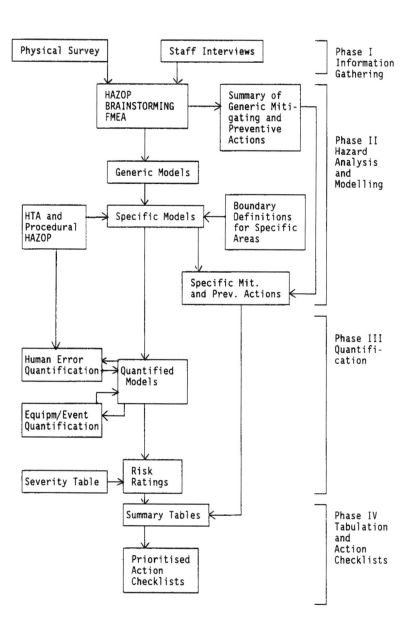

FIGURE 1 Schematic diagram of the proposed methodology

1   The quality of the floor surface (eg. slippery when dry/wet etc..).

2   Location of rubbish bins and vending machines if any.

3   The general standard of housekeeping (how speedily are spillages cleaned up, light tubes replaced, etc?)

4   Presence of protruding objects, temporary barriers etc...

5   Quality of lighting and signing.

It is also important to obtain any plans or instructions used by or security personnel staff as a reference for crowd control purposes. This information may be supplemented by staff interviews, used to gather information on what is perceived as hazardous by staff and on their attitudes, training, experience and practices (whether established in writing or not). This may involve scrutiny of personnel selection procedures where staff are employed on a temporary contract.

Phase II - Hazard Identification and Modelling. Phase II commences with the parallel application of techniques such as HAZOP (Hazard and Operability Studies), Brainstorming and FMEA (Failure Modes and Effects Analysis), and continues with the construction of high level generic models (Fault/Event Trees) and the compilation of a list of identified consequences plus a Severity Table. The generic trees are subsequently developed into trees specific to sub-areas within the system.

The following referenced tables are examples of the type produced during a crowd congestion study within the rail transport industry.

1.   Table 2 shows part of a HAZOP table for a railway station using the property word FLOW and key word NO/PART OF.

2.   Samples from a table summarising the results of a brainstorming session is shown in Table 3 and the FMEA for an entrance hall with turnstiles is shown in Table 4.

TABLE 2 HAZARD AND OPERABILITY STUDY SHEET

DESCRIPTION: Station Entrance

FUNCTION: Flow - Flow Along Hall, Flow To and From Entrance

Property Word	Guide Word	Causes	Consequences	Severity	Action	Comments
Flow	NO/ Part Of	1. Limit density reached	Crowding in entrance hall		Avoid maintenance work during peak hours.	
		2. Temporary obstruction	Crowd anxiety and frustration			
		3. Disturbance in hall (illness, fighting, etc)				
		4. Flows from different directions meet and so unable to move				

TABLE 3   Sample from the Table Summarising the Results of a Brainstorming Session for a Hypothetical Station

	PEOPLE	GEAR	STAFF	OTHERS	EQUIPMENT	STATION	ENVIRONMENT
STATES	No hands free. Aged disabled, blind, deaf. Children.	Loose clothing. Footwear, scarves.	Attitude, shortage, turnover.		Seat location. Vending stands. Photo booths.	Siting of posters, adverts. Phones in halls. Dark tunnels. Spiral stairs.	Street congestion. Rain. Hot or cold weather.
EVENTS	Loss of balance. Stepping on toes. large groups. Behaviour in crowd. Emergency drills. School parties. Falls on stairs. No ticket.	Wheelchairs. Possessions lost. Rain protection worn. Walking sticks. Food, baskets, drinks.	Illnesses, absence, strike. Racist attacks.	Buskers, pickpockets, drunks, smokers. Police presence. Contractors working.	Maintenance repairs. Vending machines/ booths under repair, maintenance, refill.	Modification to flows. Fouling, litter, vomit.	
CAUSES	Flow on/off platform. Pushing. Fall onto floor. Rushing to enter. Barrier hopping.	Ghetto blasters, personal stereos. Clothing catches.	Staff overloaded Wrong control actions.	Vandals. Emergencies in street. Bomb scares, cranks.	Sabotage. Handrail fails. Collapse of temporary barriers.	Masonry collapse. Rats, other pests present. Counter-flows. Poor route flows. Temporary route changes.	Floods, fires, smoke. Fire, emergencies in street.
RESULT/ CONSEQ.	Barging, anger. Distressed people. Fainting, sickness. Suffocation. Panic, fear of falling.		Evacuate station. Failure to reach area.	Emergency services cannot reach congested area.	Impaired access to hoses, extinguishers.	Doors open in emergencies. Wrong way flows.	

TABLE 4 FAILURE MODES AND EFFECTS ANALYSIS

ITEM	FAILURE MODES	CAUSE	EFFECT	COMMENTS
Stairs	Slippery	Very heavy rain or snow. Mud brought in from roadworks close by. Flooding from road.	People slip and fall.	Main problem - blockage/ obstruction by things or people.
	Blocked off	Repair - heavy congestion.	Crowd diverted to other/ parallel routes. Stairs not available as emergency exit route.	
	Obstructed	Obstacles pushed by crowd from temporary barriers - left by vandals - drawn in by crowd from road.	Congestion - tripping - injury.	
Turnstiles	Stile closes on person going through or irregular opening and closing.	Control failure.	Injury - barrier out of service, not available for emergency exit.	Two main aspects: closed (especially Common Mode Failure [CMF]) exit blocked. Open (especially CMF) difficulty in controlling entry.
	Blocked	By jammed object/cases/ pushchairs, etc.	Stiles out of service, not available for emergency exit.	
	Fails closed.	Control failure.	Exit/entry not possible via this channel.	
	Stile fails open.	Control failure or sides jammed.	Difficulty in restricting entry - counterflow between entering and exiting crowd.	
	All stiles open.	Control failure - spurious use of emergency button.	Difficulty in restricting entry - counterflow between entering and exiting crowd.	

3. Fig. 2 shows a Hierarchical Task Analysis for crowd control at a station entrance, whilst Table 5 shows a procedure HAZOP table for guidewords NOT DONE, LATER THAN, SOONER THAN.

Phase III - Quantification. Data for input to the fault tree are obtained in the usual way either from field data or estimates based on established methods. Data include equipment failure and repair rates, event probabilities/frequencies and human error probabilities. Note that the authors have found HEART (Human Error Assessment and Reduction Technique, [2]) to be a suitable technique for human error quantification.

Table 6 summarises the formulae and mathematical basis of the different stages in quantification. It should be noted that severity indices are selected from a severity table produced for the specific system.

The formula used for the hazard cause risk rating is more complex than the consequence risk rating due to its dependence upon :

1   The sum of all consequence risk ratings for a given hazard;

2   The frequency of the specific hazard cause to which the preventative action applies.

Note that the result should not be interpreted as a probability or frequency (in fact it has the dimensions of a frequency squared) but as an index to allow comparison between otherwise quite unrelated events.

When human errors appear not as initiating events but as contributory failures, F_{HC} is calculated as :

$$F_{HC} = F_{IN} \times P_{HC} \tag{1}$$

Where F_{IN} is the frequency of the potentially dangerous condition requiring intervention and P_{HC} is the probability of inadequate human response. This ensures consistency between risk ratings assigned to all hazard causes; the ratings are all based on hazard cause frequency rather than on frequencies and probabilities for different areas.

The risk rating formula also allows comparison of preventive actions associated to causes leading to different hazards.

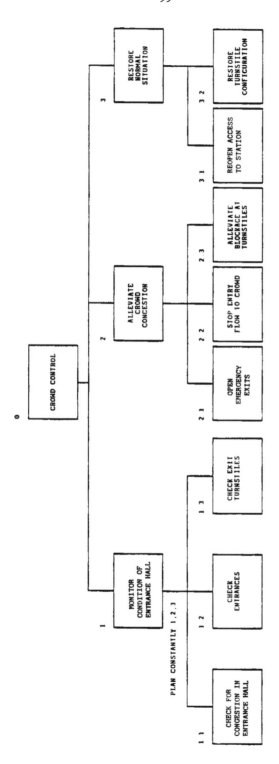

Figure 2 Hierarchical Task Analysis for crowd control procedure

100

Table 5   PROCEDURE HAZARD AND OPERABILITY STUDY SHEET FOR CROWD CONTROL

DESCRIPTION: Alleviate Hall Overcrowding

GUIDEWORD	CAUSES	CONSEQUENCES	SEVERITY	ACTION	COMMENTS
Not Done	Failure to check hall.	Passenger discomfort.		See 1.1.1 Check hall.	
		Passenger anxiety			
		Fall, injuries.		Train staff to recognise problem (use of videos).	
Later Than	Check delayed			See 1.1.1 Check hall.	
Sooner Than	Mis-diagnosis of situation.	Passenger annoyance.		Training (videos).	
		Bad publicity.		Double check with hall staff.	
	Over cautious - new staff.	Passenger anxiety (unnecessary)		Double-man hall when staff inexperienced.	

TABLE 6 Quantitative Information Organised in the Standard Tables

Type of Quantitative Information	Mathematical Derivation
Hazard frequency (F_{HAZ}).	Boolean analysis applied to specific trees.
Hazard cause frequency (F_{CAU}).	Boolean analysis applied to sub trees or direct field data (for single event).
Consequence probability, conditional probability (P_{CONS}).	Estimated or field data as available.
Severity index (S).	Estimate on the basis of severity category.
Risk rating (RR): for consequences	$RR_C = F_{HAZ} \times P_{CONS} \times 10^S$
for hazard causes.	$RR_{HC} = F_{CAU} \times \sum_{i=1}^{n} RR_{c\,i}$
	Over all consequences n with S greater than a predetermined value.

Phase IV - Tabulation and Action Checklists. The methodology proposes a specific format for tabulation of results to allow easy cross referencing between hazards, their causes, their consequences, and any suggested specific preventive or mitigating actions.

Figure 3 demonstrates the relationship between a hazard and its causes and consequences. On the left hand side of the hazard lie the most important events or event combinations from which the hazard originates (causes). The preventative actions suggested for these events should achieve the greatest reduction in the frequency of occurrence of the hazard. On the right hand side of the hazard lie all hazard consequences. The mitigating actions quoted in the tables are those most likely to reduce the overall impact of the hazard on the well being of the crowd if the hazard occurs.

Tables 7 and 8 are examples of result summary tables created by the authors during the study of a rail transport system. Note that the hazard 'overcrowding at entry turnstiles' acts as the link between the two tables. Table 7 presents potential hazards and preventive actions whilst Table 8 presents consequences and mitigating actions.

The risk ratings are used to create action checklists for hazard causes and consequences presented in order of priority; for example the higher the RR_{ci} the higher the priority of the action. The proposed risk rating formula can be used to produce a list of actions across a system for all identified hazards. Alternatively, priority lists can be compiled across separate system areas, provided that modelling assumptions have been kept consistent throughout the system.

It is however recommended that preventive and mitigating action lists are presented separately.

CONCLUSIONS AND FURTHER WORK

The methodology described in this paper provides useful guidance when applying traditional safety analysis techniques to systems handling crowd flow. This is achieved by:

1   The provision of a format which allows easy cross referencing between results of qualitative and quantitative analysis.

103

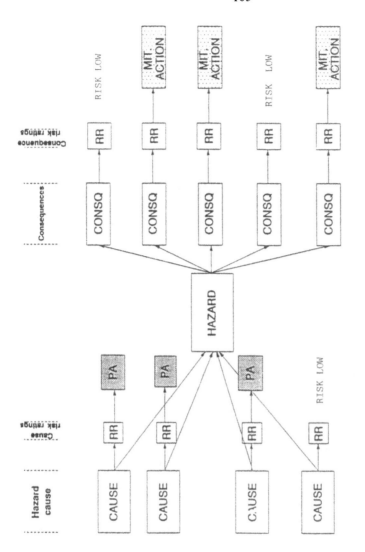

Figure 3 Organisation of qualitative and quantitative information to link events and consequences to suggested actions

TABLE 7 Example of format used for presenting sources of potential hazard

SCENARIO: Entrance hall Overcrowding   AREA - Between entrance to hall and turnstiles

TABLE 1   SHEET 1

SOURCES OF POTENTIAL CONGESTION LEADING TO HAZARDS

SOURCES OF POTENTIAL HAZARD	FREQ.	HAZARD	RISK RATING	PREVENTIVE ACTION	TREE FIG.
Turnstile blocked by object	4	Overcrowding at entry turnstiles	3.83	- Limit size/shape of objects allowed on ground - Provide automatic opening of turnstiles (fail open) - Design turnstile with smooth lines to prevent trapping of object in corners, recesses, etc.	1
Turnstile sticks closed	10^{-2}		0.0096	- Maintain stiles regularly test emergency open - Review design to increase fail safe features	1
Cross flow between entry and exit crowd	10^{-2}		0.0096	- Establish well defined one way system by separating flow streams with barriers - Direct exiting crowd via alternative route	1

TABLE 8 Example of format used for presenting potential consequences of a hazard

SCENARIO: Entrance Hall Overcrowding   AREA – Between entrance to hall and turnstiles

TABLE 1   SHEET 2

HAZARD QUANTIFICATION, CONSEQUENCE AND MITIGATING ACTIONS

HAZARD	FREQ.	CONSEQUENCE	PROB.	CONS. FREQ.	SEVERITY	RISK RATING	ACTIONS
Overcrowding at entry turnstiles	2.3E-3	Physical discomfort (breathlessness, claustrophobia			1		- Design lighting for comfort - Ensure adequate ventilation
		Bruises, concussion, cuts			1		- Provide medical assistance at hand - Ensure no sharp edges to turnstiles and objects in hall - Pad corners
		Person crushed against wall/object	0.01	2.3E-5	4	0.23	- Reduce numbers of protruding objects in hall - Provide warning signs outside hall to keep customers out
		Passenger falls down emergency stairs	0.1	2.3E-4	3,4	0.727	- Provide grab rails - Cover steps with impact absorbent surfaces
		Frustration at delay/ long wait			1		- Play soothing music - Keep customers informed - Apologise on PA
		Fainting/dizziness			2		- Ensure adequate ventilation - Provide drinking points

106

2   The use of risk rating formulae which allows the classification of mitigating and preventive actions in order of priority.

Because of its structured approach the methodology has been found to be successful in dealing with the complexity of systems in which human reliability aspects affect system flow dynamics.

The main area for further work is to improve the proposed formulae for risk ratings. These are adequate as a first approximation to represent dependency on probability, severity and number of consequences, as well as on hazard frequency.   However, these parameters could in principle be combined in several alternative ways.

One possible alternative for the hazard cause risk rating formula is to use the sensitivity of an event rather than its frequency F_{HC} to calculate the risk rating. This would enable the relative importance of events contributing to the same hazard to be taken into account. It is the intent of the authors to explore these alternative risk rating methods in order to support the methodology presented.

ACKNOWLEDGEMENTS

We wish to thank J Corrie and K Norgrove of Mott MacDonald for their useful comments on several aspects of the methodology.

REFERENCES

1   The Institution of Chemical Engineers, Risk Analysis in the Process Industries, Report of the International Study Group on Risk Analysis (European Federation of Chemical Engineering) - 1985.

2   Humphreys, P., Human Reliability Assessors Guide, SRD, October 1988.

THE DEVELOPMENT OF A HUMAN FACTORS STRATEGY FOR THE DESIGN AND ASSESSMENT OF PLANT AND EQUIPMENT

TREVOR WATERS
Safety Department
British Nuclear Fuels plc, Risley

ABSTRACT

A strategy is described that will ensure that the treatment of Human Factors in plant design and assessment is both effective and efficient. The strategy provides guidance on the depth of Human Factors analysis that is appropriate according to the consequences of operator error. In the strategy, plant is graded into one of four categories according to the potential hazards that are involved in its operation and the potential for operator induced accidents. Each category of plant has a different programme of Human Factors analysis, consisting of different techniques and different levels of detail. These techniques are described and the content of the different programmes is provided.

INTRODUCTION

It is generally accepted that Human Factors (HF) problems are a significant cause of system failure and that these need to be addressed in the design and assessment of plant. However, it is also important that the analysis of Human Factors is planned carefully, if resources are to be used both effectively and efficiently.

If insufficient HF analysis is performed then the design of the man-machine interface may not be adequate and Human Factors deficiencies may go undetected. This may result in problems such as:

o Reduced safety margins

o Reduced operability

o Reduced product quality

o Costly backfit design measures

Insufficient HF analysis may also cause overly pessimistic human
error assumptions which may result in inappropriate and expensive plant
design.

While it is important to perform sufficient HF analysis, excessive
and unnecessary analysis should also be avoided as this is wasteful of
resources and distracts away from the key areas.

The paper describes an approach that has been developed within the
Engineering division of British Nuclear Fuels plc (BNFL) to ensure that
the treatment of Human Factors is efficient, effective, consistent and
systematic. It provides guidance on the level of HF analysis that is
appropriate, according to both the potential for operator induced
accidents and the consequences of such accidents.

THE PURPOSE OF HUMAN FACTORS ANALYSIS

The purpose of Human Factors analysis is to create an environment in
which the level of human performance that is required on plant is
actually achieved.

Human Factors analysis can help improve safety margins, operability
and product quality in the following ways:

o It decreases the number of errors, eg through good display design,
   the appropriate use of interlocks etc

o It improves the speed and accuracy of the operator in detecting,
   diagnosing and responding to events

o It increases the potential for error recovery

o It improves the likelihood of successful operator behaviour in
   scenaria that involve difficult diagnostic and decision making
   skills. This is particularly important as a defence against
   operator error in highly complex scenaria caused by unpredictable
   combinations of events.

The same techniques of Human Factors analysis can be applied to the
areas of interface design, written operating procedures and training and
the philosophy that underpins this strategy is applicable to all three
areas.

METHODS OF HUMAN FACTORS ANALYSIS

Techniques of Human Factors analysis can be used to ensure that the demands that are made upon the operator are within his capabilities. A number of approaches can be adopted to achieve this and these are shown below:

The Application of a Human Factors Standard
A considerable amount of Human Factors knowledge has been developed over the years that can be applied in situations without any special analysis. For example, there is detailed information regarding the use of colour, VDU screen design, coding methods, workspace layout, alarm handling, etc. BNFL has developed a Human Factors design standard for control room design that overcomes many common ergonomics problems. This can be applied by designers without the need for specialist involvement.

Task Analysis
Task Analysis, in its widest sense, is a method in which the tasks, the demands of the operator, and the operator's capabilities are formally identified and analysed. It is an umbrella term that covers a large number of different techniques. These different techniques are used for different purposes eg for devising training programmes, for designing equipment, for developing manning policies or for human reliability assessment.

Although Task Analysis can be time consuming it provides greater insights about areas of weakness and where there is a greater potential for error.

Human Error Analysis
Human Error Analysis is a systematic technique for the analysis of operator error and can be performed separately or as part of a Task Analysis. If done separately it would typically take the form of a Failure Modes and Effects Analysis considering the following aspects:

o The step in a procedure

o Possible errors

o Possibilities for recovery

o The effects of the errors upon the system

o The criticality of the error

o The cause of the error

o Measures to reduce either the cause or the consequences.

HAZOP Study
BNFL performs HAZOP studies on all new plant and this technique provides an important contribution towards the identification and treatment of Human Factors. The method is systems led (and thereby focusses on critical errors) and uses experienced operational staff. HAZOPs are expensive however and cannot be relied upon on their own to identify and treat all problems.

Quantitative Human Error Assessment
It is usually necessary to assign probabilities to the errors that have
been identified for inclusion into the quantitative safety assessment. A
number of methods are available that require different amounts of
resources and provide assessments of differing degrees of accuracy.

Two systems are used in BNFL (1). JHEDI (Justification of Human
Error Data Information) which is a quick, auditable and defendable method
of quantification and HRMS (Human Reliability Management System) which is
applied for critical or 'sensitive' human errors.

DESCRIPTION OF THE STRATEGY

BNFL has considerable experience in the treatment of Human Factors. This
has included the development and application of many of the techniques
described in the previous section. In particular, the Human Factors
programme in support of THORP (2), has provided practical experience of
the benefits, costs, and limitations of such methods. This experience
has been used in the development of this strategy and has helped ensure
that the strategy is both practicable and effective.

The purpose of the strategy is to ensure that the level of HF
analysis that is applied to a plant (or subsystem of a plant) is
commensurate with the potential and the consequences of human error.

For example, if there is little opportunity for operator error or if
the hazards arising from operator error are low then the level of HF
analysis that is required will be minimal. At the other extreme, if a
plant contains a large inventory of hazardous material, involves
hazardous processes and requires a high degree of operator control, then
more detailed HF analysis will be required. This would involve high
quality ergonomics design, based upon a detailed understanding of the
operator's task and personal characteristics and also the analysis of
error possibilities.

Between these extremes lie plants of varying levels of hazards
requiring different levels of HF analysis. While the depth of the HF
analysis for these intermediate plants does not need to be extensive, the
analysis that is performed should prevent significant problems. This
strategy can be applied on either a plant wide basis or, (for larger
plant) on a plant area basis.

The strategy includes two major components:

o The categorisation of plant (with respect to the consequences of
   human failure)

o The implementation of a corresponding programme of Human Factors
   analysis

The following sections describe these aspects in greater detail.

The Categorisation of Plant

The categorisation of a plant (and hence the level of HF analysis that is required) is determined by two factors:

o The potential consequences of accidents/outages.

o The potential for operator induced accidents.

These combine to give four different categories as shown in table 1.

TABLE 1
Determination of the level of Human Factors analysis

Potential Consequences	LOW		MEDIUM		HIGH	
Operator Error Potential	Low	High	Low	High	Low	High
Level of HF analysis required:	1	2	2	3	3	4

The consequences of accidents can be considered in either financial or safety terms. In the latter case, the consequences would be determined by the materials, inventories and processes involved. In BNFL, the risks are mainly radiological, while other industries may be involve hazards arising from toxic or explosive materials. It is helpful to describe the categories LOW, MEDIUM and HIGH in strict numerical terms to enable the grading to be objective and scrutable. Again, the values and units of the criteria depend upon the industry and hazards involved. Within BNFL, the categories are based on specific levels of exposure to radiation of the workforce and the public. Examples of other numerical criteria that could be used in other industries include: the numbers and seriousness of injuries, the volumes of material released etc. The assessment of potential hazards can be done at on early stage, either from the HAZOP 1 study or from a scoping risk assessment. While this is not detailed, sufficient is normally known about the plant and process to give a reasonable 'feel' for the magnitude of the potential risks.

The assessment of the potential for operator error is difficult without analysis, due to the complex and insidious way in which events can combine to cause accident scenaria.

However, experience suggests that the following types of activity are particularly susceptible to operator error:

o Diagnosis of faults in complex chemical plant

o Decision making that involves financial penalties

o Operation in novel/non-routine/variable conditions

o Operation on systems with low error tolerance

Operator Error Potential is classified as HIGH, if any one of the above conditions exist and LOW otherwise.

It is likely that this procedure can be used in many different industries, as the factors that cause errors are common.

It should be noted that, if the Operator Error Potential is LOW then the significance of maintenance activities may increase. This is undoubtedly the case on plant that is widely interlocked or highly automated. In such situations design and assessment of maintenance and testing activities should be given careful attention.

Programmes of Human Factors Analysis
The content of the four different levels of Human Factors analysis are shown below:

Plant Category		HF programme
		Minimal Human Factors analysis
1	A	Application of a design standard.
		Limited specialist review
2	A	Application of a design standard.
	B	Sampled design assessment by HF specialist.
		Specialist involvement
3	A	Application of a design standard.
	B	Specalist design support and assessment
	C	Broad review of Human Factors issues by HF specialist
	D	Task analysis (outline)
		Detailed Human Factors plan
4	A	Application of a design standard.
	B	Specialist design support and assessment
	C	Full review of Human Factors issues by specialist
	D	Task analysis (outline and detailed)
	E	Human Factors input to selected HAZOP Studies
	F	Human Reliability Analysis (including accurate quantification)

Maximum benefit is gained by careful programming of the activities into the overall project timescales. For example the design standard is applied early to avoid unnecessary design modifications that would be expensive if done late.

DISCUSSION

The strategy suggested in this paper provides an overall framework within which there is flexibility to meet the specific requirements of different situations. For example, the broad Human Factors reviews that are proposed for plants in categories 3 and 4, need not be resource intensive but provide a means by which problem areas can be detected early in the project. Also, application of the strategy on a plant area basis further improves the efficient use of resources.

It is anticipated that some refinement of the strategy will be possible as experience is gained in its implementation. Particular aspects that could be developed are the assessment of operator error potential and the activities on the category 3 and 4 plants that should be subjected to task analysis.

CONCLUSIONS

The conclusions of this paper are shown below:

1. The strategy is based on the principle that the level of HF analysis that is applied to a plant should be commensurate with the potential consequences of operator error. This ensures the most efficient use of resources and ensures that resources are directed to the most important issues.

2. The method ensures a Human Factors design input on all plant at a low cost by application of a design standard.

3. Quantitative human error assessment is performed for all plant at low cost. This is achieved by using a quick and conservative method first and only where necessary is this supplemented by a more detailed and accurate method.

4. Complicated (and unpredictable) events in high category plants are defended against by high standard ergonomics design. This is achieved by reviews and specialist design support and assessment.

REFERENCES

(1) Kirwan, B., A resources-flexible approach to human reliability assessment for PRA. SARSS '90 Paper, (in these proceedings).

(2) Kirwan, B., A human factors and human reliability programme for the design of a large UK nuclear chemical plant. In Proceedings of the Human Factors Society, 33rd Annual Meeting, Denver, USA. 16-20 October 1989, pp 1009-1013.

A RESOURCES-FLEXIBLE APPROACH TO HUMAN RELIABILITY ASSESSMENT FOR PRA

B Kirwan
Safety Department
BNFL, Risley

ABSTRACT

There are many different methods of human reliability assessment, each differing in its performance according to a number of criteria such as accuracy, resources usage, usefulness and auditability. For a large Probabilistic Risk Assessment (PRA), it is necessary to be resources-efficient when applying techniques of human reliability assessment (HRA), so that the real human error problems get the attention in terms of depth of analysis which they need, and the more trivial problems (i.e. with less risk impact) are dealt with in a more expedient fashion.

In practice this is not so simple as often the perception of the severity of the problem alters as a function of the depth of analysis, and thus in real PRA's the decision as to whether to evaluate a scenario in depth or not is based on fairly broad risk sensitivity criteria and analysts judgement.

BNFL are utilising two human reliability assessment approaches. The first is a 'screening' type of approach for putting broadly accurate estimates into risk assessments, but which also concentrates on making such assessments auditable and involves limited task and human error analysis. The second approach is a far more powerful approach to in-depth human reliability assessment (which is also far more resources - intensive), which involves full task and human error analysis, Performance-Shaping-Factor-based quantification, and error reduction analysis. The way in which these two approaches are utilised in order to make the PRA/HRA process efficient, is discussed in the paper with examples of the decision criteria for deciding when to implement the more detailed approach.

INTRODUCTION

Human reliability assessment (HRA) within probabilistic risk assessment (PRA) is now more the rule than the exception. There are a range of techniques which can estimate human error probabilities (HEPs), and critical assessments of such techniques which can be used to select technique(s) appropriate to a particular assessment[1,2]. One of the major pragmatic criteria for selecting the approach to be utilised is the resources likely to be used by the technique. Differences in resources requirements on real PRA scenarios can be dramatic between different techniques and even between different assessors[3]. Since the accuracy and precision of many techniques still remains in question[1,3], the user is faced with a decision problem of which technique to use. The first and most obvious solution to such a problem may be to opt for a technique which has some apparent validity and yet is low in resources requirements. However, two problems may arise with this simple solution. The first problem can occur when it is believed by assessors and/or designers and operators that a particular scenario will be dominated in risk terms by human error. In such a situation all parties may wish to have more confidence in the risk estimates, perhaps by using an apparently more robust or in-depth tool of HRA. They may also wish to know how and where to attempt to reduce human error potential, hence requiring a HRA technique that specifies error reduction mechanisms and strategies (this assumes that the alternative action of system redesign to remove the human error impact is not taken automatically). The second problem, which can overlap with the first, is if the PRA itself shows the risk to be dominated by the HEPs and/or to be putting the calculated risk uncomfortably close to the risk criteria allocated for the system. In such situations, the analyst knows that the HEPs may have quite large uncertainty bounds associated with them, and may well prefer to have a second look at the scenario using a technique with a higher degree of accuracy, or simply one which does not make fairly crude assumptions, and models the scenario in detail.

A second solution to the question of which technique to use for HRA might be simply to use an in-depth high-resources technique all the time. As more assessments are carried out and more experience is gained and data synthesised, resources usage per assessment may be seen to decrease. However, such a solution requires a rather large initial commitment in resources terms, which may be difficult to justify in the short term (which is often the frame in which financial justification must take place on all but very large projects).

This paper looks at the third logical solution of the resources problem of HRA, which is to have a flexible approach based on a fast 'screening' HRA approach and a more in-depth powerful tool for the more human-error sensitive assessments. Such a resources-flexible approach must contain robust criteria for the selection of the appropriate resources-commitment level for individual scenarios, such that resources are not wasted, but, more importantly, so that human error sensitive scenarios gain the attention they deserve.

The resources-flexible approach to HRA has been developed by the author at BNFL's Risley Safety and Technical Department, largely during and related to a PRA for a large UK nuclear chemical plant. The PRA for this plant is based on over two hundred person years of effort, and a

significant number of safety cases contain human error probabilities.
This has therefore generated a generic need to develop a
resources-efficient approach, although the approach is only now being
finalised and implemented at the end of the PRA. Nevertheless the
approach will be utilised for certain other projects.

The approach utilises two systems, the first called the Justification of
Human Error Data Information (JHEDI) system, and the second called the
Human Reliability Management System (HRMS)[4]. Both systems are
computerised, data-based and highly auditable. This paper outlines these
methods and defines in broad terms the criteria for their application.
In addition some case study material is utilised to demonstrate the first
system (JHEDI), and some extracts are given from a HRMS analysis. Since
HRMS is highly detailed, only a sample of its functions are contained in
this paper, although a full case study is documented elsewhere[5].

OBJECTIVES

2.1 To postulate a resources-flexible approach to HRA in PRA.

2.2 To delineate criteria requirements for application of the higher
    versus the lower-resources usage technique.

SCOPE

This paper argues the case for a resources-flexible approach using JHEDI
and HRMS as possible counter-parts of such an approach, however other
techniues and combinations of techniques could be used in a similar
fashion. It is simply more easy to investigate the utility of such an
approach in the context of real systems.

METHOD

This section defines the two systems JHEDI and HRMS, and the criteria for
their respective application. It also briefly discusses human
performance limiting values[6] (HPLV's) which are utilised by the JHEDI
system in particular. The two systems are explained below.

The JHEDI System

For some time BNFL has carried out assessments based on a database system
containing HEPs and associated descriptions or categories of
performance. This database was based originally on such HEPs as found in
WASH-1400[7], and was called simply Human Error Database (HED). HED was
evaluated along with five other more standardised approaches[3], and was
found to possess a reasonable degree of accuracy in a range of PRA
scenarios, and in this sense was therefore highly resources-efficient as
it could be applied quickly. During an internal review it was found
necessary to improve upon its consistency of application and
auditability, as well as on its performance in high human reliability
scenarios. For these reasons the HED database was encapsulated within a
system which essentially made the usage of HED more traceable and
justifiable, by documenting the information available to the assessor
(hence its name : Justification of Human Error Data Information). The
system comprises the following module/functions:

Scenario description	-	free-text description
Task Analysis	-	linear 'nested loop' architecture
Human error identification	-	rapid human error identification support module available
Quantification	-	usage of HED database/alarm handling HEP algorithms
PSF/Assumptions Documentation	-	documentation of the status of various important performance shaping factor (PSF) parameters, and documentation of additional assumptions underpinning the assessment.
Approval/Audit by HRA manager	-	HRA manager checks all JHEDI's produced.

A case study example using JHEDI is given in Annex 1. This shows the assessment for the case in which operators in a plant must manually undertake a sampling operation prior to discharging the contents of a vessel to its next location (called sentencing).

The scenario descriptions, task analysis, human error output, and an example of the PSF output with the derived HEP (chosen from the database by the assessor) are shown. These are the actual printouts from the JHEDI system, all of which are input entirely by the assessor, who may have limited formal training in HRA theory/practice. The database is not shown. The assessor also has access to alarm handling performance algorithms which give the assessor an HEP for either a simple alarm response HEP (eg. failure to respond to high level alarm), or for more complex cases involving diagnosis, based on the answers the assessor gives to a number of PSF questions. An extract from the alarm quantification module in JHEDI is given at the end of Annex 1. It should be noted that the PSF questions in HRMS and JHEDI are not judgmental, ie. they do not ask the assessor to make relative judgements, but ask for the answers to factual conditions. This can be seen from the PSF questions/answers in Annex 1. Both systems ask assessors only what they can be expected to answer, and do not ask complex questions (eg. 'how would you rate the interface in this scenario on a scale of 1 to 10') which a human factors specialist would have difficulty in answering, let alone a safety assessor with no ergonomics experience. The fault tree for this sampling/sentencing scenario (not printed out by JHEDI) is also shown.

It is apparent that this approach has a good deal of auditability since an independent auditor can see the nature and structure of the task, the errors identified and the PSF parameters encompassing the scenario. In fact such information is of significant use if it is ever required to carry out the assessment again some years later when the original assessor may have moved on. The independent auditor has to judge whether the HEP selected is appropriate or not.

The JHEDI system is relatively rapid and may take up to half a day for a scenario with several HEP's. This assumes that the assessor knows the task requirements reasonably well. Answers concerning training and procedures, if not known by the assessor at the time of assessment, can be documented as assumptions and then agreed for future implementation with the operations group (ie. by inclusion in their training/procedures systems). This has in fact occurred on the major BNFL project mentioned earlier.

This system is therefore auditable, has a certain degree of validity from its database, and is relatively fast compared to several other techniques. Its database is also fairly conservative, so that it is more likely to err on the pessimistic rather than the optimistic side.

Dependence treatment in JHEDI System: The JHEDI system deals with human error dependencies in three ways. The first is the use of relatively conservative values which itself can account quantitatively for a moderate degree of dependence: the JHEDI system quantifies at a fairly aggregate level of behaviour for many scenarios, rather than decomposing down to behavioural elements, where treatment of dependence becomes arguably more essential (eg. as in the THERP system[8]). The JHEDI system contains a small number of conditional probabilities, often concerned with checking actions, eg. if the operator must check the contents of a computerised analysis report prior to sentencing (eg. with an HEP of 10^{-2}), the supervisor must in some cases act as an independent check of the operator. If the supervising check was really independent, a similar HEP of 10^{-2} might be applied, yielding a total failure probability for this error of 10^{-4}. However, this would be optimistic, ignoring the dependence between the two actions which is predictable from a psychological standpoint and is verified via incidents. Hence a conditional probability of 10^{-1} is utilised for the supervisor failure to check.

The third method of treating dependence is aimed at preventing highly optimistic estimates through the concatenation of a number of apparently independent HEPs, all of which may be required for the occurrence of an undesirable event. Such scenarios frequently involve different personnel, perhaps even at different times during the event, and multiple check-points. However, even such situations are limited in their degree of human reliability, as incident and accident frequency have shown. A set of human performance limiting values (HPLV's)[6] have therefore been derived which are used in fault trees to accompany JHEDI HEPs in high human reliability scenarios. The HPLV's are essentially human common mode limiting values, and are primarily either 10^{-4} or 10^{-5} in value. They are not used on their own to quantify human error in a situation, but set a limit on the human reliability level attainable.

The JHEDI system therefore enables relatively quick HRA which is documented, and conservative as well as justifiable. It is used by assessors on all human error scenarios, in the first instance. For many scenarios JHEDI is the only system used. In certain cases though, having carried out a JHEDI, and according to certain criteria, a more in-depth system is required.

Human Reliability Management System (HRMS)

The overall modular structure and functions/rationale of HRMS have been
described elsewhere[4]. Given the detailed nature of HRMS and its large
number of outputs, it is not appropriate to document these outputs within
the confines of this paper, although a case study is described in another
paper[5]. This section therefore describes the overall functions
achieved by HRMS, and Annex 2 gives an extract from the main module of
interest with respect to this paper, the error reduction module. Figure
2 however shows the overall HRMS framework. HRMS analysis also begins
with a nested-loop task analysis as JHEDI does. However, in addition to
this a tabular scenario analysis may be utilised if the scenario is
'event-driven', e.g. in an accident scenario, wherein operators are
responding to events as they occur in time, rather than strictly
following a routine procedural sequence. The tabular scenario analysis
(see Annex 2) takes some time to develop but gives a very detailed
picture of what is happening, and who is doing what in the scenario at
each time segment reviewed. It also documents the indications that are
available, and can be used to plot the likely current diagnosis of the
operators.

The human error analysis is achieved by a detailed and exhaustive
question-answer system which guides the assessor through a comprehensive
range of external error mode forms and psychological error mechanisms.
Because this human error identification (HEI) module is in-depth and
comprehensive, which leads to the identification of a large number of
potential errors, HRMS contains a representation module which considers
which of the identified errors need to be quantified in the fault/event
tree, as some of the errors may be recoverable, or may be aggregated with
other HEPs if data permit. The representation module allows the
auditable screening of all the possible errors, aimed at gaining a
tractable and quantifiable framework which still nevertheless assesses
the principal human error forms contributing to the undesirable event.
Recovery points and dependencies between errors are also noted in the
representation tables. Once the representation analysis is completed the
system automatically selects the errors requiring quantification for the
next module.

The quantification module is based on six Performance Shaping Factors
(PSF): time; the quality of information; training/experience/familarity;
procedures; task organisation; and task complexity. Each PSF has a range
of questions associated with it, all of which can be answered as long as
the plant has reached at least the detailed level of design. The answers
feed into algorithms to generate HEPs by comparison with HEP data
profiles stored in the system. The HEPs derived are then fed into the
PRA via fault trees and event trees in the usual manner (not within the
HRMS software itself).

Up to this point, HRMS and JHEDI have followed similar paths, albeit with
HRMS going into far greater detail via the tabular task analysis, HEI and
representation, and with more PSF parameters to consider. However, as
noted earlier in the introduction, one of the primary reasons for
selecting a more complex and resources-intensive system would be for its
error reduction potential. HRMS contains an error reduction module which
allows the user to determine cost-effective means of improving human
reliability, by one or more of the following methods:

120

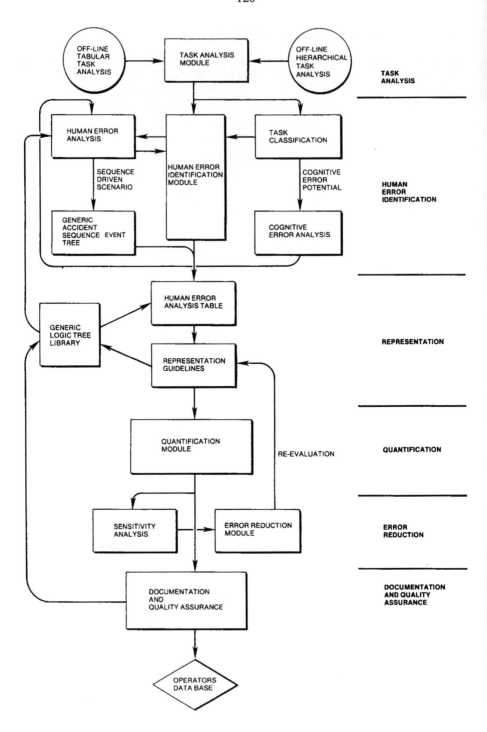

FIGURE 1: HUMAN RELIABILITY MANAGEMENT SYSTEM STRUCTURE

consequence reduction/error pathway blocking	-	a system change which removes the operator from danger, or removes the consequences of the error. Usually error reduction mechanisms (ERMs) in this module are hardware oriented, and are most appropriate in cases of a severe human reliability problem.
error recovery enhancement	-	introduction of error recovery loops into the failure logic, either via procedures or other forms of task organisation.
PSF based error reduction	-	altering of the task PSF to improve the HEPs.

The first two mechanisms affect the event or fault tree logic either by making certain paths less dominant or removing them totally from the analysis. The PSF based ERM's allow the user to consider the effects of, eg. inserting a separate hardwired alarm, or instituting refresher training, etc. on the HEP. Furthermore the PSF-based error reduction module guides the user as to the most effective PSF to consider to reduce an individual HEP by automatically carrying out a basic sensitivity analysis using the same algorithms used to calculate the original HEP. Any improvement in PSF has to be justified and it is logged. Ultimately all ERM's have to be validated in terms of, eg. noting which design drawings were changed, etc., and these changes are logged with the scenario analysis and become auditable assumptions underpinning the safety case. An example from the error reduction module is also given in Annex 2. This example shows a sampling error whose task organisation leads to a high error probability, which is resolved in this case by moving to an automatic sampling system. The error probability improves by a factor of ten. This HEP does not improve further in this scenario because although the system is now less prone to error, the fundamental error of selection of the wrong sampling location still remains possible (though less likely), even with an automatic system. The actual guidance used in the module to reduce error is not shown, but the results of the assessor interaction with the system is given in Annex 2, in the form of what must be implemented to achieve the quantified error reduction effects.

Comparison of JHEDI and HRMS

Both systems have been computerised for ease of use and are largely self-training and self-documenting to cut down resources in terms of training time and reliance on 'experts'. At present only certain personnel are designated users of HRMS. The resources usage varies dramatically. JHEDI may take up to half a day from start to finish, and in some cases may only take an hour. HRMS may take two weeks to assess a scenario within a design project, because a good deal of time is required to collect information at the task analysis phase, including discussions with designers/operators, reviewing alarm and instrumentation schedules, etc. Once the task analysis has been completed, then unless it is a trivial problem, the HEI, representation, and quantification phases will take one or two days, depending on the complexity of the scenario. Error reduction if required can involve several iterations whilst the analyst considers improvements in the PSF but must also discuss their feasibility with operational/design personnel.

From purely the resources perspective, HRMS may seem extravagant, but it offers several advantages which, under certain circumstances, may justify the resources utilisation. Firstly the scenario is analysed in great depth, and this gives a good deal of confidence that the human error inputs have been properly and comprehensively modelled. Secondly the system is based on far more specific information via the detailed PSF consideration, ie a specific scenario in a specific plant is assessed rather than a generic scenario in a general plant (the latter is usually the case in PRA). This may or may not ultimately prove more accurate in an absolute sense, but the specificity of the assessment enables a clearer recognition of the relative importances of different errors according to their PSF profiles. Thirdly, the error reduction module allows proper analysis of means of improving human performance, and allows the effects of error reduction mechanisms to be quantified in the assessment.

Selection of appropriate method

In any PRA the estimated risk is compared against formal risk (frequency and consequence-based) criteria, to decide whether or not the design is acceptable. It is therefore useful to relate the decision of which technique is required to such risk criteria. Such criteria will vary from industry to industry but will all tend to have the same format, eg. the predicted frequency of an event of particular consequences should not exceed 1E-6 per year.

The JHEDI and HRMS systems, to constitute a resources-flexible approach, require a mechanism for selection according to specified criteria which themselves should be based on the degree to which the PRA risk estimates impinge on the allocated plant risk criteria. It was noted earlier that JHEDI was utilised on all assessments in the first instance. The criteria which can call upon HRMS have been derived as follows:

(1) If the risk calculation is within a factor of 10 of the risk criteria allocation (or if it is in fact above that criterion) and a ten-fold increase in a HEP (or group of dependent HEP's) causes at least a doubling of the calculated risk

OR

(2) If the event outcome frequency is sensitive to one or more HEPs such that a ten-fold increase in an HEP would yield a ten-fold increase in event frequency, AND the outcome frequency is within two orders of magnitude of its risk criteria allocation.

OR

(3) A particular scenario requires in-depth assessment as a result of a desire to gain greater confidence in the assessment (eg. for high consequence scenarios) or means of improving its reliability, either as a consequence of assessor, designer, or operator request, or as demonstration of the integrity of an operator-machine system.

The rationale for these criteria is as follows. If the risk calculated is within a factor of 10 of the criteria, then a more in-depth assessment is warranted given potential inaccuracies or omissions of critical error pathways. Even if the risk calculation is two orders of magnitude away from its target, if the event frequency is directly sensitive to human error input, the changes in several HEP's could ultimately come too close to the allocated criterion. Lastly, but by no means least, if certain personnel believe a scenario is worth investigating in depth, then this should be reason enough: there is a danger in any 'art' such as safety assessment, of ignoring a basic feeling of lack of confidence in an assessment and hence the desire to model it more fully. Ultimately any assessment is as good as the assessor carrying it out, so it therefore follows that if the assessor has low confidence in an assessment, then it should be at least repeated by another assessor or, preferably, modelled in more depth until confidence is gained or a problem is found which undermines the original assessment and was probably underlying the assessor's concern.

These criteria can therefore be used to decide when HRMS should be utilised, and they rely on the monitoring of the impact of safety cases on their associated risk criteria, as well as sensitivity analyses of safety cases.

CONCLUSIONS

A resources-flexible approach to HRA has been postulated for dealing with human error scenarios in PRA's, based on a quick but auditable method and an in-depth HRA system. The advantage of this approach is that it allows resources to be allocated according to the severity or potential severity of human error impact. The resources-flexible approach has been demonstrated using two specific techniques, but other combinations of techniques are possible.

Two systems, JHEDI and HRMS, have been outlined and some of their aspects outlined by use of case study material, showing the relative depth of analysis achieved by the different systems, and hence the relative advantages of one versus the other.

Criteria for the application of HRMS to scenarios have been derived, based upon the concepts of risk tolerance and sensitivity to human error, as well as based upon the desire to achieve greater confidence in an assessment or the achievement of high human reliability for scenarios.

ACKNOWLEDGEMENTS

Acknowledgement is given to Bryen Martin and Bob Page (BNFL) for useful comments on JHEDI and criteria for application of HRMS, and to Mike Lihou (LLPS) who programmed both systems in such a user-friendly way that not only can they be used but they are.

REFERENCES

1. Kirwan B., Embrey D E., and Rea K. (1988) The human reliability assessor's guide, ed. by P Humphreys, NCSR Report RTS TT/95, August. NCSR, Wigshaw Lane, Culcheth, Warrington.

2. Swain A D. (1989). Comparative Evaluation of Methods for Human Reliability Analysis. Gescellschaft Fur Reaktorsicherheit, (GRS mbH). GRS-71. April.

3. Kirwan B., (1988) A comparative evaluation of five human reliability assessment techniques, in Human Factors and Decision Making, ed B A Sayers, (London : Elsevier Applied Sciences).

4. Kirwan B., and James N J., (1989). A Human Reliability Management System. Presented at National Reliability 89, UKAEA/Bradford University, June 1989, Brighton Metropole. (Proceedings available from NCSR, Wigshaw Lane, Culcheth, Warrington).

5. Kirwan B (1990: in press) Detailed Human Reliability Assessment : A Case Study. To be published in Applied Ergonomics.

6. Kirwan B., Martin B., Rycraft H., and Smith A, (1990) Human Error Data Collection and Data Generation. Paper presented at the 11th Advances in Reliability Technology Symposium, Liverpool, 18-20 April.

7. Rasmussen N., (1975) Reactor Safety Study. Atomic Energy Commission Report WASH-1400, AEC (USNRC: Washington DC 20555).

8. Swain A D., and Guttmann H E., (1983) A handbook of human reliability analysis with emphasis on nuclear power plant applications. USNRC, Nureg/CR-1278, Washington DC-20555.

ANNEX 1

Sample from a JHEDI Case Study

The following acronyms are used in the various figures and tables in
annex 1 and 2:

BUP   Back-Up Panel
CCR   Central Control Room
CRO   Control room operator
DCS   Distributed Control System (A VDU based system)
DCU   Distributed Control Unit
DOG   Dissolver Off-Gas
DP   Differential Pressure
ESD   Emergency Shutdown
GNI   Guaranteed Non-Interruptible (power supply)
MANAUTH   Management Authorisation
REC   Recovery (virtually certain)
SS   Shift Supervisor
SUB   Subsumed
UN   Uranyl Nitrate
VDU   Visual Display Unit

JUSTIFICATION OF HUMAN ERROR DATA INTERPRETATION

FILE NAME   :

ASSESSOR   :

DATE   :

SAFETY CASE :

SCENARIO:
The enrichment monitoring and blending sequence reaches the point where a daily
sample of UN is required.The CCR operator requests that a local operator goes
to the plant room and takes a sample from the correct tank.The local operator
does this and despatches the liquor to the laboratory.
        In the laboratory the analyst performs the test,records the result
and transmits it back to the CCR.Here the CCR operator inputs the result into
the DCS.If this result differs markedly from the on-line measurement then the
DCU draws this to the operators attention.The operator makes a desision on
sentencing the liquor and must obtain a Management Authorisation before
proceeding.It is assumed that this supervisors check should reveal any mistake
by the CCR op when inputting the result and any difference between this result
and a corresponding on-line measurement which has not been noticed by the CCR
op.

TASK ANALYSIS

    q1.   INITIATE SAMPLING
    1.1   CCR op recognises need for daily UN sample
    1.2   CCR op communicates with local op
    1.3   Local op goes to plant room and takes sample
    1.4   Local op despatches sample for test
    2.   ANALYSIS
    2.1   Analyst carries out test
    2.2   Analyst records results
    2.3   Analyst transmits results
    3.   INPUT OF RESULT TO DCU
    3.1   CCR op enters lab result into DCU
    3.2   DCU warns op of discrepancy between measurements
    3.3   CCR op obtains Management Authorisation
    3.4   CCR op sentences liquor

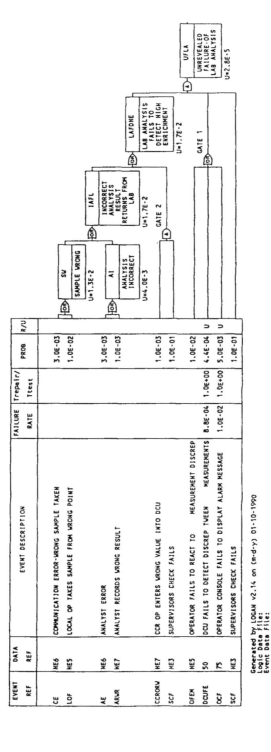

FIGURE . FAILURE LOGIC DIAGRAM FOR UNREVEALED FAILURE OF LAB ANALYSIS

HUMAN ERROR ANALYSIS

```
Assessor    :
Date        :
Safety Case :
File name   :
```

1.2	CCR op communicates with local op
03120	Communication error between CCR and local operator
1.3	Local op goes to plant room and takes sample
04120	Local operator takes wrong sample
2.1	Analyst carries out test
07120	Analyst makes mistake whilst undertaking test
2.2	Analyst records results
08120	Analyst records result wrongly
3.1	CCR op enters lab result into DCU
11120	CCR op enters wrong value into DCS
3.2	DCU warns op of discrepancy between measurements
12120	Op doesnt react to discrepancy warning
3.3	CCR op obtains Management Authorisation
13120	Supervisors check fails

PERFORMANCE SHAPING FACTORS

FILE :

STEP :   1.2   CCR op communicates with local op

ERROR: 03120   Communication error between CCR and local operator

TIME

  The time available is about 60 minutes
  The available time is significantly more than required

QUALITY OF INFORMATION

  The task step is based in the Central Control Room
  Are diverse hardware and software based alarms available ?   NO
  Is the information available in the situation clear and unambiguous ?   YES
  Is feedback given promptly of all control actions confirming that
    that they have been carried out effectively ?   YES

EXPERIENCE / FAMILIARITY

  The event is very frequent (at least weekly)

PROCEDURES / INSTRUCTIONS

  Written procedures will be available

TASK ORGANISATION

  The personnel involved is a single operator
  Do you think the operator would understand the task significance,
   ie the importance of reliable performance and the cost/consequence
   of errors ?   YES

TASK COMPLEXITY

  The task involves either a single clear step or a few well trained
    and very familiar routine steps in order to achieve a clear goal

Probability Chosen: 3E-3   HED Reference: HE6

Authorised by:

PERFORMANCE SHAPING FACTORS

FILE :

STEP : 2.1   Analyst carries out test

ERROR.: 07120   Analyst makes mistake whilst undertaking test

TIME

The time available is several hours
The available time is significantly more than required

QUALITY OF INFORMATION

The task step is based in an area other than a control room
Is information presented as a written/typed document ?   YES
Are diverse hardware and software based alarms available ?   NO
Is the information available in the situation clear and unambiguous ?   YES
Is feedback given promptly of all control actions confirming that
 that they have been carried out effectively ?   NO

EXPERIENCE / FAMILIARITY

The event is very frequent (at least weekly)

PROCEDURES / INSTRUCTIONS

Written procedures will be available

TASK ORGANISATION

The personnel involved is a single operator
Do you think the operator would understand the task significance,
 ie the importance of reliable performance and the cost/consequence
 of errors ?   YES

TASK COMPLEXITY

The task involves some interpretation or straightforward calculations
 in a basically non-complex procedure

Probability Chosen: 3E-3   HED Reference: HE6

Authorised by:

PERFORMANCE SHAPING FACTOR ANSWERS

```
Assessor    :
Task        :
Safety Case :
File name   :
Error       :
```

RESPONSE TO ALARMS - SIMPLE

Is the alarm response based in the Central Control Room ?   YES

Is the alarm an ESD signal, both hardwired and displayed on VDU,
and occurring both audibly and visibly ?   NO

Is the situation unambiguous, requiring a simple and
straightforward response ?   YES

Is the situation high priority ?   YES

Are there diverse hardwired and software based alarms ?   YES

Response time available is about 60 minutes

HUMAN ERROR QUANTIFICATION RESULTS

```
Assessor    :
Task        :
Safety Case :
File name   :
```

```
+-----------------------------------------------------------------------+
| 1         RESPOND TO ENR DISCREPANCY ALARM                            |
+-----------------------------------------------------------------------+
|           CRO fails to respond to discrepancy alarm                   |
+-----------------------------------------------------------------------+
| DATA POINT   Alarm response - simple          1                       |
+-----------------------------------------------------------------------+
| H.E.P.    0.010000                                                    |
+-----------------------------------------------------------------------+
```

ANNEX 2

Sample from a HRMS Case Study

The first figure in Annex 2 shows an example tabular task analysis page,
which is developed off-line (is not on the computer). This shows the
depth of qualitative analysis required when utilising HRMS.

The following two figures show an extract from a sampling/sentencing
error reduction exercise. The error reduction menu on HRMS (each module
has its own sub-menu) is followed by the original sensitivity analysis
table showing that task organisation would be the most cost-effective
means of achieving better human reliability (ie. 2E-2 rather than 1E-1
which is the current HEP). The final sheet shows the new sensitivity
analysis table with a revised HEP figure of 1E-2 achieved by optimising
the task organisation PSF. The actual error reduction mechanisms which
must be implemented to do this are listed below the second sensitivity
table.

Tabular Task Analysis – Dissolver Off-Gas Fan Failure (Non GNI Failure)

Task Goal	Time	Personnel	System Status	Information Available	Decision/Action/ Communications	Equipment/ Location	Feedback	Distractions/ Other Duties	Comments and Operator Expectations
1. Normal Operations		CRO, SS	Normal operating envelope			CCR		Busy on activities	Single CRO, SS not busy on another fault. Assume beginning of leach sequence (worst-case timescale)
2. Respond to first alarm	T+10s	CRO	Both DOG fans fail for a cause other than GNI failure	VDU alarm (yellow) on low DOG fan shaft speed	CRO targets 'fetch alarm'; goes to level 4 mimic.	VDU, DOG console	Flashing yellow alarm message	Previous tasks; other alarms begin to occur	
3. Respond to further alarms	T+20s	CRO	Further alarms occur due to: (i) low DP across fans (ii) low flow in DOG (iii) low leg fan DP across dissolver (iv) low pressure drop across columns	VDU alarms: - yellow - red - red - red	CRO targets 'next alarm'; goes to level 4 mimic. CRO goes to group alarm listing. Looks at printer.	VDU, printer	Flashing audible messages	Other alarms	
4. CRO identifies fan failure & calls supervisor (acknowledge alarms)	T+60	CRO/SS	Alarms on: (i) Low flow DOG duct at stack monitoring room -Hardwired (ii) Hardwired in DOG duct manifold	VDU alarm; - red -Hardwired	CRO monitors level 3 mimic. CRO acknowledges alarms, calls supervisor. SS detects alarm from own console	DOG VDU. DOG VDU & BUP. SS VDU console	Audible & visual. Audible & visual. "	SS involved in other tasks	

| ERROR REDUCTION MENU |

--> Introduction

    Selection of Error Reduction Mode

    Consequence Reduction / Error Path Blocking

    Increasing Error Recovery

    Reducing Error Likelihood

    Root Cause Reduction

    Sensitivity Analysis

    Print Error Reduction Analysis Report

    Return to Main Menu

Use Up-Down cursor keys to select... Confirm with <Enter>

PSF REDUCTION MENU

STEP :   1   Go to sample point
ERROR: 02310   Goes to wrong sample point
ORIGINAL HEP: 1.0E-001   CURRENT HEP: 1.0E-001

	ORIGINAL PSF	CURRENT PSF	TARGET PSF	HEP RESULT
TIME	4	4	2	1.0E-001
QUALITY OF INFORMATION	4	4	2	1.0E-001
TRAINING / EXPERIENCE	0	0	0	1.0E-001
PROCEDURES / INSTRUCTIONS	4	4	2	1.0E-001
--> TASK ORGANISATION	8	8	6	2.0E-002
TASK COMPLEXITY	2	2	0	1.0E-001

Initialise Current to Original
Return to Error Reduction Menu

Use Up-Down cursor keys to select... Confirm with <Enter>

PSF REDUCTION MENU

```
STEP :          1      Go to sample point
ERROR:          02310  Goes to wrong sample point
ORIGINAL HEP:   1.0E-001                        CURRENT HEP:  1.0E-002
```

	ORIGINAL PSF	CURRENT PSF	TARGET PSF	HEP RESULT
TIME	4	4	2	1.0E-002
QUALITY OF INFORMATION	4	4	2	1.0E-002
TRAINING / EXPERIENCE	0	0	0	1.0E-002
PROCEDURES / INSTRUCTIONS	4	4	2	1.0E-002
--> TASK ORGANISATION	8	2	0	1.0E-002
TASK COMPLEXITY	2	2	0	1.0E-002

Initialise Current to Original
Return to Error Reduction Menu

ERROR REDUCTION MECHANISMS

```
Assessor      :
Task          :
Safety Case   :
File name     :
Task Step     : 1       Go to sample point
Error         : 02310   Goes to wrong sample point
Error Mech.   : Topographic misorientation     Event Ref: OGTWSP
```

CONSEQUENCE REDUCTION & ERROR PATHWAY BLOCKING

```
Description   : Implementation of auto-sampling system (10)
Comments      : Removes error from tree
```

INCREASING ERROR RECOVERY

```
Description   : auto-sampling prompts operator with sample point (1)
Comments      : adds error recovery step into tree
```

REDUCING ERROR LIKLEHOOD

```
Description   : Remove distractions
Comments      : Move this sampling operation under the
                auto-sampling system.

Description   : Computerised checking
Comments      : Carry out this sampling operation within the
                auto-sampling system, backed up by error-checking
                on the information computer.


Description   : Built in self-checking
Comments      : Use a tick-sheet procedure which notes the
                sampling point to be sampled and the code of the
                actual sampling local valve.
```

IF THE ABOVE MEASURES ARE IMPLEMENTED THE H.E.P. WILL BE: 1.0E-002

ASSESSING THE SAFETY OF EXISTING PLANTS - A CASE STUDY

PETER BALL AND CHRISTOPHER LEIGHTON
British Nuclear Fuels plc,
Sellafield, Seascale,
Cumbria CA20 1PG

ABSTRACT

BNFL is preparing fully developed Safety Cases, using qualitative and quantitative techniques, for all its operating plants at the Sellafield irradiated nuclear fuel site in Cumbria. The Safety Cases include hazard and operability studies for fault identification and hazard analysis for fault quantification. Lessons learnt from the first three years of the programme are presented.

INTRODUCTION

British Nuclear Fuels (BNFL) Sellafield is undertaking a five year programme of 'fully developed Safety Cases' (fdSCs) covering all operating plants at its irradiated nuclear fuel handling and reprocessing site. The Head of Sellafield Site has retained overall responsibility for the programme, and has ensured that adequate funding continues to be available as well as the support of management. The programme costs £5 million per year, started in 1987, and will have provided each of the 50 major plants at Sellafield with a state-of-the-art fdSC by 1991. About 35 man-years of professional specialist safety assessment effort is required each year. A mixture of company employees and hired contractors is used. They are supported by a few dedicated administrative and word processing staff. In addition, substantial input is provided by plant management and operators, and engineering maintenance staff. In later years, each plant's fdSC will be updated to cover modifications, any incidents which may have occurred, changes in data or published information on radiological risks and assessment methodology.

In the mid 1980s there were a series of incidents at Sellafield. These incidents, although minor in safety terms, attracted a great deal of adverse publicity. The nuclear industry was already at that time highly regulated, principally by the Nuclear Installations Inspectorate. The public concern arising from these incidents led the regulators to involve themselves even more in the minutiae of Site operations. The effects on BNFL have been exceedingly costly, and out of all proportion to the magnitude of the hazard posed by the incidents. Experience indicates that the costs following on from even quite minor events can be far more than those from the immediate loss of production or harm to people, particularly if the events occur close together and attract publicity with a "band wagon" effect.

SAFETY CASE DESCRIPTION

Safety is assessed by both qualitative and quantitative methods. Qualitative assessments, reviews and audits are important and valuable. However, the topic of this paper is the quantitative assessment of potential fault conditions by probabilistic safety analysis (PSA).

A structured hazard and operability (HAZOP) study identifies systematically potential hazards and their causes. This leads on to hazard analysis (HAZAN) which quantifies the frequencies of occurrence and the radiological consequences of identified faults. The results are compared with Sellafield's numerical accident risk criteria.

RESULTS

The original purpose of the fdSCs was to satisfy regulatory demands for BNFL to compare the safety levels for these operating plants, some over 30 years old, with new plant standards. These are generally proving to be fully adequate. It is perhaps surprising to find that plants designed several decades ago against such safety standards as then existed are capable of meeting strict modern safety criteria. The reasons for this seem to be the robustness of the original design, and later improvements to the plant and operational practice resulting from experience.

EXPERIENCE GAINED

Organisation

Ideally, it would be preferable to divide each plant into areas and systems. Each area would be assessed by two analysts, who would act as HAZOP Team Leader and Secretary, and who would additionally carry out all of the follow-on hazard analysis. This approach has not been adopted because:

a) The time available for each fdSC is too short.

b) The number of suitable company employees available to lead such studies was insufficient.

Instead, each plant is assigned one experienced project leader. He is supported by additional, somewhat less experienced staff, if required. These people are all company staff, and lead HAZOP teams and manage the overall PSA (ie HAZOP and HAZAN). Each Team Leader is assisted by a HAZOP Secretary, a more junior analyst, who may be a company employee or a contractor. When further experienced company safety analysts are available, they carry out HAZANs on the more complex or sensitive projects. The remainder of the HAZANs are carried out by experienced contract staff. In this way, the overall PSA is managed by and is the responsibility of experienced company staff. The pressure of work on these staff is considerable. They are required to lead a HAZOP (meeting typically for 12 hours each week for several months) and manage the HAZAN work of several part and full time assessors. Nevertheless, experience has shown that this workload is sustainable in the long term.

In order to motivate permanent staff and ensure a high level of professionalism, a career structure is provided. Recent graduates act as HAZOP Secretaries or junior analysts. Promotion is available for suitable staff to the next grade as HAZOP Team Leaders or safety assessors. Staff may be promoted further to Lead Safety Assessors, with overall responsibility for larger projects or groups of smaller projects. A progressively lesser number of more senior management posts in safety complete the specialist career structure. Not everyone wants to remain a safety specialist in the long term. This work has proved an effective 'stepping stone', allowing people to broaden their careers into plant management, engineering design etc.

Progress with all the PSAs is monitored in regular progress meetings every 2 or 3 weeks attended by HAZOP Team Leaders and departmental line managers. Resource allocation may be adjusted, as a result of these meetings, to ensure that target timescales are met at acceptable quality.

Training

If coherent and consistent safety cases are to be produced, it is essential to provide those carrying them out, both company employees and contractors, with clear guidelines, supported by adequate training. HAZOP Team Leaders and Secretaries receive initial formal training, based on syndicate exercises. They and the HAZAN staff receive formal training in HAZAN, based on lectures. Each course is accompanied by a comprehensive manual which acts as a detailed reference book for later use. In this way two aims can be achieved:

a)   To introduce inexperienced staff to the basics and indicate where they can find the detail later.

b)   To acquaint experienced staff, especially HAZAN contractors, who have not previously worked on fdSCs, with the way BNFL Sellafield requires the work to be done.

Company staff acting as HAZOP Secretaries are regarded as potential future HAZOP Team Leaders : so their Team Leaders have the responsibility for training them on-the-job with this career progression in mind.

Each new member of company staff assigned to HAZOP or HAZAN receives a personal Training Schedule. This lists subjects and techniques with which he will need to become familiar, together with relevant information sources. Progress with training is monitored by departmental line managers, who endorse when each subject has been mastered. On completion of training, usually after about a year, a Certificate is formally prepared and the achievement noted in the employee's staff record.

Hazard and Operability Studies (HAZOP)

The HAZOP technique is a very effective tool for identifying hazards and causes. It is also, unfortunately, expensive in man-hours, accounting for about a third of total fdSC production costs. Worse still, it uses the time of those people whom the plant management are least able to release, for example plant managers and foremen, and maintenance engineers. So it is important that these people should see the HAZOP as worthwhile. It has been reassuring to note that these people, even if initially neutral or even hostile to HAZOP, have generally come to value it, and seek later to use it themselves when considering plant modifications : it is becoming an integral part of Sellafield's safety culture. Indeed, the process of carrying out the study changes the attitudes of plant management towards their plant, often forcing a re-appraisal.

Initial concerns that plant and engineering staff might be reticent at HAZOP meetings proved groundless. It was astonishing how honest and open people were willing to be.

Because HAZOP is expensive, it is prudent to make it do as many jobs as possible. By keeping unusually full records, properly indexed, cross-referenced and summarised, its usefulness in future work can be maximised (eg for minimising the time and effort required for assessing future modifications and other non-routine work, to assist design of similar plants). The records kept show not only where hazards could arise, and how they are controlled, but also why other deviations from normal conditions do not result in hazards. A properly recorded HAZOP may be used later for purposes which had not been thought of when the study was done. The expense of maintaining very detailed records has been found by experience to be fully justified by the benefits which later accrue. The temptation to limit short-run costs by keeping records 'by exception' or in summary is well worth resisting.

An organisation that wants to set up a major HAZOP programme when it has little previous experience faces a considerable task. BNFL Sellafield set about it by:

a) Retaining experienced HAZOP consultants to train the first batch of staff, advise on customising the methodology, attend HAZOP meetings as advisers, and monitor and assist generally, but not to run the meetings.

b)   Using suitable BNFL staff to run the HAZOPs, trained and advised by consultants.

This approach works, and BNFL Sellafield now has a pool of about 20 experienced HAZOP team leaders who are knowledgeable both of HAZOPs and of Sellafield operations.   External consultants are still retained, but only for quality control and to provide a link with developments in the methodology.

HAZOPs on operational plant rarely (but occasionally) identify operability problems that were not already understood.   The cost of the studies can be reduced by concentrating primarily on hazards, and not studying in detail those plants or plant areas with a low hazard potential (eg facilities for laundering contaminated clothing or cleaning safety equipment such as respirators).   It is clearly important to identify rigorously areas that need not be studied in depth.   Generic HAZAN based on plant inventories and worst case faults, possibly combined with an overall HAZOP concentrating on interfaces or a HAZOP 1, is one means for such facilities.   The penalty paid for this economy is the lost opportunity to identify operability problems and the like : it is not a loss in safety terms.   There are many different workable techniques such as these for assessing a new plant design at a stage when full HAZOP would be impossible, or for screening a plant to identify the likely problem areas.   The most appropriate technique needs selecting case-by-case.

Costs can be reduced further by limiting the study to routine operations and those non-routine operations where quick action is required (eg recovery from fault conditions where operator intervention is important in minimising the effects).   There is no need to cover non-routine operations which can be assessed at the time they are going to be done.

HAZOPs traditionally use line diagrams of plant pipework and vessels, together with operating and maintenance instructions as the basis of the study.   More modern plants are computer controlled : HAZOP can be applied to the software program, if it is presented in an appropriate form, in the same way as normal instructions.   Photographs and videos of plant and operations are proving cost-effective supplements, reducing time spent in meetings in describing what the plant does.   These visual aids are also of benefit for HAZAN work.

BNFL is using HAZOP primarily as a means of identifying potential hazards and their causes.   The analysis of these hazards is a separate, complementary exercise.   The interface between the two is of crucial importance.   The HAZOP must provide the hazard analyst with an understandable summary of hazards and causes, and access to the HAZOP staff to discuss it.   The normal HAZOP procedure of raising actions to bridge the gap has been found by experience to be inadequate, and was abandoned during the first year of fdSCs.   A more detailed listing has been found to be indispensable (see Appendix 1).

One of the benefits of assessing an operational plant is that historical information is available.   It is possible to use the record of deviations from normal to cross-check the HAZOP independently.

HAZOP has several incidental benefits. Two of these are being increasingly recognised as significant:

a)   Provision of up to date plant drawings.

b)   Production of improved instructions.

It is a feature of HAZOPs on operating plants that identification of major hazards which had not previously been known about is extremely rare. However, the HAZOP identifies numerous, usually minor, additional causes of hazards. It also provides confidence that the major hazards have been identified.

Hazard Analysis (HAZAN)

HAZAN is a tool for quantifying risks (consequences and frequencies of potential fault conditions). For its worth to be properly exploited, it requires safety criteria against which the results can be compared. Risks can then be judged as acceptable or not. Where risks are unacceptable, HAZAN can indicate the areas where remedial modifications to plant, process or operational practice would be effective.

Quantification against numerical criteria also allows identification of operations and equipment which are particularly important in maintaining risk at acceptable levels. These can be designated formally, so that plant operators and maintenance staff are aware of their special importance.

The efficiency of HAZAN can be maximised by adopting common databases for such things as release fractions, and item reliability, and standard methodologies for common cause failure evaluation and estimating human error probabilities.

Gathering and using plant-specific item reliability data improves the credibility of assessments. BNFL's experience is that reliability data gathered on plant shows failure rates and unavailabilities significantly lower than what would be expected from the literature. One of the qualitative elements of the fdSC is a Serviceability and Operability Audit, carried out by external consultants. This audit sometimes identifies equipment that is failed or in poor condition. Such faults are usually rectified promptly by plant engineers. However, such faults are taken into account in HAZAN when estimating time to repair, failure rates etc. Another useful qualitative element is the review of the Site's comprehensive record of incidents and minor events. The results from this review allow a crude calibration of some occurrences, particularly human error probabilities in the assessments.

Certain categories of HAZAN are found to be needed in many plants (eg loss of ventilation extract filtration leading to enhanced aerial activity releases from stacks). Over the years, the quality and realism of these standard assessments improves, provided assessors are kept aware of what was done in the best previous similar assessment. To this end, detailed notes on several standard topics have been prepared as statements of current assessment practice. These notes are frequently updated. There is no point in re-inventing assessments, but on the other hand it is important not to stifle innovation and improvement.

HAZAN can and should be subject to thorough quality control. Assessments can be checked to ensure : the data are applicable, plant and process details correct, logic and methodology sound, and, of course, calculations error-free. The HAZOP-HAZAN loop can be closed by requiring the staff who carried out the HAZOP to check that the HAZAN has dealt with everything relevant that was found at the HAZOP. Future use of the assessments is aided by insisting on a standard format. If all HAZANs are checked in this way, it adds 10-15% to the overall cost of an average HAZAN. Staff who carry out the checks have all been involved in HAZAN, even though they may have particular specialisations (eg reliability data). It is considered that the use of checkers who lack experience in the overall HAZAN would be counter-productive. It has been found by experience that checkers should have other tasks as well as checking, otherwise their effectiveness as checkers is reduced by acute boredom.

The results of HAZANs are compared to numerical safety criteria. These typically consist of primary criteria based on nationally and internationally accepted standards of risk, and more restrictive secondary criteria set to control event categories (eg large consequences) that are perceived as particularly undesirable.

Criteria have a major impact on HAZAN costs. Obviously, unrealistically restrictive safety criteria are harder to meet, and assessments of borderline cases can be particularly costly. Less obviously, the form of criteria, whether tight or not, can affect assessment costs. A step-like system of consequence-frequency criteria is currently popular because it is believed to be adaptable to reflect public perception of risk. However, as a system of criteria, it makes generic or bounding case assessments difficult, and therefore tends to multiply the number of distinct cases requiring assessment. If the steps are too numerous they can exceed the accuracy of the assessments.

FUTURE DEVELOPMENTS

As the fdSC programme has proceeded, areas of methodology where improvements may be worthwhile have been found. These are at varying stages of implementation. They include:

Sensitivity analysis to determine the important features in assessments
Extending the existing programmes for gathering data from operating plants at Sellafield (eg equipment failure rates)

Reanalysing published release fraction data, and carrying out research into fault conditions where existing data are sparse

Developing human error probability methodologies and data (for example, to investigate the link between human error probabilities and time available in diagnostic and decision-making errors).

CONCLUSIONS

This paper has outlined the approach taken for PSAs on operating plants at BNFL Sellafield. it has indicated the organisational and technical features which have allowed 30 major safety cases to be prepared to time, budget and quality.

Problems found with achieving adequate flow of information from HAZOP to HAZAN have been described, and the solutions that were adopted have been explained.

Other lessons learnt and future developments have been mentioned.

144

HAZARD 3
Table B2.1   HAZARDS AND THEIR CONTRIBUTORY CAUSES

Description of Hazard	List of Initiating Events	HAZOP Record Sheet Reference	Reference to Assessment
Abnormal Aerial Release from LA Effluent Cell	Venting of Pre-CIRC Tank via Body Feed Tank	2.2, 14.10	
	Aerosol from V.800	4.3, 4.5 17.9, 17.10	
	Release from CF	3.4, 4.5, 5.1, 5.2, 16.1	
	Release from FF	3.5, 4.5, 6.1, 13.4, 16.2, 16.4	
	Release from unprotected sludge flask	7.2, 7.3	
	Manual entry to FF cell during OPS	7.2, 8.3, 8.5	
	Inappropriate PLC over-ride	8.1., 8.5, 11.4	

'OPERABILITY STUDY ADAPTATIONS IN SELECTED FIELD WORK AREAS'

C. J. BULLOCK, F. R. MITCHELL and R. L. SKELTON

Jenbul Consultancy Services Ltd., 4a Old Market Place, Ripon,

North Yorkshire HG4 1EQ, U.K.

ABSTRACT

The basic 'hazop' method, essentially simple but rigorous if correctly used, was conceived within an industrial environment about thirty years ago. However, current adaptations frequently reveal scope for realising the full potential of the technique. This paper suggests some of the means by which this objective is achievable; firstly, practitioner efficiency can gain from an improved awareness of perceptive ability and secondly, there is evidence that departure from the orthodox guide word approach should be discouraged.

Formats are included for the reliable documentary control of changes, actioned by 'HAZOP' review, applied throughout project development.

INTRODUCTION

The operability study (HAZOP) of process systems has attracted growing allegiance over the last 20 years but many users of the method have devised variants, some of which offer novelty but dubious advantage. The objective of identifying operability problems does not always take precedence in practice over a search for hazards. In certain areas of chemical process handling 'in-house' guidance documentation has been generated wherein basic HAZOP study principles are interwoven with features of risk assessment strategy, quantified hazard analysis procedures or subjective numerical ratings. Table 1 is an example of the 'scoring' approach adopted on a site where perceptive effort did not extend sufficiently, for example, to prevent an incident of operator unreliability which resulted in disaster for want of removing a vent manifold blind after maintenance and prior to plant start-up. Insurance company premium rating strategies were intended for existing risk level evaluation and not for quantitative conversion of a qualitative analytical tool effectiveness.

Hazard 'eligibility scoring' as a corrective procedure is to be discouraged. Its use can mean that simple systems, e.g. water, air, drainage etc. may never be studied and it often happens that such systems can cause major incidents; e.g. the 'Three Mile Island' nuclear reactor catastrophe can be attributed to a prime cause of water in instrument air.

TABLE 1

'AN EXAMPLE OF ELIGIBILITY SCORING FOR A HAZOP STUDY'

The method requires a summation of points derived from a consideration of ten characteristics:-

		Maximum Points Score
1.	Flammability of materials in process and storage	4
2.	Reactivity of materials in process and storage	4
3.	Inherent fire hazard of process or storage	5
4.	Inherent explosion hazard of process or storage	5
5.	Heat content derived from tonnage of combustible chemicals	4
6.	Reaction temperature	3
7.	Reaction pressure	5
8.	Toxicological Assessment	3
9.	Skin and Eye Irritancy	4
10.	Carcinogenicity and Mutagenicity	3
		40

The aim is to divide plant and processes broadly into 4 categories of hazard:-

'Low' hazard	9 points or less
'Medium' hazard	10 - 15 points
'High' hazard	16 - 22 points
'Very High' hazard	over 22 points

Hazard and Operability studies (HAZOP) is recommended to be conducted on assessments scoring a total of more than 15 points. In practice, the 'Maximum Points Score' for each characteristic is sub-divided into materials pointing (inherent characteristics) and a separate score for "in process" where applicable. The maximum total points allocated to the former can account for 25 of the 40 so that a decision to apply HAZOP may emerge from this contribution alone. (The assessment of certain properties such as flammability is founded upon established guidance, e.g. NFPA 704 'Standard System for the Identification of the Fire Hazards of Materials').

Basic doctrines such as introduced by T. A. Kletz [1] and H. G. Lawley [2,3] have more than withstood the test of time; more recent literature [4,5] has emphasised the application of this proven disciplined approach to all stages of project development and plant condition. There is currently, however, a wider recognition of the usefulness of operability study in non-process environments and of the need to impose stricter control and approval procedures backed-up by managements devoted to the objectives of quality assurance. At the same time, it is needful to correct misconceptions, discourage 'short-cut' interpretations such as the 'WHAT-IF' hazard-seeking ploy and to constantly recognise the vagaries of human perception for which no amount of team study activity represents an antidote. There is an over-riding need in every HAZOP study to be convinced that things really are what they are supposed to be. Initial study preparation, aggressive enquiry into perceived and credible possibilities, together with a marked distrust of all assumptions are readily attainable pre-requisites for practitioners of the technique.

THE INFLUENCE OF PERCEPTION

The efficient application of operability study is very demanding of individual perceptive effort; it is acknowledged that study team activity is mentally tiring to the extent that a 3 to 4 hour 'sitting' within a working day is preferably not exceeded, although an experienced leader or chairman may be capable of supervising teams alternately.

It is not always appreciated, however, that much of a study team's potential relates to examining facets not shown on drawings or in other information normally available in documentary form within a project. The reliability of plant operating staff in following repetitive instructions involving imperfectly identified controls is one common example within the broader consideration demanded. Temperament is known to condition response, depending upon whether observations are 'synthetic' or 'analytic' [6], thereby suggesting that there is always scope for questioning the integrity of, for example, the sequenced operating step instructions for a batch-type process; e.g. an 'analytic' observer tends to break up the field of view into its constituent parts or details without necessarily considering each in the context of the whole. Perceptive ability must always strike a balance between rapidity and ease on the one hand, and the highest attainable accuracy on the other. One of the challenging objectives of safety technology training is to teach the practitioner to search carefully and study closely, although residual differences in response may have an innate cause such as intellectual efficiency. There is evidence [7] to suggest that there is no such quality as all-round perceptual efficiency or inefficiency, but rather that we tend to be relatively more efficient in some types of perception and in some perceptual situations than in others. Effective leadership of operability study teams needs to acknowledge the significance of these individual characteristics additionally to the more obvious reasons for multi-disciplinary representation.

One of the golden rules in practising the HAZOP technique is to 'break all assumptions' and yet, of many examples made available to the authors of the aforementioned commercial 'in-house' guidance documents the following inclusions are typical:

> "It is assumed that all the pipework, vessels and other equipment are fit for purpose, and will be correctly maintained, operated and inspected."

"It is assumed that a formal modification procedure is in use covering both hardware and process changes."

"It is assumed that operator actions will generally be correct and in accordance with approved operating procedures."

From a contractor's viewpoint, bearing in mind that HAZOP studies are frequently contracted-out, a disclaiming statement relating to subsequent operating consequences may become less acceptable as the terms of BS5750 "Quality systems" become more widely applied to monitoring purchased services. Part 2, para. 4.8 states "The supplier shall establish criteria for acceptable workmanship through written standards or representative samples."

THE ROLE OF PERCEPTIVE ABILITY

Enhanced perception in the context of process and plant operability can be most usefully applied in the two related areas of:

a)   property-related hazards of materials in process

and   b)   interaction with standards and codes of practice which influence hardware and plant design.

a)   Thirty or more years ago training doctrines within the chemical engineering discipline in the United States began to include the safety dimension in the form of the now well-established 'hazard workshop'. A guide [8] published by Purdue University in 1968 was usefully abstracted [9] for student guidance in the U.K., based upon the premise - "failure to anticipate the outcome---of events by application of (physical chemistry) principles". This latter paper, together with T. A. Kletz's refreshing and reciprocal treatise [10] are amongst the most useful reference sources available to a HAZOP practitioner.

At the conceptual process design stage, operability study via a 'check list' approach is capable of eliciting useful input from related physical properties such as flash point/vapour density (e.g. risk and ventilation levels) and solubility/flammability (e.g. containment strategy and inerting). Amongst other hazard-associated physical characteristics worthy of attention are the behaviour of immiscible liquids (especially on transfer between differing equilibrium states), volume changes between phases, flashing and heats of mixing. Problems possibly arising from kinetics, thermal imbalances and contact of chemicals with their containment are usually more evident and better emphasised in the literature.

b)   Codes and regulations intended essentially for guidance often become mandatory when transcribed into company documentation, thus rendering it more difficult to introduce departures for reasons of safety and reliability. Regulatory authorities are usually appeased provided that any deviations from compliance are resolved, accepted and documented. On the other hand, the authors have experienced instances wherein the influence of a regulating body has been so forceful that reputable equipment suppliers have been obliged to modify their established design practices to meet a market, e.g. in the developing technology of off-shore activities.

Another facet of compliance is the need to guard against relying solely upon relevant regulations, thereby providing perhaps the best opportunities for exercising perception backed by experience.

This can be exampled in three different ways:-

i)   NFPA requires only an overspeed trip on a diesel-driven fire pump - but it should also trip if no water is being pumped, otherwise overheating and damage could result, apart from failure of emergency duty.

ii)   API RP14C lays the stress on "recommended practice" but in many areas covered, good common sense should prevail. The code is focused upon the safety requirements of individual equipment items, whereas it is more important to consider the overall needs of a unit or section of plant as an operating entity.

iii)   Again, API recommends a gas detection unit above a housed compressor but not if installed in a well-ventilated environment. But what regard should be given to the sensitivity and response time of such a protective device?

Apart from aspects of external compliance, certain vulnerable areas of design practice are not always referenced internally. Pipework fittings and the difficulties which they present are possibly the most fertile area for any method study procedure which is required to look beyond the conventional P. & I.D. or E.L.D. All types of flexible joint are prominent in this category, with particular reference to anchors, guides and emphasis upon the impact of operating parameters. E.g. restrained bellows expansion units may require to be fitted vertically to reduce the chance of a malfunction caused by deposition in the convolutions. A recent study on a North American site failed to secure a satisfactory response regarding a skid-mounted vendor packaged pumping unit fitted with armoured flexible couplings on both suction and delivery. Were they to counter vibration or expansion, or both? The flexibles were spanned by substantial welded struts, certainly not essential in transit, but no evidence emerged of pre-commissioning instructions regarding removal.

Strainers are often specified for strategic locations, especially temporarily on start-up, to prevent debris passing into machinery. Invariably, these are not intended to withstand a high pressure drop and, if overlooked, failure can have catastrophic consequences. Such devices can also prevent other protective functions such as alarms and trips from operating in their intended manner.

On many projects, particularly smaller ones and modifications to existing facilities, design documentation is inadequate or found to be out-of-date, sometimes before commissioning. Possibly the most useful outcome in recent times of the obligatory approach to operability study has been the realisation that the field of formal documentation often warrants a thorough review in order to be effectively administered. Prominent areas for enhanced attention are relief systems (drains, valves, bursting discs, vents, flame arresters, flares and blowdown headers etc.) and the tendency for operating staff to make unrecorded changes. In one company survey [11] of relief valve reliability covering over 1600 individual tests on R.V.'s of different types and origins, only 65% lifted within -10% of set pressure, 14% failed to lift by 110% and 10% were found to leak.

It is also appropriate to question the inviolability of log sheets, especially in relation to the quality of contribution to a study team by plant operating personnel. Too often these records represent the interface between conscientiousness and familiarity; better than nothing, maybe, but not foolproof. Now that the HAZOP principle is extended to the early design stages by using check lists, these too should be subjected to scrutiny by independent individuals or teams.

CHANGE AND ACTION CONTROL FOR PROJECT DEVELOPMENT

To protect against unauthorised or ill-conceived changes, a system of control covered by recorded sanctions is mandatory.

Difficulties can start in the design office even before the plant is built. Piping and layout engineers can make what seem to them to be very simple changes in pipe plans to make their life easier. Sometimes the effects of such modifications can be catastrophic.

The difficulties continue on site where construction engineers again wish to make life easy and commissioning engineers sometimes make hasty decisions to overcome one problem and in the process create much more serious problems.

Pressures of production, shortage of spares etc. all create further problems for which ad hoc solutions are sometimes applied. The effects of such solutions are not always fully examined and Flixborough [12] alone should always remind us of this.

1. Control at Design Stage

Once a P. & I.D. has reached the "approved for design" (AFD) status it should have been through the principal HAZOP review (HAZOP 3 in I.C.I. terminology). After this point all changes must be subject to very strict control.

One tried and tested solution uses P. & I.D. change note (Figure 1). Company procedures demand that all changes to a P. & I.D. post-AFD status are made using this procedure. In this way all disciplines are aware of the change and have the opportunity to assess its implications to their activity. In addition, a safety engineer has to decide whether or not the change affects the integrity of the HAZOP.

When the P. & I.D.s are at the next status "approved for construction" (AFC) all the change notes are incorporated into the drawing and a further HAZOP review is conducted. The main objectives of this review are:-

(i)   to ensure that all actions arising out of the process review have been cleared
ii)   to check that the changes made have not breached the integrity of the HAZOP 3 review
iii)   to remove any safety-related "HOLDS" on the P. & I.D.
iv)   to incorporate data not available at the time of HAZOP

Items (i) and (iv) above are largely administrative but some actions could well involve operating instructions yet to be written. In this case the HAZOP chairman must assure himself that the

FIGURE 1

PIPING AND INSTRUMENTATION DIAGRAM
CHANGE NOTIFICATION

DESCRIPTION OR DIAGRAM OF THE PROPOSED CHANGE

REASON FOR CHANGE

CHANGE REQUIRED BY (DISCIPLINE)

	SIGNED	DATE		SIGNED	DATE
PREPARED BY			PROCESS APPROVAL		
ORIGN. DISCIPLINE APPR.			ENG. APPROVAL		
PROJECT APPROVAL					
IS HAZOP REQUIRED? YES/NO					

DRG No.   REV.

CHANGE NOTE No.   SHEET No.   OF

requirements in HAZOP 3 have been passed to the relevant people.

Item (ii) is a review based on the change notes. The HAZOP chairman must first cross-check the decisions made earlier as to whether or not a HAZOP review is needed and countersign the sheets. Items requiring review are then subject to a simplified guide word review, usually with a reduced team. This team would comprise a chairman, a safety engineer, a systems engineer and a process engineer. Actions if deemed necessary are handled in the same way as HAZOP 3.

(It is interesting to note that J. S. Fitt [13] advocates a reduced team of similar structure, even for 'full' hazard studies, including a client representative where applicable and particularly if the client is the licensor).

Item (iii) is usually largely cleared as a result of item (i) though some "HOLDS" may have to be carried forward to AFC status.

Once all the actions from the above item are cleared or it is deemed that they remain on hold, the P. & I.D. can be issued to site at AFC status. Figure 2 is a typical HAZOP Action Sheet and it will be noted that a tape-recording reference is included. Increasingly, especially on critical designs, all study team activity is recorded.

2. Control at Construction Stage

The same basic principles apply but because of the probable distance from home office the discipline of change control is more difficult to enforce. The site engineer must follow procedures which lay down that no modifications must be carried out without consent of the design office and the same paperwork system as for P. & I.D. changes is used.

3. Pre-commissioning Review

A further review is carried out at pre-commissioning. The objectives are basically the same as the pre-AFC but, in addition, a "walk through" is carried out to ensure that the P. & I.D.'s are brought up to "as built" status. This is best done by the commissioning team marking up a set of P. & I.D.'s, checking line-by-line and noting any deviations. Special attention must be paid to drains and vents which tend to be more subject to site modifications. Any unapproved modifications discovered must be referred back to the designers using the change note system as before.

The pre-commissioning review follows a similar format but the commissioning team manager must also be included in the review.

By this time the operating instructions must be available and the chairman must check that all actions from earlier reviews relating to operating instructions have been incorporated.

All remaining "HOLDS" must also be reviewed to ensure that they have no safety implications.

153

FIGURE 2

Project:

HAZOP ACTION SHEET

Project No: Action No:

Drawing No: Date: Tape Ref:

Meeting Sheet: Rev:

Action On: To be
                                        cleared by: (date)

Description:

Reply:

Please return to: Team Secretary

Commissioning must not commence until the review chairman is satisfied that all outstanding issues are resolved.

4. Control during Commissioning

Experience shows that this is a potentially hazardous stage. There is always a tendency for commissioning personnel to think they know better than the designers. If things start to go wrong during commissioning the temptation to make on-the-spot modifications must be prevented by rigorous procedures.

All changes must be referred back and subject to the same review procedures. Any modifications which are accepted must be written up and the drawings modified as built.

Particular care is needed in the case of programmable logic controllers (PLC-based systems) where the implications of ad hoc modifications can be very difficult to fully check out.

5. In Service Modifications

These can take several forms:-
a)   improvements
b)   maintenance
c)   process change

Items a) and c) must be documented in a similar way to the P. & I.D. change note procedure or plant modification procedures (PMP's) produced. PMP documentation is generally more complex, particularly if process changes are involved, and must include a safety case section.

The HAZOP technique can be equally well applied to PMP's and very often it is carried out as a completely new study if only because access to the first study may not be possible.

Usually adequate time is available for a reasonable safety review to be made and it is essential that company procedures ensure that some form of PMP is used and that it includes a safety case. The safety manager's signature must also be required as part of the approval procedure.

The dangers usually arise from maintenance and from pressures to keep production going after a breakdown. Again, rigorous company procedures are the best safeguard but a simplified form of PMP must be used for any temporary modifications. The signatures of all discipline heads, particularly engineering, process, C. & I. and safety must always be obtained.

Figure 3 shows a 'change note' as used in a European oil refinery and now being modified to acknowledge the incorporation of HAZOP reviews.

The origins of HAZOP-governed change control can be traced back to at least 1982 when a paper [14] was presented at an I.Chem.E. Symposium emphasising the need for feedback to support design concept changes to process plant on offshore platforms.

FIGURE 3

NUMBER		ENGINEERING SERVICES REQUEST DATE					
DEPT. OF ORIGIN	DATE	SECTION APPROVAL	DATE	PROJECTED COMPLETION DATE	IS SHUTDOWN REQUIRED?	PLANT LOCATION	

DESCRIPTION OF TASK/PROBLEM (inclu. possible solution if known)

JUSTIFICATION (Give precise reasons and costs)

PROBABLE CONSEQUENCES if not implemented

DEPARTMENTAL ENDORSEMENTS/PROGRESS DATES/COST ESTIMATE APPROVALS

Safety can only be assured by rigorous application of procedures, usually part of a quality assurance plan. Such systems only work if they have a full management back-up. It must be more than a paper chase; that is, people who sign must be aware of what they are signing and must accept responsibility.

It must also be remembered that, in the U.K. and most of Europe, individuals can now be personally held responsible for accidents and hence they must check very carefully before they sign any safety documentation.

EXTENDED GUIDE WORD USAGE IN OPERABILITY STUDY

So far as is known few comprehensive guide word lists compiled for application to specific process and non-process environments have been publicised. One form, with deviations expanded directly from the classical I.C.I. list and used successfully in many HAZOP training events is shown in Table 2. It was built up over a period of time from observation of features common to a wide range of case studies adapted from field experience and not from purely academic considerations.

Amongst other versions seen in use, additional guide words include such as "sooner/later than" and "changes in physical/chemical condition"; the former presumably cater for batch process conditions now more conventionally handled by operating sequence time-line diagrams tied to a given plant status, as exampled in Figure 4, and the latter should be covered by the conventional entries of pressure, temperature, composition change etc. By the same token, HAZOP report sheet column headings such as 'design intent' would appear to be superfluous. Computerised recording programmes (e.g. 'HAZSEC'[15]) of proven and time-saving attributes appear not to rely upon more than the basic preamble to "ACTION", "ACTION BY" and "REVIEW" columns. The expansion of each process step (Figure 4), whether automated or performed manually, is a useful 'aide memoire' for study team use.

Table 3 is an extract from a proposed expanded guide word list adapted for drilling systems; it follows throughout the sequence of Table 2 and it can be seen that the predominant emphasis within a mechanical system highlights the operator interface notably. One adage in this context, of course, - stemming from the familiar oft-repeated question "How will the operator know---? is not so much to ask what will happen if the operator makes a mistake but 'how often'? will he act incorrectly.

It cannot be over stressed how important it is to follow the basic methodology. From both training and consultants' viewpoints there are two main divergent paths from which, often with difficult diplomacy, it is necessary to recover a faulted approach. Either a client will have adopted a concept born of compromise between available abilities and the time factor (without necessarily deprecating the virtues of competent 'coarse' or 'mini' HAZOP studies) or a premature attempt is made to create an internal 'Company guidance document' before any hands-on experience is acquired. An example of this latter circumstance is represented by Table 4 which was incorporated in a HAZOP 'orientation programme' at one location along with hazard severity and probability scales plus other excursions across the qualitative/quantitative threshold. Comparing this with an adaptation for electrical distribution systems of the established guide words (Table 5) it can be seen that little advantage, much confusion and - above all - the risk of overlooking important deviations could be the outcomes

TABLE 2

EXPANSION OF GUIDE WORDS

NO FLOW:-	WRONG ROUTING - BLOCKAGE - INCORRECT SLIP PLATE - INCORRECTLY FITTED N.R.V. - BURST PIPE - LARGE LEAK - EQUIPMENT FAILURE (C.V., ISOLATION VALVE, PUMP, VESSEL, ETC) - INCORRECT PRESSURE DIFFERENTIAL - ISOLATION IN ERROR - ETC
REVERSE FLOW:-	DEFECTIVE N.R.V. - SYPHON EFFECT - INCORRECT DIFFERENTIAL PRESSURE - TWO WAY FLOW - EMERGENCY VENTING - INCORRECT OPERATION - ETC
MORE FLOW:-	INCREASED PUMPING CAPACITY - INCREASED SUCTION PRESSURE - REDUCED DELIVERY HEAD - GREATER FLUID DENSITY - EXCHANGER TUBE LEAKS - RESTRICTION ORIFICE PLATES DELETED - CROSS CONNECTION OF SYSTEMS - CONTROL FAULTS - ETC
LESS FLOW:-	LINE RESTRICTION - FILTER BLOCKAGE - DEFECTIVE PUMPS - FOULING OF VESSELS, VALVES, RESTRICTOR OR ORIFICE PLATES - DENSITY OR VISCOSITY PROBLEMS - INCORRECT SPECIFICATION OF PROCESS FLUID - ETC
MORE PRESSURE:-	SURGE PROBLEMS - LEAKAGE FROM INTERCONNECTED H.P. SYSTEM - GAS BREAKTHROUGH (INADEQUATE VENTING) - ISOLATION PROCEDURES FOR RELIEF VALVES DEFECTIVE - THERMAL OVERPRESSURE - POSITIVE DISPLACEMENT PUMPS - FAILED OPEN P.C.V.s - ETC
LESS PRESSURE:-	GENERATION OF VACUUM CONDITION - CONDENSATION - GAS DISSOLVING IN LIQUID - RESTRICTED PUMP/COMPRESSOR SUCTION LINE - UNDETECTED LEAKAGE - VESSEL DRAINAGE - ETC
MORE TEMPERATURE:-	AMBIENT CONDITIONS - FOULED OR FAILED EXCHANGER TUBES - FIRE SITUATION - COOLING WATER FAILURE - DEFECTIVE CONTROL - FIRED HEATER CONTROL FAILURE - INTERNAL FIRES - REACTION CONTROL FAILURES - ETC
LESS TEMPERATURE:-	AMBIENT CONDITIONS - REDUCING PRESSURE - FOULED OR FAILED EXCHANGER TUBES - LOSS OF HEATING - ETC
MORE VISCOSITY:-	INCORRECT MATERIAL SPECIFICATION - INCORRECT TEMPERATURE
LESS VISCOSITY:-	INCORRECT MATERIAL SPECIFICATION - INCORRECT TEMPERATURE
COMPOSITION CH.:-	LEAKING ISOLATION VALVES - LEAKING EXCHANGER TUBES - PHASE CHANGE - INCORRECT FEEDSTOCK/SPECIFICATION - INADEQUATE QUALITY CONTROL - PROCESS CONTROL PROCEDURES - ETC
CONTAMINATION:-	LEAKING EXCHANGER TUBES OR ISOLATION VALVES - INCORRECT OPERATION OF SYSTEM - INTERCONNECTED SYSTEMS - EFFECT OF CORROSION - WRONG ADDITIVES - INGRESS OF AIR - SHUTDOWN AND STARTUP CONDITIONS - ETC
RELIEF:-	RELIEF PHILOSOPHY - TYPE OF RELIEF DEVICE AND RELIABILITY - RELIEF VALVE DISCHARGE LOCATION - POLLUTION IMPLICATIONS - ETC
INSTRUMENTATION:-	CONTROL PHILOSOPHY - LOCATION OF INSTRUMENTS - RESPONSE TIME - SET POINTS OF ALARMS & TRIPS - TIME AVAILABLE FOR OPERATOR INTERVENTION - ALARM & TRIP TESTING - FIRE PROTECTION - ELECTRONIC TRIP/CONTROL AMPLIFIERS - PANEL ARRANGEMENT AND LOCATION - AUTO/MANUAL FACILITY & HUMAN ERROR - ETC

TABLE 2 (continued)

SAMPLING:-   SAMPLING PROCEDURE - TIME FOR ANALYSIS RESULT - CALIBRATION OF AUTOMATIC SAMPLERS/RELIABILITY - ACCURACY OF REPRESENTATIVE SAMPLE - DIAGNOSIS OF RESULT - ETC

CORROSION/EROSION:-   CATHODIC PROTECTION ARRANGEMENTS - INTERNAL/EXTERNAL CORROSION PROTECTION - ENGINEERING SPECIFICATIONS - ZINC EMBRITTLEMENT - STRESS CORROSION CRACKING (CHLORIDES) - FLUID VELOCITIES - RISER SPLASH ZONES - ETC

SERVICE FAILURE:-   FAILURE OF/DRY INSTRUMENT AIR/STEAM/NITROGEN/COOLING WATER/HYDRAULIC POWER/ELECTRIC POWER - TELECOMMUNICATIONS - HEATING & VENTILATING SYSTEMS - COMPUTERS - ETC

MAINTENANCE:-   ISOLATION - DRAINAGE - PURGING - CLEANING - DRYING - SLIP PLATES - ACCESS - RESCUE PLAN - TRAINING - PRESSURE TESTING - WORK PERMIT SYSTEM - CONDITION MONITORING - ETC

STATIC:-   EARTHING ARRANGEMENTS - INSULATED VESSELS/EQUIPMENT - LOW CONDUCTANCE FLUIDS - SPLASH FILLING OF VESSELS - INSULATED STRAINERS & VALVE COMPONENTS - DUST GENERATION & HANDLING - HOSES - ETC

SPARE EQUIPMENT:-   INSTALLED/NON INSTALLED SPARE EQUIPMENT - AVAILABILITY OF SPARES - MODIFIED SPECIFICATIONS - STORAGE OF SPARES - CATALOGUE OF SPARES - ETC

SAFETY:-   FIRE & GAS DETECTION SYSTEM/ALARMS - EMERGENCY SHUT DOWN ARRANGEMENTS - FIRE FIGHTING RESPONSE TIME - EMERGENCY & MAJOR EMERGENCY TRAINING - CONTINGENCY PLANS - T.L.V.s OF PROCESS MATERIALS & METHODS OF DETECTION - NOISE LEVELS - SECURITY ARRANGEMENTS - KNOWLEDGE OF HAZARDS OF PROCESS MATERIALS - FIRST AID/MEDICAL RESOURCES - EFFLUENT DISPOSAL - HAZARDS CREATED BY OTHERS (ADJACANT STORAGE AREAS/PROCESS PLANT ETC) - TESTING OF EMERGENCY EQUIPMENT - COMPLIANCE WITH LOCAL/NATIONAL REGULATIONS - ETC

159

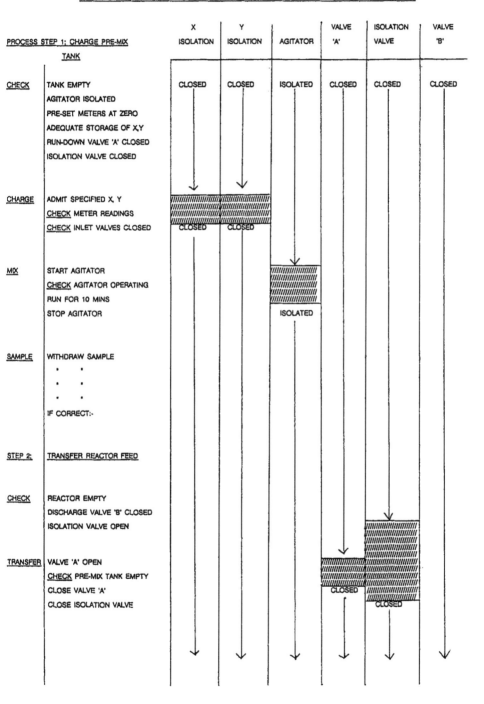

FIGURE 4

PART OF OPERATING TIME-LINE DIAGRAM FOR CATALYTIC BATCH REACTOR

TABLE 3

PROPOSED HAZOP GUIDE WORDS FOR DRILLING SYSTEMS

STANDARD 'HAZOP'	SUGGESTED 'DRILLING HAZOP'	BASIS FOR DEVELOPING QUESTIONS
NO FLOW	PROPOSED ACTION FAILS NO MOVEMENT	PIPEWORK OR HOSE FAILURE (INCLUDING FLEX. COUPLINGS) EQUIPMENT FAILURE (BEFORE OR DURING OPERATION) INCORRECT OPERATION (OPERATOR ERROR) LIFTING EQUIPMENT FAILS TO LIFT/OR FAILS DURING LIFTING
REVERSE FLOW	OPPOSITE OR DIFFERENT ACTION REVERSE MOVEMENT	FAILURE OF LIFTING EQUIPMENT (FALLING OR DROPPED OBJECTS) OPPOSITE/DIFFERENT OR UNEXPECTED ACTION OCCURS HOSE CONNECTIONS INCORRECT
MORE FLOW	INCREASED ACTIONS INCREASED MOVEMENT	OTHER ACTIONS CARRIED OUT AT SAME TIME PROCEDURES NOT FOLLOWED CORRECTLY (SHORT CUTS) LIFTING EQUIPMENT NOT CONTROLLED CORRECTLY FAILURE OR HYDRAULIC CONTROL SYSTEMS FAST ROTATION OF ROTARY TABLE FAST OPERATION OF LIFTING EQUIPMENT COMBINED DRILLING AND PRODUCTION OPERATIONS PRODUCTION INTERFERENCE X ------- X ------- X ------- X ------- X ------- X ------- X ------- X INCREASED MOVEMENT OF DRILLING RIG (DEFECTIVE HEAVE COMPENSATOR, BALL JOINT AND TELESCOPIC JOINT MOVEMENT EXCESSIVE) DYNAMIC POSITIONING, DRAGGED ANCHORS ETC. X ------- X ------- X ------- X ------- X ------- X ------- X ------- X
REDUCED FLOW	REDUCED ACTIONS REDUCED MOVEMENT	DEFECTIVE EQUIPMENT (SLOW OPERATION) OPERATOR FATIGUE (SLOW RESPONSE) RESTRICTED PIPEWORK/FILTER BLOCKAGE FAILURE OF HYDRAULIC CONTROL SYSTEMS PIPEWORK OR EQUIPMENT LEAKAGE
INCREASED PRESSURE	INCREASED LOADING	EXCESS TORQUE GENERATED BY HYDRAULIC EQUIPMENT EXCESS LOADING OF CABLES OR LIFTING EQUIPMENT EXCESSIVE HYDRAULIC PRESSURES SURGE PROBLEMS IN LIQUID SYSTEMS DAMAGE CAUSED BY FALLING OBJECTS DESIGN SPECIFICATIONS EXCEEDED INCORRECT PRESSURE DIFFERENTIAL (HYDRAULICS/DOWNHOLE PRESSURE CONTROL) WEATHER, WIND, ROUGH SEAS ETC.
REDUCED PRESSURE	REDUCED LOADING	RESTRICTED PUMP SUCTION LINE LEAKAGE FROM PIPEWORK (UNDETECTED) PASSING RELIEF OR ISOLATION VALVES DEMAND BY HYDRAULIC EQUIPMENT EXCEEDS CAPACITY OF SYSTEM INCORRECT PRESSURE DIFFERENTIAL (HYDRAULICS/DOWNHOLE PRESSURE CONTROL)

TABLE 4

HAZARD AND OPERABILITY DIAGRAM
FOR ELECTRICAL SYSTEMS

Category	Sub-items	Description
Changes In Current	High current Low current No current Reverse current	One supply, two supplies
Changes In Voltage	High voltage Low voltage Reverse voltage Potential differences (too great / too little)	Surge, spikes, transients, flash-over Voltage dips, fluctuations, variations Back emf, connecting errors High impedance, Low impedance,
Frequency	Too high Too low	Harmonics
Phase	Lead/lag Loss of	Inductive/capacitive, synchronising
Start-up/Shutdown	Switching: Common mode, Spurious, Failures Maintenance Synchronization/sequencing Testing/commissioning/spares holdings	Corrosion, water-ingress, interference, fire/explosion Too fast, too slow, never partial, erratic, too late, too soon, access, isolation, procedures, interlocks Responsibility
Interference	Electrical	AC on DC, DC on AC, transients, lightning, earth faults, insulation faults, corruption, electro-chemical, magnetic fields
	Environmental	High/low temperature, corrosion, flooding, fire/explosion, mechanical failure/damage, humidity
Emergencies	Power failure	Emergency lighting, drives, alarms, battery failure, communications
Grounding	None	

TABLE 5

"HAZOP GUIDE WORD FORMAT FOR ELECTRICAL DISTRIBUTION DIAGRAMS"

NO FLOW	Distribution Board component failure, connection failure between components, faults in overload trip system, isolation in error (Human Error)
REVERSE FLOW	Wrong connections after maintenance
INCREASED FLOW	Short circuit, leakage to ground, required load in excess of supply capacity, or in excess of design requirement, overload settings too high. Plant being operated in excess of design parameters
REDUCED FLOW	Faulty load shedding arrangement, reduced generator capacity, phase imbalance
INCREASED PRESSURE	Voltage increase etc.
REDUCED PRESSURE	Voltage reduction, voltage dip
INCREASED TEMP.	Resistance to the fire situation, effect of ambient temperature changes, overheating of distribution board components, (how detected), detection of electrical fires, cable routing.
REDUCED TEMP.	Winter conditions, effect of sub-zero temps, ice formation around electrical components, failure of 'HVAC' equipment
CONTAMINATION	Ingress of flammable atmosphere into electrical distribution rooms. Internal damage by animal, insect life. Mechanical and water damage.
INSTRUMENTS	Set points for MAX. overload condition, (how tested and adjusted), load shedding arrangements, (how controlled), ground leakage detection
CORROSION	Corrosion resistance of equipment, incorrect humidity
SERVICE FAILURE	Electrical supply failure, failure of any associated services, reliability of standby equipment, paths of communication
MAINTENANCE	Maintenance isolation procedures, work permit systems
SAFETY	Fire protection, fire fighting arrangements for distribution equipment. Personnel protection, area classification. Training of maintenance and operating staff

of this particular variant. Table 5, of course, is capable of being adapted for studying control loops associated with electronic systems.

Computer control systems represent an important current challenge for operability study. One prominent consultancy organisation has reported [16] good results using a modified HAZOP technique to carry out a failure mode analysis in conjunction with a revised set of guide words for examining the related human factors and the interface.

Finally, of relevance to the use of the HAZOP 'check-list' approach [5] to HAZOP stages 1 and 2 (in I.C.I. terminology) are the guidelines [17] published by the American Institute of Chemical Engineers. These include a "Sample HAZOP checklist" sub-divided under equipment identity headings (pumps, towers, pipework, reactors etc.). Whilst useful, but obviously not unique, some entries tend to inhibit the critical influence of perceptive scope. Under the heading, for example, "Is backflow prevented?" one directive reads "Double check valve for $\Delta P > 1000$ psi". Elsewhere, other finite values for operating parameters are also quoted which rather suggests a close relationship with established design code usage thereby, at least in the minds of the authors, tending to invalidate the unfettered use of the HAZOP study method, an objective which encapsulates its most important potential.

CONCLUSION

The use of operability study can be unwittingly abused and is sometimes quoted as evidence that a system is thereby protected against untoward events. This paper has attempted to demonstrate that the quality of human input is capable of improvement and that there is a case for viewing abbreviated variants of the orthodox method with suspicion. Competent formal induction of newcomers to the technique is vital if its effectiveness in other areas is to be fully and confidently exploited.

REFERENCES

1.	T. A. Kletz	'HAZOP AND HAZAN - notes on the identification and assessment of hazards', I.CHEM.E., 1986
2.	H. G. Lawley	'Operability studies and hazard analysis', Chemical Engineering Progress, 1974, 70, 45-56
3.	H. G. Lawley	'Size up plant hazards this way', Hydrocarbon Processing, 1976, 55 (4), 247
4.	P. C. Merriman	'HAZOP - the views of the Chemical Industries Association', I.CHEM.E. Loss Prevention Bulletin, 046, 1981
5.	J. L. Hawksley	'Process Safety Management: a U.K. approach', Plant/Operations Progress, 1988, 7 (4), 265-9
6.	J. L. Singer	'Personal and environmental determinants of perception', J. Experimental Psychology, 1952
7.	L. L. Thurstone	'A factorial study of perception', Univ. of Chicago Press, 1944
8.	W. H. Tucker	'The Chemical Engineering of Safety', Purdue University, Illinois, June, 1968
9.	A. J. D. Jennings	'The Physical Chemistry of Safety', The Chemical Engineer, Oct. 1974, 637-52
10.	T. A. Kletz	'Myths of the Chemical Industry', I.CHEM.E., 1984
11.	F. R. Mitchell	Private communication, Nov. 1987
12.	"The Flixborough Disaster - Report of the Court of Inquiry", HMSO, London, 1975	
13.	J. S. Fitt	'Sizing up process plant hazards', Health and Safety at Work, July, 1980
14.	M. A. F. Pyman and F. R. Mitchell	'Identifying major process hazards at the concept design stage', Design 82, I.CHEM.E.LDLldmemem.d!%%NXS#6ies
15.	Technica Ltd.	'In-house' computer programme for recording 'HAZOP' meetings, Feb, 1989
16.	L. J. Bellamy and T. A. W. Geyer	'Addressing human factors issues in the safe design and operation of computer controlled process systems', SARSS '88 Symposium
17.	"Guidelines for hazard evaluation procedures", Appendix A, A.I.CHEM.E., 1985	

SOME ASPECTS OF THE USE OF QUANTIFIED RISK ASSESSMENT FOR DECISION-MAKING

By R P PAPE
HSE Technology Division, St Anne's House,
Stanley Precinct, Bootle, Liverpool L20 3RA

Introduction

1   In this paper I offer some personal thoughts on the use of QRA
as an aid to decision-making, particularly for decisions by society
beyond the technical sphere. This draws mainly upon my experience
gained with the development and implementation of various approaches
for the assessment of major hazards, for the purposes of siting and
public building control nearby. HSE (and its forerunners) have been
involved in such work since 1972. They act as advisers to local
planning authorities and others whose task it is to decide on
planning permission for land-use development.

Initially HSE relied on an assessment of the consequences of the
worst reasonably foreseeable accidents, giving hazard ranges which
defined the zones where there seemed possibly to be significant cause
for concern. In the late 1970s, a seminal project to quantify the
risks from major hazards at Canvey Island gave some indication of the
potential usefulness of QRA here[1]. At the same time, HSE assessed
the risks from a cross-country pipeline to assist in decisions on the
siting[2]. Further work led to a general approach for use in-house
by HSE, based on a computerised 'shell' called RISKAT - the Risk

Assessment Tool[3]. At the same time, the Royal Society published a Study Group report on QRA, with suggestions for numerical criteria[4].

HSE's experience with its approach for specific cases, using the Royal Society criteria as guidance, allowed HSE to produce a summary of its own views on Risk Criteria for the present purpose[5]. This followed the publication of HSE's document on the Tolerability of Risk from Nuclear Power Stations[6]. Since then, HSE has completed the trilogy with a general discussion on the use of QRA in decision-making[7], with descriptions of several case-studies and an analysis of factors which affect the perception of risk.

The notice for this conference raised several questions, to which this paper will attempt to relate. The general issue of QRA as an input to high-level decision-making is illuminated by the particular usage discussed here. The question of resources to do the work is exemplified by the need to provide a practical tool for casework through RISKAT. The question of how far to go in using 'prediction' as the basis underlies much of the discussion.

Regarding our learning from incidents, I would hope that we are continually aware of the need to test our predictions against the historical record, and to review the methods in the light of new accidents. (However, where the predictions disagree with 'experience' we should not always assume that the predictions are wrong!) Finally, we must try to understand the attitudes and perceptions of the public and their representatives, so as to encourage the best use of risk predictions in society's crucial decisions. Here, a brief outline of the responses to HSE's documents may give some insights, and this may be supplemented by the practical experience of regulators and the studies of social scientists.

2   Contributions and Limitations of QRA for Decision-Making

It seems necessary to begin by noting that the results of 'quantified risk assessment' may be used in a wide variety of situations, from plant design options to high-level policy. The results may be used as they stand in comparison with set criteria, or they may be input to some form of cost-benefit assessment (CBA). As we move up the decision-level hierarchy the significance of the numbers decreases relative to other factors involving qualitative and value judgements.

Somewhere towards the top of the hierarchy comes legislation; and it is interesting to note how opinions differ on whether to incorporate QRA and criteria into legislation. Presumably a society's views on this depend on the wish to retain flexibility, perhaps to allow for the uncertainty in QRA and CBA, or to allow value-judgements to be applied. Also, familiarity with quantified risk decisions may be significant. I have heard it argued that the Netherlands government may be more willing than most to adopt QRA[8] because of their long experience in judging the height of sea-defences from the statistics of sea-levels.

In the UK, the HSE has recently set out its views with some precision, in a foreword to [7] and in evidence to the Piper Alpha inquiry[9]. It is worth quoting from the former, as it touches on several of the themes in this conference:

> "There is a general conclusion. Major disasters can and do
> happen: but the chance of any one happening must be kept
> very low. The graphs in this paper in part reflect actual
> experience: some accidents that have actually happened in
> the UK - in work-related and other situations. They also
> predict the probabilities of even larger disasters. These
> are not mere theoretical constructions: some very recent
> events at home and abroad, such as the fire at King's Cross
> Underground Station where 31 people were killed, the
> Zeebrugge Ferry disaster (184 killed), and the Piper Alpha

North Sea Oil Rig disaster (167 dead) amply prove that large-scale disaster can happen. So do the events at Bhopal, Mexico City and Chernobyl which are noted in the text. Before Bhopal, the largest death-toll known to us from an accidental release of toxic gas was 60. After Bhopal it was over 2000. But predictions of possible consequences from major disasters had already envisaged disasters on such a scale.

It is therefore important to be able to predict what could happen - and as far as possible how likely or unlikely that is - as well as recording what has actually happened, and then to see how best to control, and if possible, reduce the risks that are identified. For this QRA is an indispensable element, but one to be used with caution and not applied mechanistically to demonstrate compliance with legislative requirements."

From this, we see that HSE would not advise that risk criteria be put into legislative requirements. QRA may be used where it is helpful to do so, as part of a demonstration of compliance with the general requirement to take precautions so far as is reasonably practicable. HSE guidance on the CIMAH regulations mentions the use of QRA in a safety report[10]. In the nuclear field, the NII insists that prediction and probability analysis are used to estimate the risk of accidental releases of radioactivity from nuclear power plant [6, para 82]. QRA has also been used for safety devices in mines[11] and in various other fields where it seemed worthwhile[7]. The Health and Safety Commission's Advisory Group on Major Hazards Transport will soon publish its First Report, with a substantial input from QRA; this is seen as valuable to give a balanced impression of the main sources and levels of risk in the various transport modes. In all cases QRA is insufficient on its own to demonstrate compliance; underlying it is the necessity that appropriate standards, operating and maintenance procedures, and other quality factors are applied.

There is not space here to discuss in detail the current
controversies over certain aspects of QRA, particularly for high-
level decision-making. These include the uncertainties in absolute
risk figures, and the significance of human factors [eg 12]. Perhaps
I could just suggest that these problems vary considerably in
significance depending on the type of assessment and its context.
For example, a comparison of options for a detail of plant design may
not suffer from the broader types of uncertainty, and such
assessments are now widely accepted as aids to design. There may
also be scope for minimising the difficulty by careful specification
of the decision-problem, eg:

1   If a 'target' risk level is set, we may be confident that it is
met in practice if an assessment which is clearly pessimistic
produces a result which is well below the target level.

2   If two options are being compared, one is clearly preferable to
the other if the difference exceeds the likely uncertainty.

However, some decisions are absolute (ie yes/no) and they involve
drawing a line (ie risks above are 'no'; risks below are 'yes').
The decisions addressed in the Risk Criteria document[5] are of this
type, since we must decide what separation-distance is required
between a major hazard and, say, a new school. In practice, we might
avoid too sharp a division, by leaving a grey area between the 'no'
zone and the 'yes' zone. This is the classic three-zone approach for
criteria. Hence:

3   If an absolute decision is to be made, a middle zone may be
inserted between the two extremes. Risks which fall within this zone
may be judged from secondary factors or by the use of more detailed
assessments, or CBA etc.

Given the uncertainty in QRA and the resources needed to carry it
out, it is important to be discriminating in its use. Some pros and
cons are listed below:

Advantages	Disadvantages
Systematic, structured	Uncertainty
Basis for comparisons	Difficult to explain
Numerical results	Resources needed
Input to CBA etc	Expertise rare
Objective	Difficult to check
Reveals weaknesses	Highlights very rare events
Understanding plant	Needs criteria

It is clear that QRA should only be used where it makes a
contribution to the information available sufficient to outweigh the
disadvantages. For higher levels of decision-making, where the
uncertainties are likely to be higher and the costs of doing a QRA
are greater, it might be favoured by:

1   Lack of historical experience, either because the situation is
novel or the type of accident is rare;
2   High cost of accidents (justifying the use of all available
assessment tools if they might help to reduce the risk);
3   Lack of tried and tested engineering standards;
4   Perceived sensitivity of the decision (requiring as much
objective information as possible - with due acknowledgement of
uncertainty).

In all cases where the use of QRA is not routine, some care is needed
in deciding the approach and the degree of sophistication to be
applied. It will be apparent from the description in [5] that the
HSE methods there have been developed for the particular purpose
described below.

3   A Case-Study: HSE Philosophy and Procedures for Control of
    Developments near Major Hazards

3.1 Assessment Approach

It was indicated above that HSE advice on the location of public
buildings relative to major hazards began with a consideration
of the potential consequences of the 'worst reasonably
foreseeable' accidents. For example, with a bulk pressurised
LPG tank it is foreseeable that there might be a fireball event
with a large fraction of the tank contents involved almost
instantaneously (ie a BLEVE - Boiling Liquid Expanding Vapour
Explosion). For a 100-te tank this could produce a fireball of
about 100m radius, with potentially-lethal intensities of heat
radiation a few hundred metres beyond the fireball edge. There
might also be damage from burning liquid spray and missiles from
the bursting tank. Another possibility - although probably less
likely with LPG - is a vapour cloud explosion of the Flixborough
type, producing severe overpressure effects to ranges of several
hundred metres.

The historical record contains sufficient examples of these
types of event to suggest that they need to be allowed for in
the planning of building development near LPG installations.
The risks are quite low (no BLEVE incidents with large LPG tanks
in the UK, after perhaps 10^5 vessel-years experience), but the
consequences could obviously be catastrophic. Thus a policy has
been developed based on hazard ranges estimated from consequence
assessments along the lines above[13]. This was endorsed by the
Advisory Committee on Major Hazards in their Third Report[14]:

    "Ideally, the separation should be such that the
    population would be unaffected whatever accident
    occurs. For hazardous installations, however,
    such a policy is not reasonably practicable. It
    seems reasonable to aim for a separation which
    gives almost complete protection for lesser and

> more probable accidents, and worthwhile
> protection for major but less probable
> accidents."

This approach seems sensible for hazards where there are
clearly-defined worst reasonably foreseeable events. Apart from
the LPG example, such cases as potentially explosive substances
(eg ammonium nitrate, sodium chlorate, ethylene oxide etc) and
certain types of process, pipelines etc might be dealt with in
this way. In all such cases there is of course scope for
quantification of risks as well as consequences. In due course
it is to be hoped that such quantification will be done to give
a more balanced impression of a hazard situation, and HSE
continues to work with consultants and industry to develop
this.

QRA is particularly desirable for other types of major hazard
where the 'worst reasonably foreseeable' or consequence-based
approach is less straightforward and the historical record is
patchy or even quite inapplicable. For example, an installation
handling a liquefied toxic gas such as chlorine could in
principle produce releases ranging from a small flange weep,
through a pipe fracture, to the contents of a whole tank. The
range of effects from any release is highly dependent on the
weather conditions; and the effects are also directional, so a
nearby building may not be affected (upwind) while one much
further away is affected (downwind). It is still possible, for
any particular type of installation, to devise a list of events
and hazard ranges in particular weather conditions, and by
judgement to deduce a range which satisfies the Advisory
Committee's advice (above). However, this would be a rather
arbitrary process and very difficult to apply consistently
between types of installation. This is where the quantification
of risks may provide a less arbitrary foundation for the
decision-making process in this area.

One incentive for work to develop QRA methods is that the
greater precision may justify the use of somewhat less extensive
control zones, thus reducing the potential effect of a major
hazard on land development in its vicinity. A reduction in zone
radius from 700m to 500m around a point would halve the amount
of land caught by the zone. With some 1700 notified major
hazard installations in Great Britain, plus many kilometres of
notified pipelines, the potential reductions in land area
affected are very substantial.

Apart from these points, HSE felt encouraged to move forward by
the Canvey Island work; also by the conclusions of a major
public planning inquiry in 1980[15], and by some early work on
the assessment of an NGL pipeline in Scotland[2]. The HSE
approach was developed to suit the nature of the problem and the
resources available, making best use of existing expertise
including cross-fertilisation between the Technical, Research
and Nuclear Installations Inspectorate divisions of HSE.

3.2 Results and Criteria

The basic objective of the assessment is to show the risk to an
individual living in a house at a particular location near a
major hazard. Here, the 'risk' is the chance per year
(frequency) that a person will suffer life-threatening exposure
to a toxic substance or heat intensity, or blast overpressure
(see [16] for more detail on the definition of dose here). This
information is used by HSE to advise whether housing in that
location should be refused planning permission at the outline
planning stage, bearing in mind that the costs to society of
refusal are less then than at subsequent stages. Thus a fairly
stringent approach seems reasonable.

For developments other than housing eg a cinema, HSE works by
comparison with housing. First, an assessment is done for a
hypothetical house resident at the location in question. A view
is taken on the advice which would be given if housing were

actually to be proposed there. Then a judgement is made as to
whether the actual development is more, less, or similar in
significance to housing. This judgement may involve more
assessment, for example a consideration of the times of day when
the development is in use (eg afternoon and evening, for a
cinema, compared to the possible continuous use of a house);
the times of day may be significant where the wind/weather
pattern varies (eg low-dispersion category'F' weather is more
likely at night). Finally, a conclusion is reached on the
particular development from the view on hypothetical housing
combined with the judgement of significance compared to housing.
To simplify this process, rules are developed, either by
generalised assessments or by casework precedent. Many types of
non-housing development are significantly less vulnerable than
housing, so the advice is more relaxed; but others (schools,
hospitals, old people's homes) are more vulnerable, and the
advice is stricter.

It might seem that a more objective approach would be to extend
the assessment to include societal risk. Then numerical
criteria could be used both for the individual risk and for the
societal risk. However, technical problems arise in the
particular context discussed here, where cases arise piecemeal
around a very wide variety of major hazard installations.

HSE's advice is intended to avoid a significant worsening of the
general national situation. Now any one development case (of
the 4000 a year referred to HSE) would probably add very little
to the existing national societal risk; but if there were no
control, cases would eventually accumulate to a significant
worsening. How do we define a 'significant worsening', and how
do we deduce whether a particular type of case, if multiplied
over the years, would contribute substantially? This is a
problem of the partition of an overall societal risk criterion
among the contributing parts of the risk. Such problems are not
insuperable, but HSE felt that the present state of the art did
not justify proposals for criteria at this time. Note that the

derivation of societal risk criteria may be much less
troublesome for other types of hazard context; for example,
where there is a single source of the hazard, or a few highly-
similar sources, and there is a limit to the amount of
development likely (eg nuclear power stations).

The HSE proposals for criteria for developments near major
hazards follow the usual three-zone approach, with an inner zone
where many types of development would automatically be advised
against; a middle zone where the details of such cases would be
discussed before advice is given; and an outer zone where all
but a few of the most highly-sensitive cases would automatically
be 'low risk'. (A limit is set on the outer zone; beyond that
limit, no cases need be considered. This defines the
'consultation zone' for an installation). The most significant
detail in the middle zone is the size of a development, so that
it is here that the general concept of societal risk is taken
into account. The criteria have been summarised as follows:

"HSE will advise against developments near Major Hazards in
the circumstances shown below. The advice is stronger the
higher the calculated risk. It is for Planning Authorities
to decide whether the development should be refused,
weighing the factor of safety with other factors in the
case. Planning Authorities are advised to give more weight
to safety for higher risks or larger developments than the
limits quoted here.

a)   Housing developments providing for more than about 25 people
      where the calculated individual risk of receiving a defined
      'dangerous dose' of toxic substance, heat or blast overpressure
      exceeds 10 in a million per year.

b)   Housing developments providing for more than about 75 people
      where the calculated individual risk of receiving a 'dangerous
      dose' exceeds 1 in a million per year.

c)   For developments in Category C [leisure facilities etc] HSE
     prefers not to suggest hard-and-fast rules based simply on risk
     figures. This is a category where the factors determining risk
     vary considerably in their importance, so it is difficult to
     generalise. HSE will tend to follow a similar approach to
     Category A [ie housing]. Thus moderately sized developments
     will usually be inadvisable where the calculated individual risk
     for a hypothetical house resident exceeds 10 in a million per
     year, and large ones where the risk figure exceeds 1 in a
     million per year.

d)   The 1 in a million per year criterion may be extended downwards
     somewhat for developments which fall into Category D [super-
     sensitive], either because the inhabitants are unusually
     vulnerable or the development is extremely large.

e)   For all types of development, HSE is likely to advise more
     strongly against planning permission where the development forms
     a precedent for further growth or encroaches within any pre-
     existing cordon sanitaire, and comes within the risk criteria
     described above."[5]

3.3 Resources Used

     It is not intended here to describe the actual techniques used
     by HSE for the risk assessment; these are detailed in [5],[13],
     and other papers referenced there. It may be of interest to
     mention the process and resources used for the work. Cases
     originate when a developer applies to a local planning authority
     (LPA) for permission to develop a piece of land near a major
     hazard. The LPA will have been told by HSE what types of
     development might be of interest there. Many minor developments
     are excluded - advertising hoardings, house extensions, small
     factory developments etc. The rest, about 4000 per year, are
     referred to the local HSE Area Office. Here a large number of
     cases are dealt with using general rules suggested by HSE HQ.
     About 600 cases a year require detailed assessment, and these

are sent to the Major Hazards Unit in HSE HQ. This Unit is
responsible for assessing these cases and formulating advice to
be sent to the LPA. Other work of the Unit includes the
development of the methodology further (in collaboration with
Research Division of HSE and consultants), and to produce
general guidelines for Area Offices. The Unit (through Area
Offices) will provide detailed support for cases which proceed
to planning appeal inquiry.

The Unit work described above is done by about ten senior
specialist inspectors of HSE, mostly with chemical engineering
or related backgrounds. They receive specific training in the
assessment techniques, and considerable peer group support from
colleagues; about 25% of cases in the Unit are submitted to a
Panel discussion of colleagues before the advice is finalised.
Much emphasis is given to codifying the work, in computer
programs, specific instructions and precedent from Panel
decisions. Thus an average workload of 60 cases per year per
person (plus supporting work) is refined to ensure that
resources are concentrated on areas of novelty or marginal
decisions. Apart from technical work, Unit staff spend
substantial time in meetings with LPAs etc, since HSE advice
against a development may cause local difficulties so that face-
to-face discussion and detailed explanation are required.

3.4   The Decision Process

The decision on a particular case is made by the LPAs, not HSE.
This is a situation where the QRA results form an input into a
high-level decision, but they do not automatically determine
that decision. The question whether LPAs should be left with
the responsibility to make such difficult decisions, where
issues of safety may clash with the potential benefits of a
development, was debated by the Advisory Committee on Major
Hazards. They held firmly to the view that the safety issue
could not be treated in isolation, and LPAs were best placed to
take all the pros and cons into account.

It is my impression that most LPAs would regard firm advice from
HSE against a development as conclusive. However, a few do
attempt to weigh safety against benefits, and on occasion they
may grant permission for a development where HSE has advised
against on safety grounds. HSE's responsibility in such cases
is to ensure that the LPA has received the best possible advice
and has been given every assistance in understanding the safety
implications. (In principle, if HSE felt that an LPA had been
perverse, HSE could ask the Secretary of State for the
Environment to take the decision away from the LPA by the use of
'call-in' procedures. To do this frequently would be to
undermine the Advisory Committee's stance).

Although HSE, as the technical adviser, does not make the
decision, it is vital for it to be aware of the pressures on the
LPAs, the underlying philosophies of planning control, and the
costs to society of refusals. Advice which failed to take such
factors into account would be seen as unrealistic, and it would
soon fall into contempt.

For example, HSE used to employ three grades of advice: 'low
risk', 'marginal risk' and 'substantial risk'. Now planning
permission should be granted unless there are 'sound and clear-
cut reasons' to refuse it[17]. HSE would only give a
'substantial risk' reply if it felt that safety stood on its own
as a sound and clear-cut reason; in forming such a view, HSE
relied on its long experience in judging what is reasonable.
The 'marginal risk' reply suggested that the risk in itself was
not a sufficiently sound and clear-cut reason to refuse
permission, but it could be added to other reasons to increase
their overall weight. HSE received considerable feedback
(including a public planning inquiry finding) that such a reply
was unhelpful and even meaningless in a land-use planning
context. This type of reply is rarely if ever used now.

Regarding the costs of refusal, HSE includes in [5] its
understanding of the economics of land-use planning. This shows
that the apparent cost of a refusal to the land owner (ie the
value foregone) is greater than the cost to society as a whole,
since the development may be located elsewhere and boost the
value of the land elsewhere. One land-owner's 'loss' is
another's gain. The cost to society as a whole is that, apart
from safety, it may be sub-optimal to build on the second-choice
land. However, in principle it seems that safety should be
included in the optimising process, so that overall the 'second
choice' is optimal, provided that the significance of safety is
properly considered. These factors, among others, led HSE to
the view that risk criteria should be set at quite low levels of
risk for the present purpose, since the real costs of a refusal
at the outline planning stage are relatively small.

4   Comparison of QRA Predictions with actual Experience

It is important to consider whether the QRA results are
realistic. For the land-use planning situation, HSE uses a
'cautious best-estimate' approach, ie best-estimate where
possible but tending to the conservative where data is weak and
judgement is substantial.

In its general discussion of QRA[7], HSE compared the historical
record from UK major hazard sites with some 'rough and ready
estimates' based on the assessment approach outlined above, in
the form of societal risk F/N curves for fatalities. The
historical record was based on actual experience for N up to 28,
and above this on extrapolation assuming that F for any given N
in the UK would be 1/25 of that for the world. This ratio,
first suggested by Gill[18], is confirmed by data for 1966-86
collected by Fernandes-Russell and quoted in [5] and [7]. The
extrapolation allows the record to be extended up to
N = 2000 or so, this being the world upper limit to date (from
Bhopal). See Figure 1. Note that this F/N plot includes
fatalities on-site as well as off-site (the point for N = 28

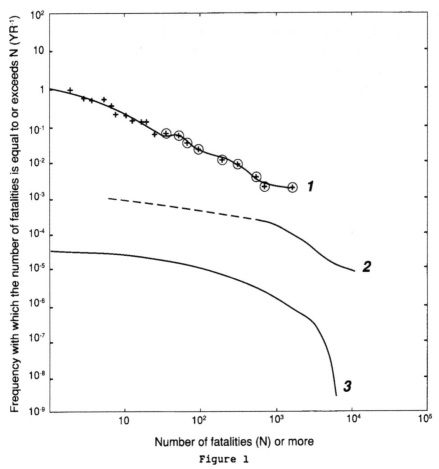

Figure 1

Key :

1 UK sites with 'major hazard' substances :

     +   :   Data points for UK

     ⊕   :   Data points from 'world', $x \, ^1/_{25}$

2 Canvey Island Study after improvements (dotted line used because there are no data points between N = 5 and N = 750).

3 Large toxic substance on site (RISKAT assessment).

Notes :

1. Lines **1** and **2** derived from (7).

2. Predictions (lines **2** and **3**) flatten out as N falls below 100-1000, due to uneven distribution of population round a specific site; neglect of effects on local workforce; and possibly neglect of smaller events in the analysis.

relates to Flixborough), but as the numbers rise the proportion of deaths off-site increases, being practically 100% for Bhopal.

Also shown on Fig.1 is an F/N plot for the Canvey Island study, and a plot for a large toxic substance site in the UK. Given the acknowledged trend to conservatism in the Canvey work, I suggest that it is not inconsistent with the historical data. The toxic substance site F is about 0.1% of the total UK F for any given N above about 100. This seems reasonable agreement as there are some 1700 notified installations in the UK, the example chosen is one of the largest, but the off-site buildings are not very close.

The 'rough and ready estimates' referred to above were based on assessments of a few typical sites including those with liquefied flammable gases, liquefied toxic gases, potentially explosive substances etc. These results were then simply multiplied by the total number of such sites to give a crude aggregate for the UK[19]. The final figures were remarkably close to the historical record line in Fig.1 - usually within half an order of magnitude on F for any given N.

The details of this last work have not been published, mainly because the gross simplifications are difficult to justify. Nevertheless it shows that even a basic broad-brush approach may produce reasonable results, at least at the level of generalisation to which it relates. This also provides an interesting example of reinforcement in decision-making, where the piece of evidence on its own may seem dubious but its consistency with other pieces of evidence lends weight to the whole. In general, I believe that it is often possible to find collateral evidence for predictions even of very rare events, given some ingenuity and willingness to consider partial or indirect evidence.

5   Discussion

HSE published the 'Tolerability'[6] and 'Risk Criteria'[5] documents
to explain its own thinking and to invite comments. 'Tolerability'
was specifically requested by Sir Frank Layfield in his report on the
Sizewell PWR, where he states that 'the opinion of the public should
underlie the evaluation of risk; ... there is at present insufficient
public information to allow understanding of the basis for the
regulation of nuclear safety'. 'Risk Criteria' was produced at HSE's
initiative, but in the context of several years' experience with
casework and a growing need for a definitive general statement to
supplement advice on future specific cases.

Many comments have been received. Those relating to 'Tolerability'
have been published[20]. HSE is still considering the comments on
'Risk Criteria'.

An obvious question raised by these responses is the extent to which
they represent the views of 'the public'. Many responses are from
representatives of professional or industrial organisations; others
are from private individuals who have had professional interests in
the field; and some are from public authorities and associations
(county or district councils etc). The last may perhaps be regarded
as representative of the public, while the others are representative
of various social partners or expert personal opinion. There has
been surprisingly little interest by the general media, and little
feedback from that source.

Should we be surprised at the lack of broad public reaction to these
publications? Perhaps not, in the light of recent comments on the
public acceptance of science and technology. The Director General of
HSE, John Rimington, has argued that the British public are not
averse to technical innovation, provided that they have confidence in
the mechanisms of regulation which are applied[21]. This was echoed
in a recent article by Brian Wynne on the public understanding of
science[22]. He suggests that there are various levels of
understanding:

1   The technical content;

2   The methods (including limitations as well as powers);

3   The forms of institutional embedding and control.

The last may be most readily understood by the layman - and it is at least as important as the others to the safety of the enterprise.

Perhaps those of us who focus on the numbers are missing the points which are seen as most relevant by the public. Some commentators on [5] may be touching on this when they suggest that we should be very cautious in our approach, and not lose sight of the potential worst-case consequences; in other words, do not be hasty in moving from the more generous distances accepted earlier.

Our experience with the published documents, and also with many discussions and presentations to planning inquiries etc, bears out the suggestions made above. We come under scrutiny for the manner in which we go about our work, the degree of professionalism, independence from interested parties and awareness of all aspects of a situation. Whether we use prediction or experience, the objectives of safety and public acceptance will depend on these broad factors as well as increasing sophistication in assessment.

REFERENCES

1   Health and Safety Executive, "Canvey: An Investigation of Potential Hazards from Operations in the Canvey Island/Thurrock Area", publ. HMSO, London, 1978.
    Also: "Canvey: A Second Report", publ. HMSO, London, 1981.

2. Health and Safety Executive, "A Safety Evaluation of the Proposed St Fergus to Moss Morran Natural Gas Liquids and St Fergus to Boddam Gas Pipelines", publ. HSE, London, 1978.

3. Pape R P, Nussey C, "A Basic Approach for the Analysis of Risks from Major Toxic Hazards", IChemE Symposium Series No. 93, 1985.

4. Royal Society Study Group, "Risk Assessment", publ. Royal Society, London, 1983.

184

5. Health and Safety Executive, "Risk Criteria for Land-Use
    Planning in the Vicinity of Major Industrial Hazards", publ.
    HMSO, London, 1989.

6. Health and Safety Executive, "The Tolerability of Risk from
    Nuclear Power Stations", publ. HMSO, London, 1988.

7. Health and Safety Executive, "Quantified Risk Assessment: Its
    Input to Decision-Making", publ. HMSO, London, 1989.

8. Netherlands Ministry of Environment (VROM): Annex to the Dutch
    National Environmental Policy Plan: Second Chamber of the States
    General, Session 1988-1989, 21 137, No.5.

9. Ellis A F, Pape R P: In Evidence to the Piper Alpha Inquiry,
    Day 133, publ. Palantype Reporting Service, 2 Frith Road,
    Croydon, Surrey, UK, 1989.

10. Health and Safety Executive, "A Guide to the Control of
    Industrial Major Accident Hazards Regulations 1984", booklet,
    HS(R) 21, publ. HMSO, London, 1985.

11. Johnston A G, McQuaid J, Games G A L, "Systematic Safety
    Assessment in the Mining Industry", The Mining Engineer, pp 723-
    735, March 1980.

12. CONCAWE: "Quantified Risk Assessment", Report No.88/56,
    publ.CONCAWE, The Hague, 1988.

13. Pape R P, in "Safety Cases" ed Lees F.P., publ. Butterworths,
    1989.

14. Health and Safety Commission, "Advisory Committee on Major
    Hazards: Third Report", publ. HMSO. LONDON, 1984.

15. Department of the Environment, Report of the Public Planning
    Inquiry into the Pheasants Wood Development, Thornton Cleveleys,
    1981.

16. Turner R M, Fairhurst S W, "Assessment of the Toxicity of Major
    Hazard Substances", HSE Specialist Inspector Report No.21,
    publ.HSE, 1989.

17. Dept. of the Environment Circular 2/80, publ. 1980.

18. Gill D W, "Approaches to Criteria in the Chemical Industry",
    Proceedings of Safety and Reliability Symposium, Stockport,
    September, 1983.

19. Davies P C: Unpublished work.

20. Health and Safety Commission, "Comments Received on 'The
    Tolerability of Risk from Nuclear Power Stations'", publ. HMSO,
    London, 1989.

21. Rimington J D, paper to Conference on "The Public Acceptance of Innovation ", University of East Anglia, 1990.

22 Wynne B, "The Guardian", 13 April 1990.

THE APPLICATION OF ROOT CAUSE ANALYSIS TO INCIDENT INVESTIGATION TO REDUCE THE FREQUENCY OF MAJOR ACCIDENTS

RB WHITTINGHAM
Principal Engineer
Electrowatt, Consulting Engineers and Scientists.
Grandford House, 16 Carfax
Horsham, West Sussex, England.

ABSTRACT

This paper describes a method of retrospective analysis of safety significant events to identify the Root Causes. A classification of causes is proposed which enables a systematic approach to Root Cause Analysis to be developed which can be applied to accident investigations. The approach is tested using a number of recent major accident inquiry reports as case studies for analysis. The results are discussed in terms of the nature of Root Causes of accidents and the implications of this for risk reduction.

INTRODUCTION

Risk Assessment methods, whether deterministic or probabablistic in nature, essentially rely on prediction of risk using a systems approach. The various possible initiating events which may lead to an unacceptable major hazard, are developed or modelled to indicate the possible outcome. The resultant hazard indicator may be an event frequency but will always include a consequence evaluation in order to assess the risk. The effectiveness of protective systems will also be taken into account to arrive at the overall risk estimate. Whilst predictive methods have made, and will continue to make, a valuable contribution to the control and reduction of risk, accidents continue to happen. It is not, of course possible to judge the effectiveness of risk prediction in terms of reducing major accident frequency.

Mercifully, major accidents are statistically infrequent events although public perception does not always recognise this. The correlation between prediction and incidence cannot be made, partly because of statistics, but mainly because risk prediction is itself not widespread in industry. For this reason, it is proposed that complementary risk reduction techniques may beneficially be employed. One of these techniques may be to apply the retrospective lessons of past accidents to the prevention of accidents in the future.

LESSONS FROM THE PAST

It is often stated, usually in the aftermath of a major accident, that the accident could have been prevented, if only the lessons of the past had been learned. There is usually an element of truth in such statements which are always made with the benefit of hindsight. The important question is whether it is possible to apply such insights prior to the occurrence of an accident. Two difficulties are apparent:-

1. Most accidents seem to be caused by the chance occurrence of a number of diverse events coinciding at a single point in time.

2. It is rare that precisely the same events coincide to cause accidents and hence similarities are never immediately apparent.

Unless it is possible to pinpoint common factors in accidents then it will always be difficult to learn from the lessons of the past. However, this difficulty often arises from a tendency to take a superficial view of the causes of accidents. It is clearly necessary to examine accidents in considerable depth if the common factors are to be identified. It is also beneficial that such in-depth examination be carried out in a structured and systematic way so that it can be applied to any accident. Then, it may be possible to express greater confidence that the lessons of the past can be learned and applied to prevent the next accident from occurring.

A SYSTEMS APPROACH

Definition

The identification of common factors in accidents which are apparently unique in terms of causation and effect is likely to yield to a systems approach. The systems approach is well known to safety and reliability practitioners, and a "system" has been defined as [1] :-

An assembly of components connected together in an organised way for a specific purpose such that the components are affected by being in the system and the behaviour of the system is changed if they leave it.

Retrospective Analysis

An accident may, in retrospect be considered as a sequence of interconnected events, usually comprising equipment and human failures linked together by cause and effect relationships. Although these links may not be organised (by definition, an accident suggests that they were not) they may well be amenable to analysis using a systems approach to determine the nature of the relationships. The more usual methods of predictive analysis, with which we are familiar, perform precisely this function by utilising various methods of modelling the system interactive logic. These models are able to describe the system and create an understanding of the cause and effect relationships between the system components thus enabling a prediction to be made of its future performance capabilities. The predicted capability can then be compared with the specified system requirements. In principle therefore, there is no reason why a similar approach may not be used to analyse an accident retrospectively, instead of attempting to predict its occurrence.

Objectives

The objectives of retrospective analysis are not, however, quite the same as those of the predictive assessments which are conventionally carried out by safety and reliability practitioners. First of all, of course, predictive analysis is attempting to postulate all the possible ways in which an accident may occur in the future due to the failure of a system. An accident which has already occurred has, presumably, been caused by one specific failure scenario which may or may not be easy to identify. In practice, even an accident which is well understood can only be explained in terms of <u>postulated</u> causes. In this sense, a court of inquiry into an accident will only be able to establish the causes of an accident "beyond reasonable doubt" rather than definitively.

The purpose of retrospective analysis of accidents is therefore seen to be twofold :

- To establish the causes of the accident beyond reasonable doubt

- To identify those particular causes which, if corrected, will prevent a recurrence of the accident.

Approach

One of the main activities which is necessary to carry out a systems analysis is that of classification of the components of the system. Accident analysis is very much concerned with the influence of human activities prior to, during and immediately following the accident. Even in the common situation of accidents which appear to have been caused by failures of equipment to perform correctly, it is necessary to investigate whether human failures have in some way contributed to the equipment deficiencies. Corrective measures derived from the lessons of the accident will almost always involve the recommendation of changes to associated human systems.

Human reliability analysis (HRA), increasingly used by safety and reliability practitioners, is highly dependent for its successful use on the adoption of a strict method of classifying human activities. It would therefore seem productive to adopt an appropriate classification scheme for the causes of accidents, bearing in mind the intention of concentrating the investigative effort on the underlying human failures.

CLASSIFICATION OF ACCIDENT CAUSES

Definitions

A comprehensive analysis of the investigation and inquiry proceedings of recent serious major accidents has revealed a variety of definitions of types of accident causation. The most common form of "classification" which can be found, comprises:

- Direct or Immediate Causes

- Underlying or Root Causes

It must be stressed here, that such a classification of causation types made from an examination of a number of diverse accident reports, does not necessarily point to common definition or usage. Such a definition must, however, be attempted if a systematic approach is to be used.

Direct causes

The direct or immediate cause of an accident usually seems to be identified with two types of events :

- Trigger events

- Latent failures

These types of events have been described in detail elsewhere [2].

Trigger events: In the present context, the trigger event of an accident will usually be the occurrence which sets off the accident sequence i.e. prior to the trigger event, the system was, to all intents and purposes, operating normally. It needs to be borne in mind, however, that "normal" operation may not correspond to the system operating "as intended" (this distinction will be taken up later). Furthermore, the trigger event will not necessarily be manifested as a failure per se. It is found quite frequently in accident investigations that the trigger event is a routine operation carried out in the correct manner at the correct time.

Latent failures: These are unrevealed failures of components of a system which remain undetected and uncorrected until a demand occurs on the failed component. They are frequent contributors to the direct cause of accidents when coupled with trigger events. Although they may be manifested as unrevealed equipment failures, these failures may also take the form of deficiencies in human components, particularly deficiencies in routine or proceduralised activities which, for various reasons which will be addressed later, have been allowed to lapse. Whatever the nature of the latent failure its most important characteristic is that it will have maintained an unrevealed potentiality to cause an accident over a period of time, which could be of considerable duration, before the accident occurred.

It is the coincidental or chance occurrence of a trigger event together with one or more latent failures which, in most instances, comprises the direct or immediate cause of an accident. This will in fact, for the purpose of this paper, suffice as an adequate definition of **Direct Cause**. It has been observed by others [3] that efforts to reduce risk by addressing latent failures should be concentrated on reducing the duration of these failures rather than their frequency. It is highly unproductive, of course, to attempt to identify and reduce the frequency of possible trigger events both because of their proliferation and their inevitability.

Root Causes

Many reports of accident investigations refer to the root cause or the underlying causes of the accident. An examination of such reports fails, however, to reveal any accepted common definition of "root cause". This term, used interchangeably with the term "underlying cause", is given a variety of different meanings according to the particular application where it is found. It is appropriate therefore at this stage to assign a definition to the term Root Cause which can then be adopted as the basis for a strict classification.

The **Root Cause** of an accident can in essence be defined as the most basic reason(s) for the accident which, if corrected, will prevent a future recurrence of the accident.

The activity which sets out to identify and define Root Causes of accidents, with the objective of specifying measures to reduce accident frequency, will be referred to as **Root Cause Analysis.**

Now it is immediately apparent that such a definition could, in theory, embrace any of the identifiable causes of an accident, including the Direct Causes as defined above. In practice it is found that the Root Cause of an accident very rarely coincides with any of the Direct Causes, although the definition cannot completely exclude this possibility. However, if it is possible to identify Root Causes as defined above, then it follows that retrospective analysis of accidents to prevent recurrences is indeed a feasible and worth while activity. This is especially the case if, of course, the Root Cause Analysis of one accident can be used to prevent other accidents occurring in different circumstances.

Influencing factors

The study of accident investigations reveals a further sub-class of the general class of "underlying causes" which cannot fit into the rather important classification of Root Cause as defined above. There are many circumstances surrounding accidents which, rather than being classed as causes which contributed to the accident, can only be described as having an influence on the course of the accident or its outcome. These influences will often have an adverse effect in exacerbating the accident but the possibility must not be ignored that the influence could also be beneficial in terms of mitigating the consequences of the accident. In either case, such factors are an important aspect of retrospective analysis since they will be of relevance, if not for accident prevention, then certainly for mitigation.

METHODS OF RETROSPECTIVE ANALYSIS

Patterns of reasoning

This section will attempt to define a systematic method of analysing past accidents so that Direct and Root Causes can be identified. It will draw on a number of case studies of recent accidents and the corresponding inquiry reports, which will be referenced. It must be stated that none of the investigations studied as the basis for the systems approach described in this paper appear to employ any such approach. It is, however, possible to discern certain patterns of reasoning which are employed in accident investigations to establish the causes of an accident. Some of these patterns are common to different accident reports and it is beneficial to draw on them as far as possible.

In the case of accidents which have received major publicity in the media, a great deal of speculation regarding the cause of the accident will have taken place prior to any proper investigation. It is probable that the cause will have already been decided in the public mind. This may also be the case to a lesser degree with most accident investigations and it is probably rare that an accident investigator will approach an investigation with a completely open mind as to the cause. It is all the more important therefore for an effective approach to accident analysis to employ a properly structured method which, as far as possible, is able to eliminate bias.

Three patterns of reasoning employed by investigators in the determination of accident causes have been identified, although this is not to exclude other approaches which may be used.

1. Hypothesis approach

This is an approach which, in theory, could result in a suppression of initial bias in the investigator. Rather than conducting the investigation with the primary objective of determining a single cause for the accident, this objective is reserved until later. In the first instance, a number of alternative hypotheses are advanced to explain how the accident may have been caused. In many ways this is not dissimilar to the group brain-storming sessions which were in vogue a few years ago in problem solving applications. The objective of these also, is to ensure that the widest possible range of solutions to the problem are explored.

Whilst not suggesting that such brain-storming techniques be employed, it does appear that the discipline involved in selecting alternative accident scenarios will ensure that the first, and perhaps most obvious cause of the accident, will not necessarily carry more weight until other possible mechanisms have been explored. Later in the investigation, the weight of evidence in support of the alternative hypotheses can be considered and a measured selection of the most likely accident cause can be made and properly validated.

An example of such an approach which, although not employed with the primary objective of elimination of bias, can be found in a study of the interim investigation into the Piper Alpha accident [4]. This particular investigation, carried out by HSE and Department of Energy inspectors, is being severely hindered by a lack of physical evidence due to the loss of the platform onto the seabed. As a result of this it has been necessary to advance alternative hypotheses to explain the cause of the accident. None of the hypotheses can, in this case, be fully proven, although in terms of the balance of evidence available, one hypothesis is more highly favoured than the others. Irrespective of the reason for this approach, it does seem to have major benefits in reducing initial bias, although in most cases it is hoped that sufficient evidence is available to eventually establish the cause beyond reasonable doubt.

2. "What if" approach

This approach has certain similarities to the Hypothesis approach described above but in most respects is subtly different. The approach is essentially one in which, the probable cause of the accident is known with some certainty. This leads the investigator to generate slightly divergent scenarios from the one originally selected. In practice, this involves "testing" the effect of fairly subtle changes in the circumstances of the accident. The result is a process of temporarily eliminating certain important components of the "system" which constitutes the accident. In essence, this involves the construction of a series of slightly divergent models of the real world followed by an assessment of how these models will now behave compared with the original sequence of events comprising the accident.

The objective of this approach is to allow an assessment to be made of the influence of the various components on the course of the accident. For instance, it may be speculated that a certain component or event is a possible direct cause of the accident, although this cannot be established with certainty from the evidence available. By removing the influence of this component in an appropriate way and constructing a "model" of the new scenario, it can quickly be determined whether this component is a possible cause. It will not be able to determine, however, whether the component is the actual cause.

An interesting example of this approach may be found in the inquiry report into the incident involving Aluminium Sulphate pollution of the public water supplies in the Camelford area of Cornwall in 1988 [5]. In this report, the direct cause of the accident had been established with some certainty in the very early stages of the investigation. It was unclear, however, whether the occurrence of a similar but unrelated incident earlier the same day had caused the control personnel to misdiagnose the accident. In fact by eliminating the earlier incident from the line of reasoning which had established the cause, it was possible to show that the outcome of the accident scenario would have been the same.

3. Change Analysis approach

This approach, whilst not defined as such in any of the accident investigation reports studied, is implicitly or unconsciously adopted by many accident investigators in the determination of causes. It is based on a common approach to problem solving and diagnosis which has been used by a number of diverse organisations, mainly in the U.S.A, in past years [6].

The principle, stated simply, is that a decline in a formerly acceptable standard of performance suggests that something has changed. The method sets out an effective means of sorting through the array of numerous and diverse changes which might have occurred, some of which may have given rise to the problem it is required to solve. The identification of the change which has caused the problem will suggest a solution to the problem. It is believed that this approach, developed specifically for accident analysis provides a systematic basis for identifying and analysing the causes.

A METHODOLOGY FOR ROOT CAUSE ANALYSIS OF ACCIDENTS

Modes of system operation

In the case of accident analysis, the problem which it is required to solve using Change Analysis is the cause of the accident. The method of approach is slightly different to that used in general problem solving, as described above. The area of change to be considered in an accident is usually closely identified by the nature and circumstances of the accident itself. In this sense, an accident is a problem which has resolved itself in an undesirable way. The areas where change is to be identified may be summarised as :

1. Design/Intent. The mode of operation of the system as designed or intended.

2. Normal Practice. The normal operating mode of the system.

3. Actual Practice. The mode of operation of the system just prior to the accident.

The study of past accidents reveals that the differences between these three states of a system, reflected in the changes that have brought about these differences, will indicate to the investigator, the direct and root causes.

Investigating the effects of change

Treating the previous statement in a simplistic fashion, it is possible to state two propositions:

A.   The changes that have resulted in differences between Design/Intent and Normal Practice will indicate the Root Causes of the accident.

B.   The changes that have resulted in differences between Normal Practice and Actual Practice will indicate the Direct Causes of the accident.

A. Design Intent versus Normal Practice = Root Causes

The basis of this proposition is, that the operation of a system in strict accordance with the principles set out in the original design intention will not be expected to lead to an accident. However, in the course of time, it is known that modified modes of operation tend to become adopted which were not anticipated in the original design. This is what is referred to above as Normal Practice.

Whilst in most situations, the normal ways of operating the system will be in accordance with the design intention, there may be certain aspects of operation which are not. These discrepancies from design intention will have become, to all intents and purposes, part of Normal Practice. The study of accidents reveals that these uncorrected discrepancies, which have become accepted as normal practice, provide the setting for accidents to occur. Some examples of this will be given later.

Normal practice, the state of affairs which exists for some period prior to an accident, has within it the roots of an accident, when this practice includes operations not anticipated by the design intent. This is not say of course, that these operations may not have been envisaged by the original designer. If they were, then it must be presumed that it was either not possible or believed unnecessary to provide any protection in the design against them causing an accident.

It is the differences between Design Intent and Normal Practice therefore, which will indicate the Root Causes of an accident. It is clear from this that the Root Causes of an accident may well have been present, possibly in an evolving form, for a considerable period of time before the accident. It is this which makes Root Cause Analysis such a potentially powerful tool for reducing risk in hazardous operations. Not only will the Root Cause, when removed, prevent the accident recurring, in accordance with the definition given above, there will be adequate opportunity for accident prevention long before it takes place.

B. Normal Practice versus Actual Practice = Direct Causes

The mode of operation corresponding to Actual Practice just prior to an accident will nearly always reveal the Direct Causes. However, in order to precisely identify the changes which inevitably occurred at this time, it is beneficial to make a comparison between Actual and Normal Operation. The justification for this is that, in Normal

Operation, for a period before the accident occurred, the conditions necessary to initiate the accident were not present i.e. the accident did not happen. In Actual Operation, just prior to the accident, these initiating conditions were present i.e. the accident occurred.

As defined earlier, the Direct Causes are characterised by the presence of a trigger event combined, usually, with one or more latent failures. It is therefore logical to compare Normal Practice with Actual Practice to indicate the Direct Causes of the accident. The question to be asked when making this comparison will always be - "what changed?" - and ideally the question should be asked using the Change Analysis approach in a structured way, as applied to general problem solving [6].

Relationship of Root Cause to Direct Cause

The identified changes between Normal and Actual Operation will, usually, lead the investigator to easily identify the Direct Causes of the accident. The Root Causes are identified from the changes which took place such that Normal Operation differed from Design Intent in the way described above. In practice, Root Causes are nearly always found to be occurrences which allowed the precursor conditions of the accident (i.e. the Direct Causes) to come into existence. The trigger event is essentially a chance occurrence, and in some cases may be a routine activity. The Root Causes therefore, are generally found to be deficiencies which allowed latent failures to occur and to be perpetuated over a period of time.

APPLICATION OF ROOT CAUSE ANALYSIS

The uses of Root Cause Analysis

Root Causes of accidents are, adopting the approach described above, associated with conditions or modes of operation with have come to be accepted as normal yet which differ in some way from the design or intended conditions. In many circumstances, this situation may well continue to exist for considerable periods of time without causing any noticeable abnormal effects. In fact, such situations almost certainly exist to some degree, undetected or uncorrected, in most organisations at some time. Where the activities of the organisation involve hazardous operations, then the situations where divergencies exist between design/intent and normal operation will contain within them the seeds of an accident waiting to happen. It is in these situations where the insights of Root Cause Analysis of previous accidents will be particularly valuable.

The approach has been tested using, as case studies, the published reports of a number of recent accidents. Published case studies have been used for reasons of availability and the fact that the material is already in the public domain. Choosing, as case studies, accidents which have already been thoroughly investigated is not, however, the ideal method of testing the approach. i.e. the test is necessarily biased because the causes have already been established. Thus the opportunity is sought to use the approach for investigation of safety significant incidents where it is important to establish the root causes definitively in a systematic way. The objective of the approach in this situation, will not be to establish blame or for litigational purposes, but to reduce future risk levels by learning the lessons of the past.

Case studies for Root Cause Analysis

A study has been made of seven serious accidents which have occurred in the U.K. in the past five years. The approach described above has been applied to each accident using the published reports of the inquiries into these accidents to provide the necessary information. The seven accidents are listed in Table 1 with the appropriate documentary references.

Modes of operation

In Table 1, the three modes of system operation i.e. Design/Intent, Normal Operation and Actual Operation, have been identified for each of the accidents. The area of operations from which these modes have been selected is of course highlighted in the accident inquiry report. However it has been necessary to identify which particular modes within these areas of operation are in fact important. This is very similar to the situation which applies when conducting an investigation into an accident for the first time. The method which has been employed to select the operational modes of interest, involves tracing the accident sequence back through the cause-effect relationships which can be seen to exist. There is nothing particularly novel about this procedure, which is routinely adopted for fault tree construction by safety analysts.

Classification of causes

The classification of causes of the accidents used as case studies can be found in Table 2. The Direct and Root Causes have been developed from a change analysis of the operational modes summarised in Table 1, using the methodology described above. The influencing causes have been identified from a study of the inquiry reports.

In each case, the discrepancy between Design/Intent and Normal Operation indicates the Root Cause. Similarly, the discrepancy between Normal Operation and Actual Operation indicates the Direct Cause. It is not intended to describe here the detailed way in which this process was conducted for each case study. The accidents which have been studied are well known in outline and the patterns of reasoning can be easily followed by reference to the published reports as necessary.

Characteristics of Root Causes

The most important outcome of this study is the insight gained into the nature of Root Causes of accidents. In every accident case study which has been analysed, it apparent that the Root Causes lie in the higher level organisational systems which carry the responsibility for the control of day-to-day operations.

TABLE 1. COMPARISON OF OPERATIONAL MODES

Ref	Accident	Design/Intent	Normal Operation	Actual Operation
8	Herald of Free Enterprise. Ferry Capsize.	Communications from bridge to bow-door operating area for control of doors prior to departure.	Word of mouth communications used without feedback.	No communication took place between bridge and bow-door area.
7	B.P. Grangemouth. Fire and Explosion.	LP Separator relief valves to discharge maximum flow from HP Separator.	LP separator relief valves sized for fire relief only.	Manual control of level in HP separator at start-up allowed gas to be released.
9	Clapham Junction. Rail Collision.	Signal works during WARS programme to be checked by a supervisor according to Departmental Instructions.	Department Instruction SL-53 not being followed.	No wire count or checking of work was being carried out.
4	Piper Alpha Fire and Explosion.	Equipment not to be put into service prior to Work Permit being signed off.	Suspended Work Permit not available for night shift to be aware of work in progress during day shifts.	Equipment re-commissioned with temporary blanks installed in place of pressure relief valve.
10	Bhopal Toxic Gas Release.	Piping to MIC tanks designed to prevent any ingress of water to tanks.	Plant modified to incorporate a "jumper" line.	Water used to wash out pipeline blockages without isolation of "jumper" line to MIC tanks.
11	Therapeutic Radiation Overdose.	Calculations to recalibrate Radiotherapy Treatment machine should be checked by a Senior Person.	Calibration calculations not checked due to staff shortages.	Standard factor omitted from calculations by Physicist.
5	Pollution of Public Water Supplies.	Water Authority personnel to be present during unloading of chemicals at water treatment plants.	Chemical supplier's tanker drivers have unsupervised access to the Water Treatment plant.	Relief Tanker Driver unloads chemicals unsupervised into unmarked tanks using a "universal" key to open tanks.

198

TABLE 2. CLASSIFICATION OF CAUSES

Accident	Direct Cause	Root Cause	Influencing Causes
Herald of Free Enterprise Ferry Capsize	Bow-doors were open on departure.	No adequate systems were installed to ensure bow-doors shut before departure.	Ship overloaded. Ship badly trimmed. "Safety Culture".
BP Grangemouth Fire and Explosion.	High pressure separator low level trips de-activated.	Pressure relief valve on LP separator was inadequately sized.	Operator error. Faulty instruments.
Clapham Junction Rail Collision.	Short circuit in signal relay due to loose wire not cut back by technician.	Instructions for checking/testing wiring work were not implemented.	Re-organisation of Regional Testing Teams. "Safety Culture". Poor staff morale and training.
Piper Alpha Fire and Explosions.	*Equipment inadequately isolated during maintenance.	*Permit-to-work system was not providing adequate protection against misuse.	Fire Protection inadequate. Emergency shut-off valves not installed in risers.
Bhopal Toxic Gas Release.	MIC tank inadequately isolated from water supply.	No provision was made for HAZOP studies on plant modifications.	Vent scrubber shutdown. Refrigeration plant shutdown. "Safety Culture".
Therapeutic Radiation Overdose.	Machine incorrectly calibrated.	No provision was made for formal checking of calibration calculations.	Shortage of qualified staff. Inadequate response.
Pollution of Public Water Supplies.	Aluminium Sulphate was unloaded into wrong tank.	Requirement for presence of authorized personnel during unloading of chemicals was not implemented.	Inadequate monitoring of water quality. Inadequate response.

* Postulated cause not yet validated by an inquiry.

In only one case study was the Root Cause of the accident found to be with the original design of the equipment. This occurred in the case of the fires and explosions at B.P. Grangemouth [7] where the defective design of the L.P. Separator pressure relief system was the Root Cause. This is a particularly good example of a case where the Root Cause definition given above is clearly seen to be justified. i.e. that cause, which, if rectified, would prevent a recurrence of the accident. In this particular case, the provision of a proper relief system design would have prevented the disintegration of the L.P. Separator and the subsequent fire and explosion. The integrity of the low level trip system on the H.P. Separator (Direct Cause), whilst important, is seen to be less effective in preventing a recurrence.

In all of the other case studies, the Root Causes are deficiencies which are found essentially in management systems. This conclusion can be justified by a study of the accident inquiry reports on which the case studies are based. The judicial findings of two of these reports in particular indicate directly that management systems are largely to blame for the accident (Herald of Free Enterprise [8], Clapham Junction [9]). In the other reports it is implicit that management systems are to blame, although in some cases the outcome is still subject to completion of the investigation (Piper Alpha[4]) or to litigation in the courts (Bhopal [10], Therapeutic Radiation Overdoses [11] and Pollution of the Public Water Supplies [5]).

CONCLUSION

Root Cause Analysis, carried out in a structured fashion, enables the lessons of past accidents to be applied more confidently to prevent future accidents. The important issue is, of course, prevention and the definition of a Root Cause given in this paper ensures that if the Root Cause can be truly established, then prevention can be achieved and future risk levels can be reduced using retrospective analysis.

The outcome of the case studies which were used to test the proposed method of Root Cause Analysis has revealed that, in these accidents, Root Causes are frequently found in the area of the management systems used to control day-to-day operations. This is borne out in many cases by the findings of the accident inquiry reports. The management and organisational influences on risk are also the subject of much current concern in the safety community. These concerns have, to some degree diverted attention from the risk impact of human error at the operational level, which, in these case studies, are frequently identified as the Direct Causes.

The problems which are associated with applying the lessons learned from Root Cause Analysis are not addressed in this paper. Where Root Causes are associated simply with equipment design deficiencies, then the remedies to be applied may be quite obvious, although, of course, the way in which these deficiencies came to exist may require investigation also. Where the Root Causes are associated with deficiencies in management systems, then practical remedies may, again, be easily identified and implemented. When these deficiencies are attitudinal in nature and concern highly diffuse issues such as "safety culture", then the remedial measures required will not be so easily understood and could be equally difficult to implement.

The method of Root Cause Analysis which is described above provides a structured approach to accident investigation. Although elements of this approach are apparent in certain investigations, there does not seem to be available any systematic method of classifying accident causation such that the results can be used to reduce future risk levels. It is believed that a retrospective analysis tool, such as that described in this paper, will complement current predictive methods of risk analysis and thus ensure that the valuable sources of past data resulting from accidents is put to a useful purpose.

REFERENCES

1. Bignell, V. Fortune, J. Understanding Systems Failures. Manchester University Press, England, 1984 p154.

2. Reason . J, The Tchernobyl Errors and their Implications. Dept. of Psychology. Manchester University.

3. Rasmussen, J., Safety Control and Risk Management: Topics for Cross - Disciplinary Research and Development, Preventing Major Chemical and Related Process Accidents Symposium, I Chem E Symposium Series No 110, 1988.

4. Piper Alpha Technical Investigation Interim Report, Petroleum Engineering Division, Department of Energy, 1988.

5. Report of an Inquiry into an Incident at Lowermoor Water Treatment Works of South West Water Authority on July 6th 1988, South West Water, August 1988.

6. Kepner, C.H., Tregoe, B.B., The New Rational Manager, John Martin Publishing Ltd, London, 1987.

7. The Fires and Explosion at B.P. Oil (Grangemouth) Refinery Ltd, A Report of the Investigations by Health and Safety Executive, HMSO, London, 1989.

8. MV Herald of Free Enterprise, Formal Investigation, Department of Transport, Report of Court No 8074, HMSO, London, 1988.

9. Investigation into the Clapham Junction Railway Accident, Antony Hidden QC, The Department of Transport, HMSO, London, 1989.

10. Bowonder. B, Miyake, T., Managing Hazardous Facilities: Lessons from the Bhopal Accident. Journal of Hazardous Materials, 19(1988)237-269.

11. Incident in Radiotherapy Department of Royal Devon and Exeter Hospital, The Report of Committee of Enquiry and Summary Report, Exeter Health Authority, 1988.

MANAGEMENT OF SAFETY THROUGH LESSONS FROM CASE HISTORIES

NICK F. PIDGEON
Department of Psychology
Birkbeck College (University of London)
Malet Street, London WC1E 7HX, UK

JOHN R. STONE and DAVID I. BLOCKLEY
Department of Civil Engineering
University of Bristol
Queen's Building, University Walk, Bristol BS8 1TR, UK

and BARRY A. TURNER
Department of Sociology
University of Exeter
Amory Building, Rennes Drive, Exeter EX4 4RJ, UK

ABSTRACT

Large-scale technological accidents are invariably socio-technical phenomena, involving technical, human, and organisational preconditions. In order to learn the lessons of past failures we must take account of this full range of preconditions. The paper describes work to develop an intelligent knowledge based system for safety management in the construction industry. Case history material on past incidents is acquired by a process of knowledge elicitation, and the information derived is represented in a knowledge base using Event Sequence Diagrams. A dictionary of concepts, expressing preconditions typically present in accident case histories, has been developed. The eventual aim is to use the knowledge based system to manage safety in ongoing projects, by comparing patterns of preconditions in ongoing projects with those associated with past failures.

INTRODUCTION

The consequences of engineering failures range from minor inconvenience to catastrophe. It is crucially important, but immensely difficult, to consider the implications of previous accidents for current projects. The study of failures is of interest to social scientists as well as engineers, since human and organisational as well as technical factors are significant. The problem is not just technical, but occurs at the social-technical interface. It is now clear that human and organisational factors can be systematically analysed, and Turner observes that '...many disasters and large scale accidents display similar features'[1]. One of the most important tasks of the engineering profession is to study the past and attempt to learn from failures. Indeed, an engineered system can be viewed as an expression of current practice - as a conjecture - and its failure as a partial refutation of that conjecture.

Our approach to tackling this need to learn from and use past events, both technical and social, was outlined by Blockley at a workshop on modelling human error in structural design and construction. The aim is '... to produce a knowledge based computer system which might be an aid in the management of the safety of a project'[2].

The central features of the approach are:

(1)   The examination of case histories of incidents and failures in structural engineering, as well as 'successful' case histories.

(2)   The acquisition and representation of these case histories in the computer.

(3)   The definition of a set of key concepts in which these case histories can be expressed; i.e a vocabulary of terms describing the technical, human and organisational preconditions to failure.

(4)   The use of 'machine learning' to extend the knowledge base from additional sets of case histories, and the development of new terms in the vocabulary.

(5)   The development of an advisory knowledge based system, based upon the case histories, to aid in the management of the safety of structural engineering projects.

Each of these items will be reviewed briefly and a set of key concepts, which have been so far isolated, described.

CASE HISTORIES

The first problem to be addressed is that of obtaining accurate details of projects, both failed and successful, past and present. Information is combined from two sources - published reports and recorded interviews. Both have their advantages and disadvantages. Written reports commonly follow inquiries into major failures, e.g. Tacoma Narrows Bridge or Ronan Point. Whilst generally providing a wealth of detail, they represent only one extreme of the continuum of failures. Much more significant in terms of numbers of occurrences are the small to medium failures, and 'near-miss' incidents, which may not undergo detailed and public investigation. An approach which we have used with some success is to interview all of the key individuals, such as the design engineers and construction staff, concerned with a failed structural engineering project [3] [4]. This enables detailed information, on the preconditions contributing to failure, to be obtained for cases which may not otherwise have been reported. However, as Blockley has noted [2] there is a narrow window of time during which suitable case histories are available. This is because recent events may be surrounded with litigation whilst those more than about ten years old are difficult to research accurately.

KNOWLEDGE ELICITATION

Having obtained interview data from key individuals the investigator is then faced with the difficult task of analysing what is initially relatively unstructured and complex material. Dealing with such material raises significant methodological and theoretical issues associated with the need to analyse systematically qualitative data. The transcripts of semi-structured interviews are normally long, complex and in non-standard format. It is far from clear when the information is, for all practical purposes, complete. The testimony of the interviewee will typically rely on unspecified assumptions, background meanings, and what the philosopher Michael Polanyi has termed 'tacit knowledge' [5].

Dealing with semi-structured interview material is essentially a problem of knowledge elicitation. Knowledge elicitation refers to the processes of obtaining, analysing, and confirming with one or more domain experts an account of the knowledge deemed relevant to the development objectives of a knowledge based system (KBS). The acquisition of knowledge from human experts is now recognised as a crucial problem area for KBS and 'expert' system development [6]. A method for the conceptual analysis of information derived from case history interviews has been developed based on the technique of Grounded Theory. Grounded Theory was initially proposed for the analysis of qualitative data in the social sciences [7]. However, it can be readily used to analyse semi-structured case history interview material for KBS and expert system applications [8].

The task of analysing interview material is not a simple one of transcribing or translating data from one form to another; rather it is an interpretative process in which the analyst, or knowledge engineer, must take responsibility for perceiving and creating order in the data. The principal steps in conducting a Grounded Theory analysis are shown in Figure 1. There is insufficient space in this paper to discuss the method in detail, and a full account is provided by Pidgeon et. al. [8]. At the core of the method is an indexing system which is constructed by asking of each paragraph in the transcribed interviews 'What categories, concepts or labels do we need in order to account for the phenomena of importance in this paragraph?' When a concept is identified it is recorded (coded) on a file with a precis and a reference to its position in the transcript. Relationships with other concepts which can be identified from the text are also noted. In an iterative process, working though the interview data in systematic detail, a core set of refined theoretical concepts and relations indexing the critical features of the data are built up.

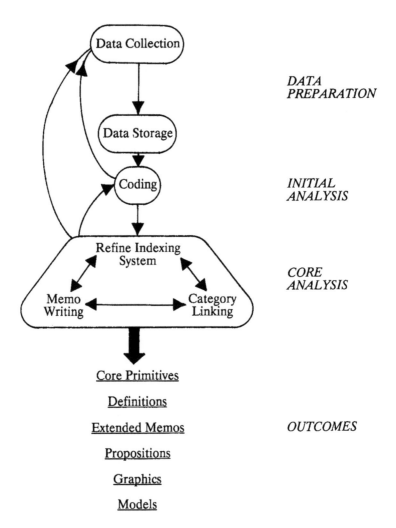

Figure 1. Principal steps in Grounded Theory analysis.

The Grounded Theory analysis typically entails the generation of a set of core primitive concepts and their definitions, a set of theoretical memos describing relevant features of the data and concepts, and one or more models of the ways in which important groups of concepts are linked. The constructed indexing system provides a detailed and comprehensive intermediate representation of the core knowledge derived from the case histories. This forms the foundation for subsequent operations to transform the knowledge into a representation suitable for machine implementation.

KNOWLEDGE REPRESENTATION

The basic case histories can be conceived of as 'stories', which need to be converted into a structured representation suitable for manipulation in a computer. The form chosen for this has been the Event Sequence Diagram (ESD), a representation that is similar in basic philosophy to the event tree technique. The diagram provides a powerful means of representing and accessing information about the sequences of events preceding a failure or near-miss incident.

The ESD format shows the temporal order and relationship of events leading up to a particular outcome [9]. The events are described in terms of the key concepts derived from the knowledge elicitation phase. The ESDs are simple tree structures, with each concept being a node of the tree. Each node may be classed as a 'head' node or 'tail' node when viewed from another node. A group of tail nodes connected to a head node in a given case history are the preconditions for that head node in that case history. Figure 2 shows a typical diagram representing the collapse of a factory roof under snow loading.

Any case history can be represented at different levels of generality. In this way a series of hierarchically arranged diagrams can be generated for any particular case history, as illustrated in Figure 3. The lowest (deepest) level contains the most detailed information, specific facts from the stories of individual case histories. At the other extreme the highest (shallowest) level represents the accumulated story of a case history in more general terms. This hierarchical structure serves two important purposes. First, the high level concepts can be used to describe (and hence link) several case histories in the computer. As noted earlier, at a general level of analysis patterns of preconditions are known to recur across incidents, and it is this aspect of failures which the system is designed to capture. Second, the hierarchical structure enables an appropriate level of problem solving to be conducted within the KBS. In some situations very detailed information on a particular failure may be required, whilst in other cases the general concepts (describing more systemic features) are more meaningful. For example, a

206

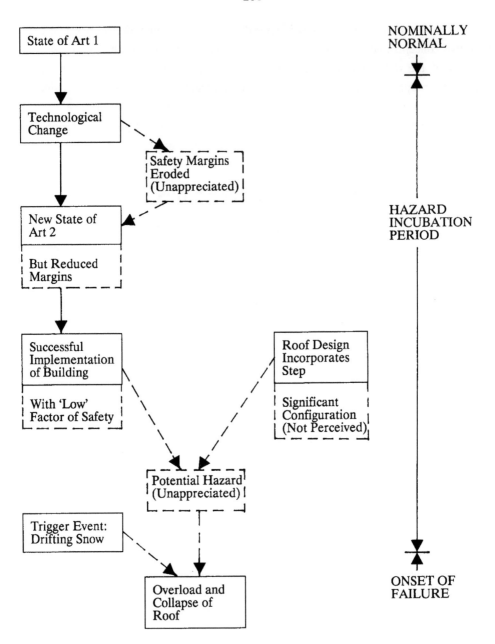

Figure 2. Event Sequence Diagram depicting the collapse of a roof under snow loading.

detailed concept such as **REINFORCEMENT BAR OMITTED** may be needed when making detailed comparisons between an ongoing construction project and a project that is known to have failed. Conversely, a high-level concept such as **POOR SITE SUPERVISION** may be utilised, in combination with other general concepts, for evaluating the overall safety characteristics of an ongoing project.

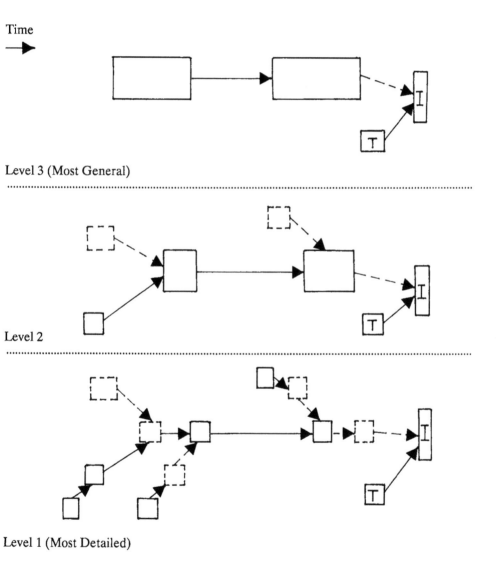

Figure 3. Hierarchically arranged Event Sequence Diagrams.

UNCERTAINTY

A number of different methods have been proposed for handling uncertainty in intelligent knowledge based systems (IKBS). These include certainty factors, fuzzy logic, Bayesian probability, and various combinations thereof [10].

The particular method used in the IKBS reported here is one which has been developed to overcome some of the major objections to fuzzy and Bayesian methods. It is an approach based upon support logic and interval probability [11]. The evidential support for a proposition or event is expressed by two numerical values, the necessary and possible support. These two values represent lower and upper bounds respectively to the uncertainty associated with any proposition in the knowledge base. This approach embodies an open-world model of reasoning in the sense that it is possible to represent propositions as true, false, or unknown, in addition to degrees of uncertainty between these extremes. Each proposition in the KBS described in this paper is given a support pair, which expresses the evidential support for the truth or dependability of the proposition within a given case history.

KEY CONCEPTS

To develop a useful knowledge base it is necessary to collect a large number of case histories and refine them into a series of linked Event Sequence Diagrams. This raises the problem of defining a suitable vocabulary of key concepts for describing events that recur across cases. Later it will be shown that a machine learning method depends on having propositional concepts (words) which occur in more than one diagram, enabling a linkage (relationship) to be established. As an initial step towards this end, a dictionary of concepts is being developed which is sufficiently large to cover the richness of the range of cases held in the knowledge base, yet small enough to ensure repeated use.

Figure 4 shows a part of the dictionary derived from previous work, and Table 1 gives a set of definitions for each of the concepts. Referring to Figure 4, at the highest (most general) level is the concept of **PROJECT SAFETY**, which defines the ultimate goal for the system user. Two lower level concepts are shown directly below this; **TECHNOS**, and **PROJECT HISTORY**. Each group of lower level concepts can be viewed as the facets of evidence which would ideally be needed to evaluate a project in terms of the higher level concept. For example, an **UNSAFE PROJECT** can be the result of **POOR TECHNOS**, of a **POOR PROJECT HISTORY**, or of some weighted combination of these two. Each intermediate concept is then further characterised in successively greater detail.

As noted earlier, the hierarchical structure of the dictionary allows groups of case histories to be conceptually linked within the knowledge base. Clearly, all accidents could be viewed as 'unique' if only a very detailed level of analysis is adopted. However, a prerequisite to effective learning from accidents and incidents is the ability (and motivation) to identify lessons at several levels of generality [12]. Thus, a failure due primarily to **INADEQUATE SAFETY ORGANISATION,** and one due to a combination of **POOR SAFETY NORMS** and **LAX SAFETY ATTITUDES,** can both be characterised at the higher level of analysis as being due to a **POOR SAFETY CULTURE.**

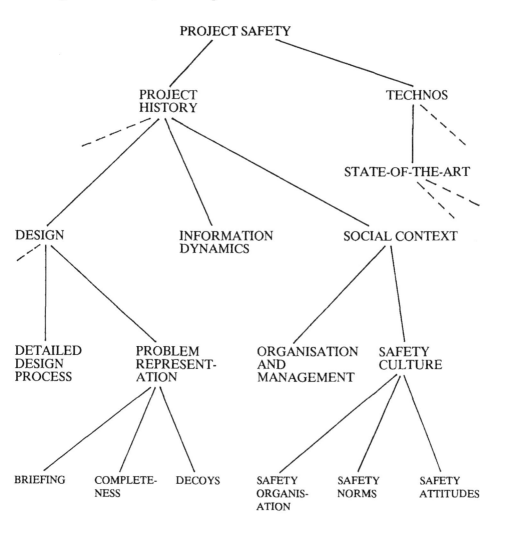

Figure 4. Section of hierarchical concept dictionary.

TABLE 1
Definitions for knowledge base concepts shown in Figure 4.

TECHNOS: The range of knowledge, information, skills, and techniques available at a given point in time to direct intended transformations of the socio-technical world.

PROJECT HISTORY: The particular assemblage and sequence of events associated with the people, groups, organisations, and artifacts of a given project or project failure.

STATE-OF-THE-ART: The collectivity of shared knowledge, skills, techniques, and information used or potentially usable in current common practice (i.e. in the process of identifying 'appropriate' problems, and providing potential solutions).

SOCIAL CONTEXT: The set of social arrangements and obligations set up to manage the problem, together with the concomitant organisational culture(s) and individual attitudes to the project and to safety.

INFORMATION DYNAMICS: The system of information handling arrangements associated with the project, aimed at effective communication between individuals and organisations involved. Under some circumstances individuals will be ignorant of key information which resides in other parts of the system.

DESIGN: The collection of relevant information, and its transformation into a form which entails it to guide a successful construction. Design outcomes will typically be a set of drawings and notes, together with the knowledge, tacit and otherwise, for the interpretation and application of the drawings as guides to action.

SAFETY CULTURE: The set of norms, beliefs, attitudes, and social and technical practices within an organisation, which are concerned with minimising exposure of employees, managers, customers, suppliers and members of the public to hazards.

ORGANISATION AND MANAGEMENT: The organisations, groups, and individuals involved in the project. Their relationships (informal and formal) and the arrangements for decision-making.

PROBLEM REPRESENTATION: The structuring of the task to be faced, including the goal(s), elements modelling relevant features of the environment, relationships between the features, and (possibly) potential solutions.

DETAILED DESIGN PROCESS: The day-to-day process of collecting, generating, interpreting, transforming, and checking the specific structural and other information necessary to guide a safe construction.

SAFETY ORGANISATION: Structural arrangements within an organisation or group for dealing with actual and potential hazardous situations.

SAFETY NORMS: The appropriateness of the organisational rules, either explicit or tacit, that provide guidance for the tackling of problems regarding hazards.

SAFETY ATTITUDES: Individual and group beliefs about hazards and the importance of safety, together with their motivation to act upon those beliefs.

BRIEFING: The communication of the problem owner's requirements to the professionals involved in the project, serving to define the goals to be met during the course of the project.

TABLE 1 (continued)
Definitions for knowledge base concepts shown in Figure 4.

COMPLETENESS: A problem representation that is too narrow to enable the problem to be effectively tackled. A relevant aspect may not be recognised at all, or not seen as urgent and/or significant.

DECOYS: The selective focus of attention on specific safety problems, leading to oversight of similar hazards.

The dictionary that we have been developing closely reflects the fact that many failures have complex, socio-technical origins. For example, the concept of safety culture is a key one within the dictionary. This term was originally used in a rather narrow sense, to describe the human failings underlying the Chernobyl accident [13]. It is clear, however, that the question of what constitutes a poor, or good, safety culture raises a number of important issues, particularly in the context of recent prominent international, as well as United Kingdom, disasters [14] [15]. In this respect many of the concepts within the dictionary are not limited in applicability to construction incidents: it is increasingly becoming clear that the human and organisational factors underlying accidents are constant across otherwise technologically dissimilar events. Therefore, in addition to its use in the KBS described here, the dictionary might ultimately serve as the basis of a check-list for carrying out a safety management audit for a complex socio-technical system, such as an industrial plant or transport network.

MACHINE LEARNING

The search for patterns in data is commonly undertaken by the use of a variety of different types of cluster analysis [16]. These methods employ a number of different heuristics to determine groupings of elements of data.

The methods of discrimination and connectivity used in our work were developed initially by Norris, Pilsworth and Baldwin [17] who wished to investigate the relationship between medical symptoms and diseases from a number of patient case histories. Discrimination entails the search for a single feature of an object which, by its presence or absence, gives evidence for the belief that an object belongs to one class rather than to another. It is therefore a serial approach. By contrast, the connectivity algorithm adopts a parallel approach to the data. This entails the search for groups of features which by their presence or absence give evidence for the belief that an object belongs to one class or another. They are those features which have been found commonly to occur together and are associated with a given object classification. The algorithm is therefore a method for pattern recognition. Each outcome is considered in turn and a search is made for groups of features which commonly occur.

Figure 5 shows an outline of our proposed development of a KBS to fulfil the objectives stated in the Introduction. The upper section of the diagram, concerned with building the knowledge base, has been implemented in C on an IBM PC AT [18]. The lower section illustrates the use of the system as currently being implemented. The learning loop in the upper part of Figure 5 involves the active participation of the builder of the KBS. At a given hierarchical level of definition the discrimination and connectivity analysis will cluster sets of key concepts which seem to be preconditions (tail nodes in the ESDs) for another key concept (a head node) across a range of case histories. These groups of concepts are then presented to the builder as hypotheses for new concepts at a higher level of definition. If the groupings make sense to the builder then he or she will name them and hence define a new key concept. For example, if the concepts **CALCULATION ERROR MADE IN DESIGN, CALCULATION ERROR MISSED BY CHECKING ENGINEER** and **WRONG DRAWING ISSUED** are found to be strongly connected then the builder may wish to define the new concept **POOR DESIGN OFFICE ORGANISATION** to represent them. If however the grouping has no meaning to the builder then that grouping can either be rejected or it can be researched in order to try to find an explanation for its occurrence.

This process is repeated for all head nodes in the Event Sequence Diagrams which are preceded by any tail nodes or preconditions. When each connected set has been re-examined and renamed as appropriate, a new analysis is carried out where the old terms in each event sequence diagram are replaced by the new concepts. Further renaming and substitution is repeated until no groups are found or until those that are do not suggest new concepts. At any stage of the iteration the builder may merge any of the newly formed ESDs which are considered to be sufficiently similar. In this way the information in the KBS is compressed as the higher level key concepts are formed.

CONSULTING THE KBS

To fulfil its desired purpose as a management tool for the control of safety and risk, the KBS, to be created from the 'learned' case histories, must be consulted by a user and advice offered. This process is illustrated in the lower half of Figure 5. The typical user envisaged may be a project engineer, or safety auditor, who wishes to make some assessment of the proneness to failure of an ongoing project. The advice provided by the system is therefore a measure of how closely the details of the project match those in

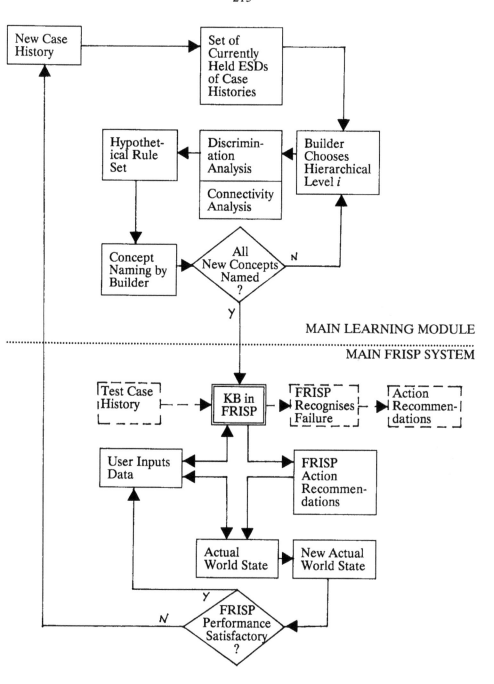

Figure 5. Outline of knowledge based system.

the KBS. The system will therefore require the user to assign supports to those concepts which are recognised as applying to the project. The evidential support for the safety of the project will then be calculated by using a measure of similarity between the proposed project and the knowledge learned from the case histories.

The hierarchical structure of the KBS enables the user to ask for an explanation of any answer given at an appropriate level. If more detailed explanations are required, then a lower level of knowledge can be used. The bottom level concepts are the observed events in the actual case histories. At the most detailed level of explanation, the user might be advised that this particular pattern of events is very similar to that observed in Bridge X shortly before it collapsed!

The advice from the KBS will necessarily be deficient in particular circumstances. It is absolutely clear that the system is not intended to replace the user but rather to help and provide clarification. If the performance of the system is poor then that suggests that the data available to it (i.e the case histories) are not rich enough to cover that particular set of circumstances. Another learning loop from the lower half to the top half of Figure 5 is therefore closed so that the case history can be used to update the knowledge base.

CONCLUSIONS

The paper describes progress on an interdisciplinary research project, involving collaboration between engineers and social scientists. The ambitious goal of this project is the production of a knowledge based system for drawing lessons from past case histories, and for the management of safety in future projects. The work integrates theories and methods from several disciplines in order to elicit case histories, to represent this information in a computer knowledge base, and to extend the knowledge base by forming links between groups of cases displaying similar characteristics. An important feature of the work has been the development of a dictionary of concepts, for expressing the common preconditions to failures. The methodology outlined in this paper is being developed for use in managing the safety of construction projects. However, the general method may be used for any situation where safety lessons need to be drawn from case histories.

ACKNOWLEDGEMENTS

We wish to thank Brian Toft for many helpful discussions. Work described in this paper was supported by grants from the Science and Engineering and Economic and Social Research Councils of the United Kingdom.

REFERENCES

1 Turner, B.A., Man-Made Disasters, Wykeham, London, 1978.

2. Blockley, D.I., An AI tool in structural safety control. In Modelling Human Error in Structural Design and Construction, ed. A.S. Nowak, American Society of Civil Engineers, New York, 1986, pp 99-105.

3. Pidgeon, N.F., Blockley, D.I. and Turner, B.A., Design practice and snow loading: lessons from a roof collapse. The Structural Engineer, 1986, 64A(3), 67-71.

4. Pidgeon, N.F., Blockley, D.I. and Turner, B.A., Site investigations: lessons from a late discovery of hazardous waste. The Structural Engineer, 1988, 66A(19), 311-315.

5. Polanyi, M., Personal Knowledge: Towards a Post-Critical Philosophy, University of Chicago Press, Chicago, 1958.

6. Neale, I.M., First generation expert systems: a review of knowledge acquisition methodologies. The Knowledge Engineering Rev., 1988, 3(2), 105-145.

7. Glaser, B.G. and Strauss, A.L., The Discovery of Grounded Theory: Strategies for Qualitative Research, Aldine Publishing Company, New York, 1967.

8. Pidgeon, N.F., Blockley, D.I. and Turner, B.A., The use of Grounded Theory for conceptual analysis in knowledge elicitation. Int. J. Man-Machine Studies, 1990, In Press.

9. Toft, B. and Turner, B.A., The schematic report analysis diagram: a simple aid to learning from large-scale failures. Int. CIS J., 1987, 1(2), 12-23.

10. Saffiotti, A., An AI view of the treatment of uncertainty. The Knowledge Engineering Rev., 1987, 2(2), 75-97.

11. Cui, W.C. and Blockley, D.I., Interval probability theory for evidential support. Int. J. Intelligent Systems, 1990, 5, 183-192.

12. Pidgeon, N.F., Risk assessment and accident analysis. Acta Psychologica, 68, 355-368.

13. OECD Nuclear Agency, Chernobyl and the Safety of Nuclear Reactors in OECD Countries, Organisation for Economic Cooperation and Development, Paris, 1987.

14. Turner, B.A., Pidgeon, N.F., Blockley, D.I. and Toft, B., Safety culture: its importance in future risk management. Position paper for Second World Bank Conference on Safety Control and Risk Management, Karlstad, Sweden, November 1989.

15. Horlick-Jones, T., Acts of God? An Investigation into Disasters, Association of London Authorities, London, 1990.

16. Everitt, B., Cluster Analysis, Heinemann, London, 1980.

17. Norris, D., Pilsworth, B.W. and Baldwin, J.F., Medical diagnosis from patient records: a method using fuzzy discrimination and connectivity analysis. <u>Fuzzy Sets and Systems</u>, 1987, **23**, 73-87.

18. Stone, J.R., Blockley, D.I. and Pilsworth, B.W., Towards machine learning from case histories. <u>Civil Engineering Systems</u>, **6(3)**, 129-135.

AIRCRAFT CRASH STUDIES IN AEA TECHNOLOGY

J P BYRNE and J JOWETT
Hazard Assessment Department,
AEA Technology, Wigshaw Lane, Culcheth,
Warrington, Cheshire WA3 4NE, UK

ABSTRACT

AEA has been involved in aircraft crash studies for almost 10 years. Over that time techniques have been developed for both frequency assessment for any site in the UK, and to assess the consequences of aircraft crash. To take account of changes in flying activity, in terms of flying patterns, levels of usage and types of aircraft flown, aerial features databases are regularly maintained. In addition, a recent revision of the background crash rates for the UK has been carried out, and aircraft impact parameters reviewed.

INTRODUCTION

Interest in the hazards posed to ground structures by crashing aircraft was initiated by the concern for the safety of nuclear power plants. Early in the development of a commercial nuclear power programme in Britain, a siting policy was developed which recommended that nuclear power stations should not be sited near airfields close to the direct path of runways [1]. Within a decade, these early ideas had found sufficient recognition for legislation to be enacted that prohibited, or otherwise restricted, flying activity near certain nuclear installations [2], although it was recognised that the probability of an aircraft crashing was, for most sites, very small. For sites experiencing average levels of flying activity (ie sites away from airfields or other significant aerial features), crash rate figures were quoted in the range from 10^{-6} to 10^{-7} per year for the area occupied by a nuclear power plant [3]. The statistics were considered again in a study which was used as supporting material for the public inquiry into the proposed Sizewell 'B' nuclear power plant [4,5]. In their response to these matters for Sizewell 'B', HM Nuclear Installations Inspectorate stated their satisfaction with the methods used to evaluate crash frequencies, but sought further assurance on crash consequences [6].

Throughout this period of interest in, and concern about, aircraft crashes in the nuclear industry in Britain, little was published about the views of the non-nuclear industries. This changed with the publication by the Health and Safety Executive of the 1985 Guide to the Control of Industrial Major Accident Hazards Regulations (CIMAH), where in paragraph 113 the following statement appears: "A safety case may also perhaps say that the risk of an aircraft crashing on the installation is insignificant in comparison with other causes of a major accident, because the site is well separated from the nearest airport and air traffic lanes" [7].

Over a number of years AEA has been interested in the assessment of both the frequency and consequences of aircraft crash. This has culminated in the development of the computer codes PRANG [8] for the determination of crash frequencies, and ACCESS [9] (Aircraft Crashes onto Civil Engineering StructureS) for the assessment of the consequences of aircraft impact on civil engineering structures.

Within the the past 12 months, as a result of a contract placed by the CEGB, much of the data used in both the crash frequency and consequence analyses has been revised. In the following we outline the basic methods of analysis and give details of the latest enhancements to the data employed in the analyses.

THE COMPUTER CODE PRANG

PRANG is used to evaluate the ground impact rate at any given site, taking into account all features of the aerial environment that may affect the crash rate. The program makes use of historical data regarding crashes away from airfields to give background crash rate information, and uses a reliability approach to assess specific features such as airfields and airways. PRANG calculates the distribution of aircraft crashes over a mesh surrounding the location of interest. The mesh is composed of 1km square cells. Past analyses have used a square mesh of side 50km, with the site placed at the centre, as any features outside this area were found not to significantly affect the crash rate at the site under study. However in the latest treatment of military combat aircraft accidents, as described later, high accident concentration zones up to 60km from the site may need to be taken into consideration.

PRANG calculates the crash rates associated with 5 different categories of aircraft for each cell within the grid. The categories of aircraft are described in Appendix 1. After ascribing a background crash rate to each cell for each aircraft category, the program deals with the main aerial features that may be included in the input data by the user. These include airfields, airways and danger/restricted areas.

The Background Crash Rate

As noted earlier, background crash rates for the UK were revised in 1989 [10] as part of a study for the CEGB. The background rate was derived from all accidents involving uncontrolled impact more than 5 nautical miles from landing or take-off. Following previous analyses, accidents to all aircraft types except military combat aircraft were excluded if a forced landing had occurred, since in such circumstances the pilot would have retained sufficient control to avoid large ground structures. As the crew of military combat aircraft generally eject prior to impact, the element of pilot control is difficult to judge, so all uncontrolled impacts have been included in the background crash rate for this aircraft type. Table 1 presents the results of this background crash rate analysis.

TABLE 1
The background aircraft crash rate

Aircraft type	Crash rate[1]
Private light	4.3
All helicopters	0.7
Small transport	0.22
Airliner/military transport	0.17
Military combat	0.5^2

[1]Units are crashes per square kilometre per year, x 10^{-5}.

[2]This figure is the crash rate for military combat aircraft more than 60km from the edge of a zone of high crash concentration. Approaching a high crash concentration zone the crash rate climbs continuously according to the function shown in Figure 3. The maximum crash rate, which occurs within the high crash concentration zone, is 1.6 x 10^{-4} km^{-2} yr^{-1}.

An important finding from the 1989 study is that
military combat aircraft crashes are largely concentrated in
4 areas of England and Wales, as shown in Figure 1. The
military combat crash rate in these areas of high crash
concentration is about 30 times that of the background rate
for areas more than 40 to 60km from the boundaries of the
high crash concentration zones. The boundary of one of the 4
areas, over the Wash, is shown in more detail in Figure 2.
Detailed maps of the other areas, over Northern England,
Dorset and mid-Wales, can be found in reference [10].

It would be physically unrealistic to have a step
change in crash rate at the boundaries of the high crash
concentration areas, and so a transition zone of roughly
40km width has been suggested. For these transitional zones
between the areas of high and low crash concentration, the
military combat background crash rate, $f(x)$, should be
calculated according to the following prescription:

$$f(x) = 1.55 \times 10^{-4}\ e^{-x/10} + 5.0 \times 10^{-6} \quad km^{-2}\ yr^{-1} \qquad (1)$$

where x is the distance from the boundary of the high crash
concentration zone in km.

The concept of high and low military combat background
crash rate zones and the transitional zone is illustrated in
Figure 3. Full details of the background crash rate
calculations, with all accident data, are given in reference
[10].

For aircraft categories other than military combat, the
background crash rate as given in Table 1 was found to be
uniform across the UK. The uniformity of background crash
rate can be explained by the following factors. Well defined
airlanes are only operated for the civil and military
transport category of aircraft. All other types are free to
choose flying routes of their convenience, albeit under the
supervision of an air traffic controller when flight paths
enter certain congested areas. Such an arrangement leads to
a widely spread distribution of flight paths and hence crash
locations. As for the civil and military transport category
of aircraft, given the high speed and high cruising altitude
of these aircraft, long distances are often travelled
between the airway location at which an accident was
initiated and the final position of ground impact. This
feature again leads to a wide spread of impact locations.

Airfields

Airfields are perhaps the single most important aerial
feature which add to the background level of crash risk. Two
reasons for this importance can be identified. Firstly,
because all normal flights by fixed-wing aircraft start and

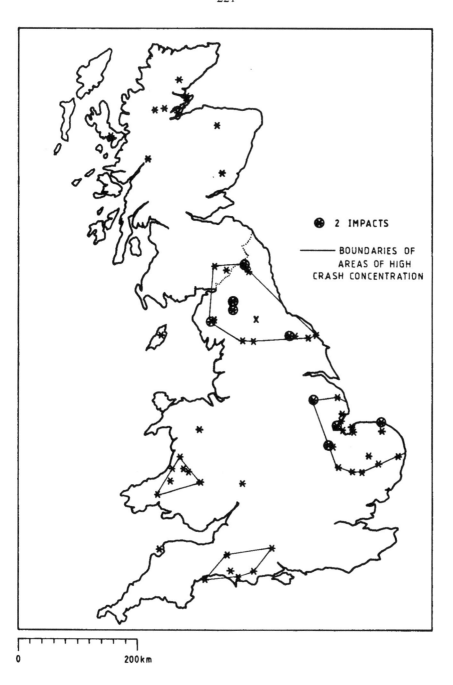

FIG. 1 ALL MILITARY COMBAT ACCIDENTS 1980-1988

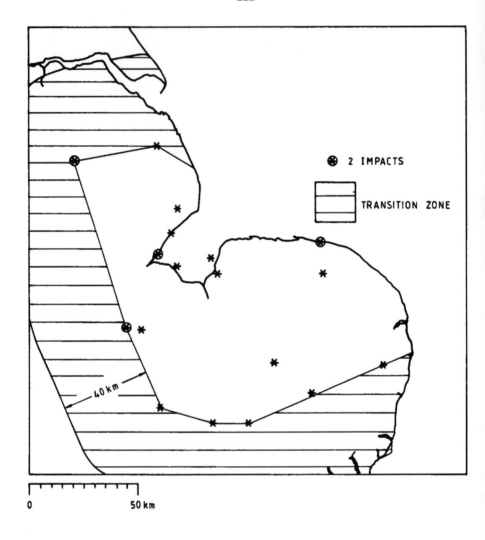

FIG. 2 THE WASH AREA OF HIGH CRASH CONCENTRATION

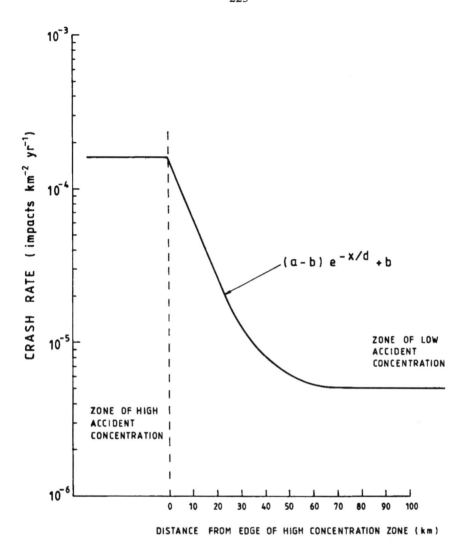

FIG. 3 VARIATION OF CRASH RATE BETWEEN ZONES OF
HIGH AND LOW ACCIDENT CONCENTRATION

end at an airfield or airport, the concentration of flying expressed in terms of flight paths per square kilometre reaches a maximum in the vicinity of airfields. Secondly, landing and take-off are the most hazardous phases of flight - a factor exemplified by the fact that approximately half of all aircraft accidents occur within 5 miles of an airfield.

Plots of the crash locations associated with near airfield accidents show that most aircraft impact the ground along the extended runway centreline. This concentration of crash locations is particularly marked for movements by military combat and transport aircraft, and is attributable to the considerable distances flown along the extended runway line during the approach to land and take-off flight phases.

Airfields are dealt with in PRANG by allowing the user to specify the details of up to 20 runways. For each runway the length, runway number, centre position (defined on a simple grid numbering system) and annual number of movements for each category of aircraft must be specified. The annual frequency of accidents associated with each runway is determined, for each aircraft category, as N x R where N is the annual number of movements along the runway by that aircraft type, and R is an aircraft category dependent landing and take-off reliability. The probability F(r,t) that an airfield related accident will result in a ground impact at a location (r,t) is given by the following probability distributions [8]:

a. For commercial/military traffic:

$$F(r,t) = 0.23 \ e^{-r/5} \ e^{-36t/\pi} \qquad km^{-2} \qquad (2)$$

b. For light fixed-wing aircraft:

$$F(r,t) = 0.08 \ e^{-r/2.5} \ e^{-3t/\pi} \qquad km^{-2} \qquad (3)$$

where r is the radial distance in kilometres from the runway end, and t is the angle in radians between the runway centreline and a vector parallel to r.

c. Helicopters: Studies of military helicopter take-off and landing crashes have revealed that 93% occurred within 100m of the helipad, and the remaining 7% within 200m. Beyond 200m from the helipad the in-flight reliability is applicable to helicopter movements. To calculate the crash rate at a location greater than 200m from a helipad, PRANG assumes that all flight paths radially around the helipad are equally likely.

For further information on the derivation of the near airfield probability distributions, see reference [8].

Airways

Airways are specified in PRANG by defining any two points along the airway centreline, usually the points at which the airway intersects the edges of the grid. An airway that changes direction within the grid can be treated as two separate airways. For each aircraft category the flying altitude is specified and the annual number of movements. The spatial distribution of airway crashes is calculated in PRANG by assuming a normal distribution about the airway centreline, with a standard deviation equal to the airway altitude.

Because airways are used almost entirely by transport aircraft whose in-flight reliability is very high, unless they are very heavily used, airways normally make only a second order contribution to the total crash rate at a site.

Areas of Intense Military Flying

AIAAs (Areas of Intense Aerial Activity) and Military Training Areas (MTAs) are well-defined regions of airspace which are subject to above average levels of military aircraft flying. They are used for activities such as pilot training and low level flying practice. The current locations of AIAAs and MTAs are shown in Figure 4. As stated earlier, the past and present AIAAs and MTAs correspond roughly to the 4 areas of high crash concentration found in determining the background crash rates for the UK. The crash rates for these areas and their associated transition zones are therefore as described earlier. For areas designated as an AIAA or MTA which are not covered by the 4 areas of high crash concentration, the situation is less clear. Some increase in crash rate above the lower level for military combat aircraft is recommended, but the exact figure to be used must take into account the type(s) of aircraft using the area, the level of usage and the nature of the activity. Each such area must therefore be considered individually as necessary.

Restricted Areas

Restricted areas include Provost Marshal Prohibition (PMP) zones and danger areas, and are areas of airspace in which flying is either continually restricted or prohibited, or restrictions exist between specified times. PMP zones exist permanently for all civil nuclear installations, usually being in the form of a cylinder of airspace of radius 1 nautical mile and height 2000 feet. The flying of both civil and military aircraft is prohibited within a PMP zone. The

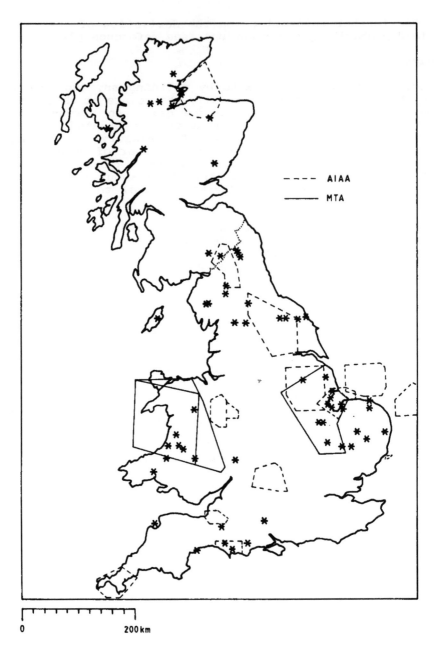

FIG.4 CURRENT AIAAs AND MTAs, AND MILITARY CRASHES

effect of a restricted area on crash rates is difficult to ascertain, and in previous PRANG assessments a conservative approach has been taken. The likely beneficial effect of a restricted area in reducing the crash rate has therefore not been claimed. Work is currently underway at SRD to quantify the effect of PMP zones on the crash rate at nuclear installations.

Using PRANG

To support PRANG, AEA has assembled a database of information on aerial features for the UK and has written associated plotting programs. Information is held on all the aerial features described above, and civil and military crash locations. By simply specifying the locations (either in longitude/latitude form or in Ordnance Survey grid co-ordinates) of the bottom left and top right corners of the grid, all the aerial features that may affect the site can be viewed graphically. The output of a PRANG run can also be presented graphically as a crash rate contour plot. An example contour plot is given in Figure 5, which shows the crash rate in the vicinity of an airfield as predicted by the PRANG methodology. Descriptions of crashes and aircraft reliability data have also been collected to provide additional data for the assessments.

THE COMPUTER CODE ACCESS

ACCESS is a computer code which has been developed to predict the structural consequences of aircraft crash. At present its capabilities are limited to the assessment of slab-sided reinforced concrete structures, although an extension to treat steel shell constructions would be possible in future.

Given a set of PRANG ground impact rates for the location in question, ACCESS determines the frequency with which internal compartments, or cells, of a structure would suffer local perforation by crashing aircraft. To make a probabilistic assessment of the structural failure frequency resulting from aircraft crash, the methodology must account for the wide spread of impact characteristics which may be realised during an accident. For example, aircraft within the PRANG military combat class may have impact masses between 2.3 and 50.0te, and impact velocities between 50 and 300 ms^{-1}. Furthermore, the aircraft's crash trajectory could vary from a near vertical dive to a shallow glide or even a skidding approach along the ground. All approach bearings might also be possible.

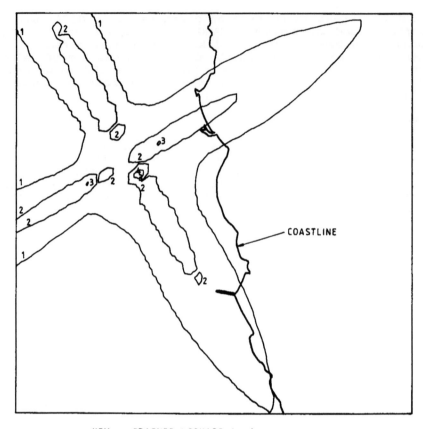

COASTLINE

KEY - CRASHES / SQUARE km / year

1 = 7,0000E - 05
2 = 2,0000E - 04
3 = 1,0000E - 03
4 = 1,6000E - 03

FIG.5 EXAMPLE CONTOUR PLOT

To account for all these variables and the range of impact locations that are possible, ACCESS uses a Monte-Carlo method to simulate a large number of crash histories. The initial parameters for each history are sampled from impact parameter distributions. After initial contact with the target has been established, structural failure models are used to follow the trajectory of the crashing aircraft as it passes through the structure. A history is only terminated when the aircraft is brought to rest or leaves the region of interest. Using such a method the cell perforation frequencies provided by ACCESS can be regarded as values obtained by sampling from the whole spectrum of crash events.

Representation of the Structure

In ACCESS a single building or series of buildings can be studied. A series of cuboids are defined, either exterior 'blocks' or interior 'cavities', by specifying for each a position vector and dimension vector in cartesian co-ordinates. In this way it is possible to build up a fairly accurate model of a structure with correctly defined wall and roof thicknesses. Subterranean cavities may also be included. A 'base-mat' must be defined which is 2-3 times the length and breadth of the site and several metres deep. This allows the simulation of skidding impacts onto the site.

The smallest cell dimension should not be less than about 5.0m. Some judgement may be required in determining which cavities are to be modelled and which not if a structure has much interior division on a small scale. Blocks must not overlap, but may lie adjacent to one another; cavities must not overlap, but may intersect more than one exterior block. To aid the visualisation and accuracy of the structural representation a graphics program 'GRAFPADS' has been developed. This program displays on screen the block and cavity geometry of the buildings specified in an ACCESS input file. GRAFPADS presents plan and/or elevation views of the buildings, and allows vertical or horizontal sections to be taken at arbitrary locations. GRAFPADS provides an efficient means of checking ACCESS input data, and can provide hard copy output of report quality. An example plot produced by GRAFPADS is shown in Figure 6.

The Aircraft Crash Parameters

It was stated earlier that a Monte-Carlo technique is used in ACCESS to generate the aircraft crash parameters of impact velocity, descent angle, mass and compass bearing. The user is required to specify distributions, which are aircraft category dependent, in the form of histogram

230

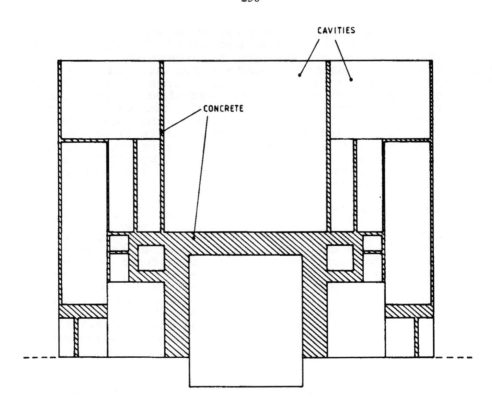

FIG. 6 EXAMPLE 'GRAFPADS' OUTPUT
(TYPICAL MAGNOX REACTOR BUILDING)

functions for each of these parameters. As this information
is part of the input data, it can be altered and separate
runs performed to examine the perforation rates from
different types of aircraft, such as military combat or
large civil transport aircraft.

Again as part of a study carried out on behalf of the
CEGB, impact parameters for non-airfield related accidents
have recently been reviewed [10]. Detailed accident reports
were studied to provide distributions for descent angle,
impact velocity and impact mass for each of the 5 categories
of aircraft used in AEA's aircraft crash studies. Full
details are given in reference [11].

Example distributions are shown in Figures 7 and 8 and
Tables 2 and 3. The descent angle distribution of Table 2 is
applicable to category 1, 3 and 4 aircraft, and also to
military combat aircraft accidents initiated at heights
greater than 2000 feet.

TABLE 2
Descent angle distribution for
category 1, 3 and 4 aircraft

Descent angle (degrees)	Fraction
10	0.077
20	0.029
30	0.149
40	0.149
50	0.149
60	0.149
70	0.149
80	0.149

For military combat accidents initiated at heights less than
2000 feet, the descent angle distribution shown in Table 3
is applicable. (A distinction is made for military combat
aircraft in terms of the height of accident initiation
because of the presence of PMP zones around nuclear
facilities). The velocity and mass distributions found for
all military combat aircraft accidents are shown in Figures
7 and 8 respectively.

232

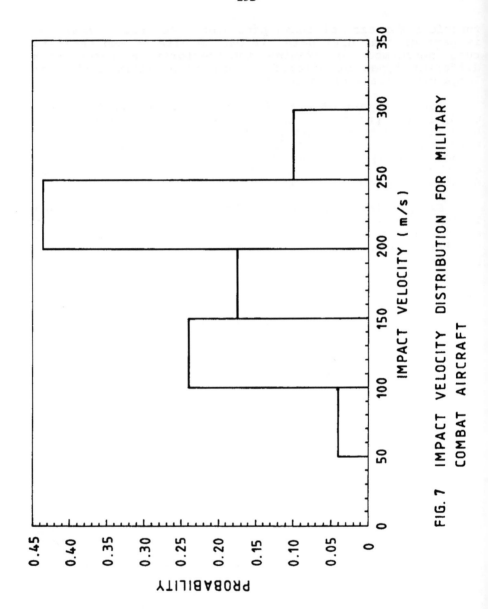

FIG. 7   IMPACT VELOCITY DISTRIBUTION FOR MILITARY
COMBAT AIRCRAFT

233

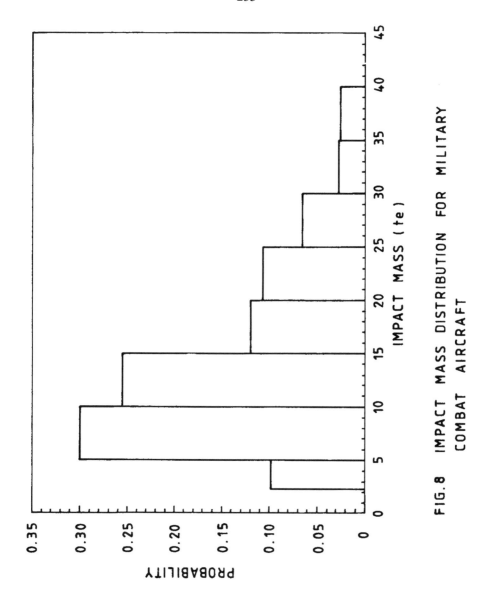

FIG. 8 IMPACT MASS DISTRIBUTION FOR MILITARY
COMBAT AIRCRAFT

TABLE 3
Descent angle distribution for military combat
aircraft accidents initiated below 2000 feet

Descent angle (degrees)	Fraction
5	0.364
15	0.204
25	0.250
35	0.068
45	0.114

Structural Failure Modelling

On impact with a reinforced concrete structure typical of
that employed in a nuclear power plant, an aircraft will
suffer severe deformations as it is decelerated. The forces
exerted during this deceleration are primarily caused by the
arrest of the aircraft's momentum. Tests conducted with
thin-walled tubular steel 'soft' missiles designed to
simulate at small scale the effects of crashing aircraft,
have revealed that when impacted by such objects reinforced
concrete slabs fail in a localised punching mode. Structural
failure is therefore assessed in the program using the local
perforation criterion described in reference [12]. The
aircraft is treated as a 'soft' missile for its first impact
with the structure, and as a 'hard' missile for any
subsequent impacts. It is assumed that compaction of the
aircraft occurs following an initial impact to form a hard
missile. If an aircraft skids prior to impact with a
structure it is also assumed to have become deformed to a
hard missile. For oblique impacts, the component of the
aircraft velocity normal to the impacted panel is used to
determine whether or not the aircraft perforates the panel.
The aircraft is assumed to leave a perforated panel in a
direction perpendicular to the exit face, whatever the angle
of incidence on the impact face. Experimental evidence
indicates that some rotation takes place following an
oblique impact, due to the large in-plane forces involved.
The exact extent of rotation is not known, therefore the
assumption used in ACCESS is conservative. The aircraft
exit velocity is found from an energy balance assuming a
frustum of concrete is ejected from the perforated panel at
the same velocity as the exiting missile.

If an aircraft perforates a panel, it is assumed to
have 'free-flight' until impacting the next panel. Any
possible obstructions to the aircraft's path are ignored,
which should be borne in mind when considering the results
of ACCESS. In previous assessments it has been assumed that

all equipment in a perforated cell is lost, either through the direct effects of the missile or through fire. This is a reasonable assumption as in most aircraft impacts onto structures an aircraft fuel fire is generated. The effects of such a fire can spread beyond the impact zone and may require further consideration.

CONCLUDING REMARKS

PRANG has been used over the last 5 years or so in assessing the ground impact rates at various sites. Studies have been performed for the CEGB's proposed PWR sites, such as Hinkley Point in Somerset and Wylfa in Anglesey. Certain MoD sites have also been assessed. ACCESS has been used to assess aircraft perforation rates for various facilities, including the Thermal Oxide Reprocessing Plant (THORP) at Sellafield, and a typical Magnox nuclear power station.

The results of PRANG assessments carried out so far for nuclear sites in the UK have shown aircraft crash rates which are generally no more than a factor of 2 larger than the background crash rate. This is largely due to the fact that nuclear stations are not sited in the immediate vicinity of airfields, or directly beneath extended runway centrelines. For these nuclear targets, the crash rates estimated have not lead to the need to provide physical protection against aircraft crash to achieve an acceptable level of risk.

The position concerning aircraft crash in the UK major hazards chemical industry is less clear, since the PRANG methodology has been applied to few such sites. However it is reasonable to assume that if such installations were located in close proximity to an airfield, the risk associated with aircraft crash may not be made acceptable without the provision of physical protection. Such a conclusion was drawn for LNG storage installations at Montoir-de-Bretagne and Fos-sur-Mer in France. The former installation is located 3km from the moderately busy airport at St Nazaire, whilst the latter is located 12km from the centre of a small airfield at Marseilles. In both cases the operators provided reinforced concrete containers of sufficient strength to withstand the effects of an aircraft impact [13].

REFERENCES

1. Farmer F R and Fletcher P T, 'Siting in relation to normal reactor operation and accident conditions', Rome Conference, 1959.

2. Air Navigation (Restriction of Flying) (Atomic Energy Establishments) Regulations 1983 (1983/640).

3. Farmer F R, Symposium on safety and siting, British Nuclear Energy Society, 28 March 1969, p130.

4. Phillips D W, 'UK aircraft crash statistics - 1981 revision', SRD Report R198.

5. CEGB, Sizewell 'B' PWR, Supplement to the PCSR on external hazards - aircraft crash, Report no. CEGB/S/724, 1982.

6. HMNII, Sizewell 'B' - A review by HMNII: Supplement 3, External hazards - aircraft crash, PWR/NII 01, 1983.

7. HSE, 'A guide to the control of industrial major accident hazards regulations, HS(R)21, HMSO, 1984.

8. Roberts T M, 'A method for the site-specific assessment of aircraft crash hazards', SRD Report R338, July 1987.

9. Byrne J P, Jowett J and MacFarlane K, 'ACCESS - a program to simulate aircraft structures onto civil engineering structures', SRD Report R398, in publication.

10. Byrne J P, Jowett J and Parry S, 'Background aircraft crash rates for England and Wales - 1989 revision', SRD Report R534, May 1990, in publication.

11. Jowett J, 'Impact parameters for aircraft crash analysis (non-airfield related accidents)', SRD Report R535, May 1990, in publication.

12. Barr P (ed), 'Design and assessment of concrete structures subjected to impact', Part 1 - guidelines for the designers, UKAEA Design Guide DG3, September 1987.

13. Chayrezy M, 'Examples of construction and design of LNG storage tanks', Proc. 1st. Int. Conf. on Cryogenic Concrete, Newcastle 1981, Paper 2, p21-37.

Appendix 1

Definition of Aircraft Categories

1   Light civil aircraft - fixed wing aircraft generally
     falling into the Civil Aviation Authority (CAA)
     classification of less than 2.3te mass. This category
     also includes military light aircraft used for
     training, such as the Bulldog and Chipmunk, which are
     less than 2.3te in mass.

2   Helicopters - all civil and military helicopters up to
     a maximum of about 10te, although some heavier
     helicopters are used on specific duties.

3   Small transport - fixed wing aircraft covering the CAA
     classification from 2.3te to 20.0te mass, including
     civil and military transports such as the Skyvan,
     Heron, Jetstream, etc.

4   Airliners and military transports - any other fixed
     wing aircraft not covered in categories 1, 3 and 5.

5   Military combat and jet trainers - all military fixed
     wing aircraft with masses up to 40 - 50te used for, or
     capable of, aerobatic style flying (ie excluding
     category 3 aircraft).

These categories are exclusive.

All masses referred to above are maximum take-off masses.

THE SELECTION OF FIELD COMPONENT RELIABILITY DATA
FOR USE IN NUCLEAR SAFETY STUDIES

BRYAN A COXSON, Nuclear Electric
MANSOUR TABAIE, Electrowatt

ABSTRACT

The paper reviews the user requirements for field component failure data
in nuclear safety studies, and the capability of various data sources
to satisfy these requirements. Aspects such as estimating the population
of items exposed to failure, incompleteness, and under-reporting problems
are discussed.

The paper takes as an example the selection of component reliability
data for use in the Pre-Operational Safety Report (POSR) for Sizewell
'B' Power Station, where field data has in many cases been derived from
equipment other than that to be procured and operated on site.

The paper concludes that the main quality sought in the available
data sources for such studies is the ability to examine failure narratives
in component reliability data systems for equipment performing comparable
duties to the intended plant application. The main benefit brought about
in the last decade is the interactive access to data systems which are
adequately structured with regard to the equipment covered, and also
provide a text-searching capability of quality-controlled event
narratives.

STUDIES ON FACTORS INFLUENCING FAILURE

The selection of field component reliability data, whether for use in
nuclear safety studies or for other applications, should ideally be an
iterative process with selection of failure metrics (e.g. calendar time
to failure, cycles to failure) proceeding hand in hand with reviews of
the reliability data. This ideal is rarely attained except in intensive
studies in controlled test environments, and normally the assessor of
field reliability data is constrained to information which was believed
to be relevant at the time the data was collected.

Similarly affected is the quantification of factors influencing failure; if information considered relevant to these factors is not available, then the ability to correlate factors with failures to determine the main contributors will be correspondingly limited. This situation is acceptable if the primary interest is to assess the historically achieved reliability on existing plant to establish that its continued operation will be satisfactory under current safety requirements.

For the design of new plant, however, and for the associated design assessment activities, novel design features, advances in equipment, or use of equipment in new situations mean that field data is frequently not comparable to the application or duty of the component under consideration. At this point the reliability assessor may be faced with either

- the use of data corresponding to the plant items of interest, but performing different duties
- the use of data corresponding to the duties of interest, but for items of plant differing from those destined for installation.

It is in this situation that the reliability reviewer has most need of not only validated failure metrics, but also quantified factors that enable him to relate the component applications or duties and associated environments, to the failure rates appropriate to the equipment to be installed, operating to some reference or datum case.

In recognition of this problem the PWR Project Group had previously participated in a research study to investigate and review factors influencing failure for an acknowledged area of particular concern. This study was on valves, where considerable variability in failure rates has traditionally been observed, and the need to quantify factors influencing failures is therefore important.

The results of this study have been reported in Reference [1]. Exploration of the factors influencing failure was carried out using

- Exploratory Data Analysis (EDA)
- Proportional Hazard Modelling (PHM)

It is evident from the paper that the choice of the data source used was pivotal to the ability of the study to identify factors influencing failure and it was selected on the following basis:-

- engineering data relating valve duty to environment was available, in terms of temperature and pressure (HP or LP)
- being a field study on a particular plant, the valve population and exposure times were capable of quantification over a $3\frac{1}{2}$ year period
- the reporting system was event based, enabling failure, repair and maintenance activities to be tracked for each valve.

The conclusions of the study regarding the factors influencing failure were unsurprising, in that the Exploratory Data Analysis and PHM both identified pressure and fluid medium (water or steam) as the predominant significant factors.

This provided confirmation of factors already known to be significant, but the more important lessons learned from the study, corroborated by other data extraction exercises carried out on various US reliability and event data banks to be referred to later, highlighted the directions for future work. These lessons included:-

-       Operational aspects such as valve function and role of the valve actuator are not as readily handled as environmental factors.
-       The usefulness of root cause analysis as a way of identifying variables influencing failure, in the absence of more proximate information such as that obtainable by condition monitoring.
-       That, in situations where comprehensive information which could quantify factors influencing failure was lacking, it was preferable to select for further study data sources oriented towards the specific applications of interest.

The last point is particularly important regarding the derivation of component failure rates for use in nuclear safety studies. The main focus is on the performance of equipment following a demand on safety systems. This involves changes of state (e.g. pumps must start, valves must change position), frequently on equipment otherwise in a standby state. Although modes of failure such as leaks are important as potential contributors to initiating faults which may challenge safety systems, they do not assume the same significance in terms of safety systems response; their occurrence will normally be revealed, and consequently their contribution to safety system unavailability is minor in comparison with unrevealed faults.

Following on from the interest in the operational aspects of failure is the requirement for event narratives, which provide important information about failure modes and root causes. The importance of root causes for determining the propensity of a component to failure for given environments has already been referred to. Equally important is the insight failure modes and root causes give to

-       how to improve design/operations so as to reduce contributions to failure from given root causes, and to determine the degree of credit to be attributed to an improved design
-       establishing if a failure is safety-critical or not (i.e. did the component fail in its safety function at the time the demand was made upon it).

The value of failure narratives to the safety/reliability assessor cannot be over stressed; experience with a variety of data banks has shown that although the use of coded fields on many aspects of failure events can be valuable in data searching, no great reliance should be placed on them. They are often missing or inaccurate, and rarely provide comparable useful insights into the exact nature of failures, and the implications of them for the component's safety duty, that is to be obtained from reviewing the narratives.

These lessons were borne in mind when assessing the available data sources for Sizewell 'B', which are covered in the next section.

ASSESSMENT OF AVAILABLE DATA SOURCES FOR NUCLEAR SAFETY STUDIES

Three main types of reliability data are required for use in Probabilistic Safety Assessment (PSA):

-        Initiating fault frequencies
-        Component failure rates
-        Dependent failure data

Due to considerations of time and space, only the second type of data is to be considered in this paper, although there are strong interactions between all three.

In comparison with the reliability data derived in 1982 at the time of the Pre Construction Safety Report (PCSR) for Sizewell 'B' [2], the station design is largely fixed, and many of the suppliers known; there is a requirement for the data intended for the safety assessments to match the known design and equipment, bearing in mind the following:-

-        Data should be best estimate, in the sense that they should be defendable and non-optimistic with minimal uncertainty.
-        Considerations of factors influencing failure, rather than aspects of statistical uncertainty, is more important for the vast majority of components.
-        Data should be based on relevant field experience. A target was set of identifying a data source based on field experience for every component item.
-        The requirement for auditability, so that revisits to the data could be made in the event of changed failure criteria for the more important items.
-        The data should be consistent. The object of a PSA is to identify relative weaknesses in a design in terms of the main contributors to the top events in the fault trees. Viewed from this perspective, it is preferable to derive all data from one data source, where reporting has been carried out to a consistent standard. For Sizewell 'B', there is also interest in the absolute value to be achieved, as in the relative contributors, however.

In many practical cases these requirements are in conflict; improvements in data sources may resolve some issues but once consent was obtained and a timescale for construction fixed for Sizewell 'B', a necessary change in emphasis towards the use of whatever data sources are readily available and accessible was required.

PPG has been supportive (technically and financially) in the development of databases such as the AEA DEFEND [3] and has developed expertise on the use of databases such as the CEGB's version of the US Licensee Event Reporting (LER) file. The scope for such research or development activities influencing the selection of component reliability data for use in the datum case Sizewell 'B' PSA has now passed, although the expertise is still available as a resource in support of other areas such as the operational phase, or for long term studies.

For component failure rates in particular, the need is to obtain estimates of failure rates for modes of failure which are fairly common. The identification of a main data source for equipment and function comparable to that intended for the station, with sufficient cumulative equipment exposure time to reduce statistical uncertainty, is the main requirement. Once such a data source, having the additional qualities of good failure narratives and minimal (or quantifiable) under-reporting problems has been identified, there will be little extra to be gained from extending the survey to poorer quality data sources - the possible additional information on equipment failure performance will be cancelled out by the degraded situation for estimating the equipment exposure times. Such supplementary sources do have a role in providing corroborative information, or for use generally as a source of operational feedback experience.

Three main operating data sources with corresponding reliability systems were identified as candidates for deriving data, on the basis of the above considerations, and are given below.

Operating Data Sources	Reliability Data Systems	Owner
US PWR Plant	NPRDS	Institute of Nuclear Power Operators (INPO)
French PWR Plant	SRDF	Electricité de France
UK Nuclear and Conventional Plant	PR/A plus specific data campaigns	CEGB successor companies

The following sections of this paper review these sources, noting their merits and drawbacks, the experiences gained in their use, and the applicability of the data derived from them to the Sizewell 'B' PSA.

US PWR PLANT

The US was surveyed as a source of PWR specific data in view of the large number of reactor years of accumulated experience.

Of the various data sources available, covering component reliability, station availability and event reporting, two were pre-eminent for use in component reliability derivation; NPRDS and the LER's. It is worth briefly restating their inter-relationship and history, as it illustrates the progress made in the last ten years, both within the nuclear industry (Three Mile Island accident provided a main impetus), and generally with information technology systems.

Nuclear Plant Reliability Data System (NPRDS)

NPRDS [4] was initiated by the Edison Electric Institute (EEI) in the early 1970's. The objective was to collect component reliability data for the nuclear industry's own use, but was not mandatory.

The software and procedures were developed by the Southwest Research Institute of Texas State. Data collection was started in 1974, so it was one of the first component data systems to be set up. As a result of the limitations of 1970's computing technology that hindered the development of non-mandatory schemes (e.g. no interactive data entry, and consequential limited feedback to data collectors), it suffered from a lack of support, and did not collect component data to a consistent standard.

Following the Three Mile Island accident and the establishment of the INPO, a general industry decision was made that NPRDS would be run as a mandatory scheme by the nuclear industry itself, and its management and technical direction were transferred to INPO in 1982.

NPRDS itself was transferred to INPO in July 1983, co-inciding with the introduction of interactive data entry from the US stations. Direct use of the system is only available to US nuclear utilities or to INPO overseas participants. Nuclear Electric participates as an operator of nuclear power stations, and obtains access to NPRDS at no additional charge to its INPO subscription.

NPRDS was subsequently upgraded in 1989 to improve access and user-friendliness. Whilst this was of debatable advantage to an experienced and intensive user of the system, the improvements have certainly eased the production of auditable interrogations which can be revisited as necessary.

The system sets a standard by which other component reliability data systems can be assessed, though this is based as much on its ready accessability by the international nuclear industry, and the comparative openness of the US regarding freedom of information, as on its intrinsic merits.

The scheme, as now currently operated, has a number of features of interest to derivation of component reliability data. In particular:-

- Mandatory reporting for the major safety-related systems for nuclear plant.
- Coding by application (e.g. Main Steam Isolation Valve) for the more significant plant items within these systems, allowing - for these items only - the contribution to failure modes such as "fail to open" and "fail to close" for valves from the valve itself, the valve operator and circuit breaker to be assessed.
- Quality control of input data by INPO staff, which is generally done to a high standard. In addition, site visits by INPO as part of their plant evaluation programmes mean that estimates can be made of the level of under-reporting. INPO initially concentrated on auditing the reporting of mechanical items. Reporting levels allow data searches to be made for 1984 onwards, and estimates of failure rates can be made with the use of prudent under-reporting factors.
- Text searches of the failure narratives can be carried out.

- In view of the diversity of plant designs and equipment in use in US PWR's (3 basic PWR Nuclear Steam Suppliers, and the BWR's are covered) NPRDS has an interactive file of engineering data. This is particularly useful for locating equipment with an engineering specification comparable to a specific Sizewell 'B' plant item for which an individual failure rate is required.

In fact an important feature of NPRDS is its stand-alone capability, as it provides a service to a large number of US PWR's which exhibit variations in their design features - though not to the same extent as UK Magnox and AGR stations. However, now that Sizewell 'B' PWR - for the forseeable future - will be unique to the UK rather than the lead station for a family this aspect of NPRDS should be seen as a virtue, and it will be beneficial to maintain links with NPRDS through INPO, or the World Association of Nuclear Operators (WANO) as appropriate, into the station's operational phase.

The confidence built up in the use of NPRDS for single component failure reporting has also led to an assessment of its use as a data source for the estimation of dependent failure frequencies. The more important of such failures should be reported as LER events, but not all dependent failures are identified as such at the time of data collection, and the considerations of estimating equipment exposure time still apply. Although the data on dependent events resulting in system failure is naturally sparse, the population of "precursor" dependent failures is sufficiently large to support some failure rate estimates.

Component failure rates for use in PSA have been derived from NPRDS for the majority of valves, and some pumps, on the basis that for these items the operating environment of a PWR is more likely to prove predictive of Sizewell 'B' experience than the environment of a UK station. Some additional mechanical items have also been covered, such as fans, dampers and filters.

In general, the failure metric chosen has been active-based failure rate, rather than frequency of failure on demand.

NPRDS does not record the number of demands on components, and the operating times are only estimates based on the number of hours in each month for which the reactor is critical. These limitations can normally be overcome for the more major items. Problems remain for items which are not coded by their application or duty.

Demand-based failure rates can be obtained by making assumptions on test intervals for standby equipment at Sizewell 'B'. Although this assumes that a standby failure rate model is applicable rather than a failure-on-demand model, this can be modelled in the PSA studies so as not to be optimistic.

The Licensee Event Report (LER) File

The reportable occurrence file maintained by the Nuclear Regulatory Commission (NRC) in the US dates back to 1969. It was computerised in 1973, by the NRC's predecessor, the AEC. A standardised LER form was introduced in 1984 [5].

More importantly the LER scope was revised at the same time to greatly restrict the reporting of single component failures. This change was instigated by the NRC on the basis that INPO had brought NPRDS up to a standard where the US nuclear industry was reporting single safety-related component failures to its own industry system, and only events with a wider significance needed to be brought to the attention of the NRC. The two schemes NPRDS and the revised LER file are, from 1984 onwards, essentially complementary, though the nature of some events will mean that they are reported in both systems.

The revised LER file is available to INPO subscribers as a database with text-searchable summaries via an interactive telecommunications system, but in addition the entire reports have been loaded into the TEXT system set up by the CEGB, and are fully text searchable.

Extensive use has been made of the CEGB's LER text file for estimating initiating failure frequencies, which is not covered in this paper. However, the LER file is also used as a check on the events reported in NPRDS in two respects:

- relevant events may not have been reported in NPRDS; in some cases this may be due to problems with definitions of component boundaries
- additional information in LER text may help to clarify the nature of failures, and assist in distinguishing safety critical failures from non-safety critical ones. An example is the reporting of failures of pressuriser and steam generator safety valves.

An event reporting system such as LER, loaded into a text handling system so as to be fully searchable, can be a considerable enhancement to the derivation of component failure rate data, provided that it is additional to a component reliability data system which provides the engineering data and estimates of equipment exposure times. Without such a component-based system the value of an event reporting system in providing component failure rates is greatly diminished.

FRENCH PWR PLANT

Electricité de France has now accummulated a significant amount of operating data on its PWR stations, of which the two main computerised data systems of interest as sources of reliability data [6] are:

- Systeme de Recueil des Données de Fiabilité (SRDF)
  (System for collecting data on reliability)
- Fichier des Evenements (FdE)
  (File of events)

These are broadly comparable to the US NPRDS and LER file respectively, although with some important differences.

SRDF contains more information about the actual operating times for the components that it covers, as a result of recording counters which are extensively employed to capture this information. Following trials at Bugey and Fessenheim, SRDF has collected information on some 1100 components from the 900 MW units since 1983, and on the 1300 MW units from commercial operation.

The events file (FdE) is operated by EDF itself, and covers a greater range of events than those reported by US plant to the NRC via the LER's. This became operational as an interactive system with remote terminal entry from the power stations around 1985. Events reported are those considered by the stations themselves to be important, according to some general criteria agreed with the safety authorities.

Notwithstanding that the amount of general operational experience available from EDF from 1984 onwards is broadly comparable to that in the US following the revisions in NPRDS and the LER file, US data has been preferred for deriving fault rates over EDF data, for the following reasons:-

-       The EDF design for both 900 MW and 1300 MW plant displays a
        considerable homogeneity, as a result of a well managed
        programme of follow-on stations, with an incremental approach to
        design changes. This same attitude is reflected in choice of
        equipment suppliers. This implies an efficient operational
        experience feedback loop to aid EDF in identifying problem areas
        in either design or equipment performance. Trends from data are
        detectable more quickly, and corrective action put in hand with
        more confidence, than for the typical US station which is not in
        a comparable homogeneous technical and managerial environment.
        Although Sizewell 'B' is considered to have a number of unique
        design features which will enhance its safety system
        performance, it will only have its own operational experience
        and data to draw on in demonstrating the merit of these
        advances. Its situation is therefore more comparable to a US
        station in terms of feedback from its own operational
        experience, and drawing lessons from the experience of other
        related stations that are nevertheless different in many
        respects.
-       The main route for obtaining information on EDF's operational
        experience, apart from publicly available material, is through
        meetings, and EDF internal reports to which confidentiality
        conditions may be attached. The reports in the reliability data
        area were drawn up for use by EDF in PSA studies on its own
        plant, and it is not always possible from the information
        provided to assess the appropriateness of these data for
        application to a different design.
-       The meetings and exchange of reports take part within the
        general framework of a technical exchange agreement with EDF,
        which is financed on each side to the extent to which
        information and services are accessed. Nuclear Electric has
        not, to date, sought to have direct remote access to either SRDF
        or the French events file, and consequently limits its requests
        to areas where it is evident that EDF experience provides
        valuable lessons for nuclear plant operations. Requests are not
        made to obtain French data where there is adequate UK or US data
        which can stand in its own right.

Areas where EDF data are being used in support of the Sizewell 'B' safety case, either in deriving fault rates for use in the PSA or corroboratively, include the following:-

- Valves and pumps. EDF data supplied is quantitative rather than qualitative, and in this area reliance is placed on EDF's technical assessment of the data. In particular, it is not possible to verify that the coding of failure modes and the assignment of failures into critical and non-critical failures will also be applicable to Sizewell 'B'. However, the EDF methodology document is available, and the quantitative analysis provides insight into the "factors influencing failure", which was a reason for seeking reliability data on the main fluid system components.
- Component data handbook. EDF's data handbook covers the totality of components used in their PSA study and is comparable with other data handbooks - of which the most well-known one is the publicly available Swedish Reliability Data Book [7].
- SEBIM valves. In this case EDF made available the operational experience (including failure narratives) and safety analysis of the SEBIM valve, which is the French designed pilot-operated pressuriser safety valve to be used at Sizewell 'B'.
- Other areas where particular note was taken of EDF's experience were the Main Steam Isolation Valves (the design is similar to that for Sizewell 'B', but by a different manufacturer) and the reactor coolant pumps (EDF has accumulated substantial experience with the Westinghouse 100D design intended for Sizewell 'B', but with a different manufacturer).

UK NUCLEAR AND CONVENTIONAL PLANT

The CEGB never operated a reliability data system specifically aimed at collecting component data on a national basis, and therefore comparable to NPRDS or the French SRDF. The situation is consequently more complicated, with a number of systems that can be accessed to identify suitable component data.

Plant Reliability/Availability (PR/A)

Set up in 1977 the PR/A system [8] ceased to be fully operational after 1989. Reporting of events resulting in a loss of station output was mandatory for all nuclear stations, and for stations with outputs at 500 MW and above. Reporting of other (trained) plant items was optional, but rarely undertaken, excepting for some feed pump studies. The system is therefore useful only for obtaining failure rates for items, the unavailability of which would result in a loss of output.

For such items (e.g. governor valves, condensers) the system provides a useful indication of the order of losses. For more detailed information regarding nature of the failures, it is usually necessary to turn to one of the sources of data discussed below.

Nuclear Plant Event Reporting (NUPER)

The NUPER system [9] is a file of events on UK power stations, and is
therefore a close parallel to the LER file and the French events file.
Although it is fully text searchable using the CEGB TEXT system, the
usefulness of the file is limited by the absence of searchable
information on engineering data, and by the criteria for reporting being
oriented to unusual events rather than component failures. Useful
qualitative information on the nature of governor valve failures on UK
nuclear stations has been obtained from this source.

UK On-Site Station Sources

In addition to the two national systems referred to above, station
maintenance management systems such as KISMET can be accessed on site, or
other data derived from specific data campaigns at sites, and held to the
REGCARD format, are available at Nuclear Electric's HQ in Gloucester.
Neither of these sources involved direct access by remote terminal from
the PWR Project Group. The majority of the data on high voltage
equipment used in the Sizewell 'B' PSA is derived from campaigns at
specific sites, such as Wylfa power station. A notable example is the
generator circuit breakers, for which extensive operational experience,
in terms of demands to open and close, had been accumulated at the
Dinorwig pumped storage scheme in North Wales.

Other UK Sources of Data

For some low voltage electrical equipment, notably batteries in d.c.
systems, relevant field data exhibiting an enhanced performance over
"generic" data has been obtained from British Telecom. There is no
reason to consider that the power station environment for batteries
should be significantly different from those installed for British
Telecom - always provided that a similar maintenance contract to a
battery supplier is taken out. However for PSA purposes the generic
estimates have been retained, although it is confidently expected, as is
the case for a number of other areas, that actual performance will better
these estimates.

REVIEW OF DERIVED DATA

The review briefly covers some of the requirements for component
reliability data, and the extent to which the derived data satisfy these
requirements.

Best Estimate (Defendable Non-Optimistic)

The majority of mechanical components have been based on US plant
operating experience for 1984-86. It is expected that, drawing on the
experience of already operating plant, Sizewell 'B' should in the long
term equal or better the average US Westinghouse PWR over this period.
Since the data derivation exercise began, there is encouraging evidence
of improving trends in US equipment reliability.

In situations of scarce data, conservative assumptions have been
made. These situations are relatively few, and are normally associated
with cases where component failure modes were requested as potential
contributors to Initiating Fault Frequency estimates.

Historical Field Data

Except for a small number of minor cases, the failure rates have been derived from actual field data, or from accepted generic sources which have been corroborated by field data. Most of the data are derived from operating experience in the 1980's, excepting where - for some limited UK data sources - an extended period was used to limit uncertainty in the derived failure rate.

Consistency

The effect of reporting scheme contributions to inconsistency in the data has been minimised by the use of US nuclear plant the majority of mechanical components, and UK data sources for the electrical ones. French data has been used where clearly applicable (e.g. SEBIM valves) and to corroborate the main data sources.

Auditability

The preference has been for data sources where access to the original failure event reports is possible. For the key components, derivation of failure rates for each mode of failure by examination of the failure narratives (when available) has been carried out. This is preferred to reliance on coded fields. NPRDS in particular must be singled out as a system where fully auditable data extractions could be generated with comparative ease. In addition the INPO back-up staff were attentive to this aspect of INPO and investigated any anomalies promptly.

Reporting Level

Under-reporting factors were used in deriving component failure rates where appropriate, bearing in mind other pessimisms that may be introduced in the course of the failure rate derivation process.

CONCLUSIONS

The ability to examine failure narratives in the data sources, and identify critical failures is an important requirement for the reliability data derivation process. The selection of datasets most appropriate to the Sizewell 'B' design is considerably aided by interactive access to a data system which provides both engineering data as well as failure data.

The data derived for supporting the Sizewell 'B' safety case are primarily based on historical field data, minimising situations where recommended failures rates involve a predictive element when applied to Sizewell 'B'.

The main benefit that has been brought to the reliability data assessment area in the last decade is the ability to access the source data with comparative ease, brought about by the use of interactive computer interrogation systems operating upon data banks dedicated to specific aspects of plant performance.

These safety analysis activities can now not only pinpoint significant safety areas, but can ensure that the qualitative experience relating to these areas is promptly brought into focus.

Future directions for development should consider the inclusion of more systematised information relating to underlying failure causes, and mechanisms of failure. This would offer the prospect of bringing to component and materials design areas the same level of information that is now readily available to the systems and equipment area.

ACKNOWLEDGEMENT

This paper is published by permission of Nuclear Electric.

REFERENCES

1. McIntyre, P.J., Gibson, I.K. and Witt, H.H., Addressing the problem of the relevance of reliability data to varied applications. Proceedings of the Euredata Conference 1989 on Reliability Data Collection and Use in Risk and Availability Assessment.

2. Sizewell 'B' Pre-Construction Safety Report, SXB-IP-771001, Issue C (P), November 1987, Central Electricity Generating Board.

3. Games, A.M., DEFEND - A dependent failures database. Euredata Conference 1989 on Reliability Data Collection and Use in Risk anᵤ Availability Assessment.

4. Nuclear Plant Reliability Data System (NPRDS). Programs Description. Institute of Nuclear Power Operations, INPO 86-010, February 1986.

5. Licensee Event Reporting System: Description of System and Guidelines for Reporting. US Nuclear Regulatory Commission, NUREG-1022.

6. Aupied, J.R., Lannoy, A., Meslin, T., Zanetti, H., Operation Experience Feedback Contribution to Probabilistic Safety Analysis. Proceedings of the International ENS/ANS Conference on thermal reactor safety "NUCSAFE88", October 1988, Avignon, France.

7. Reliability Data Book for components in Swedish nuclear power plants, Nuclear Safety Board of the Swedish Utilities, RKS 85-25.

8. Taylor, D.T., A Reliability/Availability Reporting Scheme for Power Station Plant. Proceedings of the Second National Reliability Conference, March 1979, Birmingham.

9. Spencer, A. and Walker, C.W., The CEGB's use of international and national unusual event reports for improving nuclear power plant safety. IAEA/OECD Seminar, May 1990. IAEA-SR-169 - Paper No. 44.

PROCESS VISIBILITY - THE KEY TO SOFTWARE RELIABILITY ASSESSMENT

CHRIS DALE

Cranfield IT Institute

Fairways, Pitfield, Kiln Farm, Milton Keynes MK11 3LG, UK

ABSTRACT

Software reliability and safety assessment should be viewed as confidence building activities, especially when one is dealing with ultra-high levels of reliability or safety-critical systems. Confidence can be built by examining various kinds of information drawn from the entirety of the development cycle of a given system, and making judgments regarding the compatibility of the information gathered with the level of reliability or safety required. This applies to all aspects of systems, not just software, but software and its development process are in general far more abstract than their hardware counterparts. This means that the information necessary to the confidence building activity can be difficult to find, unless special care has been taken to ensure that the software and its development process are made sufficiently visible. The main focus of this paper is process visibility: strategies are described for ensuring availability of the information necessary for confidence building.

INTRODUCTION

A number of categories of people have an interest in being able to assess software reliability, including the customer or procurer of the software (or the system of which the software is a component part), the project manager responsible for delivering the product, and the software developer's quality manager.

Superficially, the customer is interested only in the reliability of the delivered product or system, and whether this delivered reliability meets his requirements. On closer analysis, however, the customer's interest often goes much further back in the lifecycle than the point of delivery. When placing a contract for a bespoke system with critical reliability requirements, it is wise to have confidence that the chosen supplier will be able deliver the required reliability. It is equally important that this confidence is maintained as the development of the system proceeds.

The project and quality managers' needs for confidence from the earliest stages of development are more apparent. The project manager must be able to establish targets, formulate a plan to meet those targets, and monitor progress against the plan. The quality manager will want to be able to inspect measurements which demonstrate quality achievement.

There is then a clear need to be able to establish confidence in the reliability of a system not only when it is delivered, but during and even before its development. This will come as no surprise to those familiar with the reliability analysis of engineering systems; many techniques exist for examining various kinds of information drawn from the entirety of the systems development cycle, and making judgments regarding the compatibility of the information gathered with the level of reliability or safety required [1].

The problem with software is that it is abstract in the extreme, and its development process also tends to lack the tangibility provided by the physical nature of traditional systems. These factors mean that unless special care is taken, the information necessary to the confidence building activity simply does not exist.

The study of software reliability assessment has in the past tended to focus on measurement of reliability of completed or almost completed products [2]. The need for some form of assessment earlier in the lifecycle, to aid the confidence building discussed above, has received scant attention. The proper management of the reliability of systems containing software, from the viewpoints of both supplier and customer, demand the ability to carry out assessments from the earliest stages of product development.

This paper introduces a five stage approach to the problem of building confidence in the reliability of software-based systems. The first three of these stages are concerned with process based assessment: building confidence by ensuring that a *development strategy* appropriate to a given *achievable target* is being properly *carried out*. The final two stages relate to product based assessment, building confidence by examining the *product* and its *behaviour* when executed.

These five stages are described in the sections which follow. There are also discussions of the use of software metrics, the impact of education and training, and the role of management.

TARGET ACHIEVABILITY

The first of the five stages is an assessment of the achievability of any reliability targets. Typically, reliability targets will be expressed at the system level, possibly based on some specified safety requirement. In decomposing the system into its component parts, a reliability apportionment exercise will result in the assignment of reliability targets to different aspects of the system. It is important to make an initial judgement of feasibility of each of these component reliability targets at this stage, so that any necessity for reapportionment or redesign can be identified at the earliest possible - and thus cheapest - time.

Ultimately, once the top level system design has been determined, there should be some statement of reliability requirements associated with the software component or components. It is important that this statement is quantitative, and that it is made in terms which can be measured by observation of software execution. Thus statements such as *the software shall be fault-free* or *the software shall have fewer than x faults per 1000 lines* must be avoided - the first cannot be measured, and the second can be measured only in ways which are not directly related to the user-perceived failure behaviour, ie the execution of the software in time.

The stated reliability requirement should reflect the various demands on individual functions of the software. Thus a protection system of some sort may be required to have a reliability greater than p for its *shut down when required* function, and a failure rate less than λ for its *keep running if safe* function. This example illustrates that a mixture of metrics can sometimes be necessary in the requirements statement, and leads to the important point that the conditions under which the measurements are made must be specified carefully.

It is vital that any statement of requirements should be for an appropriate level of reliability. The reliability must clearly be sufficiently high for the application in question, but must be kept reasonably low because of considerations of achievability and testability. Thus demands for perfection must be avoided - perfection is (at least) difficult to achieve and impossible to demonstrate. Similarly, high levels of reliability are difficult (and costly) to achieve and can be demonstrated only be extensive (and expensive) verification, validation and testing activities.

The software reliability assessment process should begin by looking at the achievability of the stated software reliability requirements, so as to establish initial confidence in the feasibility of the target. There is little point proceeding to subsequent phases of development if the software reliability target is judged to be unachievable at this stage.

PLAN CREDIBILITY

Once a target has been set which is adjudged to be achievable, a development plan can be drawn up to show how the product will be developed to meet its reliability (and other) requirements. It is important then to evaluate the planned development strategy, so as to generate confidence in successful achievement of the target, at the early stages. This avoids the danger of embarking on development with a high risk of failure to meet targets.

In the safety-critical and ultra-high reliability areas there is an additional problem: it is impractical to demonstrate target achievement on the basis of test results alone [3], so that a high level of *a priori* confidence is essential to any reliability or safety claims made later.

The main aim of this second stage of assessment is to establish whether the planned development is likely to meet the target. There is a need to exercise a healthy degree of conservatism at this stage, because in many software-related areas, the current state of knowledge and understanding concerning which techniques and methodologies are likely to achieve which levels of reliability is meagre. In most cases, it is not possible to say definitively that method X is likely to meet target Y. Thus, extreme claims and targets must be treated with scepticism.

There are no well established techniques which can be used to make this assessment of likelihood of the planned development meeting the established target. Instead, an appropriate review process should be applied to the plan, and the experience of those involved in the review used to make the assessment as objective as possible. Any data on the way previous comparable projects were carried out, and the levels of reliability achieved in practise, will be invaluable in this context. Such data is a very scarce commodity at the current time.

The result of an exercise to establish the credibility of a plan to achieve a given level of reliability may lead to recommended revisions to the development process, such as additional or more formal verification activities. Once again, recognising the need for this at an early stage is much better than trying to put the problem right later.

DEVELOPMENT QUALITY

Having established an achievable reliability target, and produced a plan of how the target is to be achieved, the software development itself can commence with a healthy degree of confidence that the target will be met. There is now a need to continue to build this confidence, as the product is built. To achieve this, the development process must be monitored to ensure that it is being carried out in a way consistent with the target and the plan.

This third stage of the assessment process is, then, concerned with determining whether the actual development is of the standard adjudged necessary for the given reliability target. In managing any kind of activity, the project manager needs to be able to monitor and control the critical success factors. In this case, the developing reliability characteristics must be monitored and controlled; it is far from obvious how this may be done. The target reliability should have been expressed in a way which enables measurement of the final product reliability, such as probability of failure on demand, or failure rate. How can the project manager hope to monitor this quantity at a point in the lifecycle when the software is not even executable?

The answer is of course that he cannot measure reliability in any direct sense until very late in development, when it will probably be too late to take the necessary remedial steps. The challenge for the project manager is to identify factors which can be measured during development, which can serve as indicators of the level of reliability likely to be achieved, based on the current stage of development. In a well managed project, it should be possible for the project manager to report progress against objectives on a regular basis, and to have supporting evidence for his report. Problems are identified, and remedial action taken. In principle, the management of the reliability aspect should be treated no differently.

There are two kinds of evidence which the project manager can use to help him in managing the reliability. The first kind of evidence pertains to the software development process: this is simply evidence that the project is being carried out in accordance with the plan, which was earlier adjudged to be capable of delivering the required reliability. This provides evidence that the job is being done properly - this gives some support to an argument that the target will be met, as it provides indirect evidence about the developing product.

The key to providing this sort of evidence is to ensure that accurate and up to date records are maintained of the activities carried out, in at least as much detail as they were originally planned. The plan was the basis of an argument of how the target would be achieved; this needs now to be reinforced by evidence that the development really was carried out in that way. Inevitably, there will be aspects of the development which do not happen exactly as planned. In these cases, the impact of the difference on reliability must be identified, and if the impact is such as to undermine confidence in achievement by simply sticking to the plan, remedial action must be planned.

The necessary records are not difficult to produce and maintain, provided there is proper discipline in documentation, configuration control and quality assurance. These things help to ensure the necessary degree of visibility of the development process, as do accurate records of verification and validation activities. The collection, analysis and presentation of data concerning the development of the product help the manager and others to understand and communicate the progress towards the stated reliability requirements.

Verification and validation activities especially need to be made visible. These are the times at which the intermediate products are subjected to maximum scrutiny, and their adequacy determined. The activities may be carried out by the programmer, within the team, within the organisation, or by an independent outside body: these are listed in increasing order of independence from the development of the object in question, which is usually associated with increasing visibility of the reliability achievement.

PRODUCT QUALITY

The second kind of information which the project manager needs must be about the product, so that there is some direct evidence of reliability-related achievement in the various intermediate products, which may include specifications, documents and other non-executable objects. This information is needed to determine whether the process is actually delivering the level of quality required: the process was planned in the way it was because it was believed that it would deliver the necessary quality, and process monitoring has been used to ensure that it has been carried out in that way; now confirmation is sought that the process is delivering as anticipated. The principle is that measurement of various aspects of the software and other intermediate products of the development enables assessment of the adequacy of the outputs. The details of the information used for this purpose will vary from project to project, depending mostly on the development methodology and techniques adopted, but it will often take the form of fault data, which can arise in a number of ways.

Formal review activities, such as Fagan inspections [4], give rise to lists of faults identified. These lists may be categorised, showing the numbers of faults of each of a number of different levels of severity. In planning the achievement of reliability, targets may be set for these reviews, determining the intermediate acceptance criteria for the objects under review. For example, a detailed design document may be releasable for coding to commence only when review indicates no major faults and less than three minor faults. There may also be criteria which determine whether a further full review is required following rework, or if review and approval of fault fixes will suffice.

Another way of examining certain kinds of product, especially source code, is static analysis. This is normally a tool aided activity, aimed at analysing the syntax of the developed code to identify such things as unreachable code and various other non-preferred features. These do not always indicate the presence of specific faults, but do serve as indicators of potential problem areas, and can show that programming standards have not been adhered to - itself indicative of the level of process quality. These tools often provide various measures of code complexity, which can be a further indication of problem areas. A related tool-aided area is that of symbolic execution, in which formulae are derived to show the relationships between code inputs and outputs; these can be inspected to determine whether they are in accordance with the specification, and so potentially indicate the presence (but not the precise location) of faults. Static analysis and symbolic execution are reviewed by Smith and Wood [5].

Dynamic measures can also be made. These include test coverage measures [6], which quantify the extent to which the software in question has been exercised by the testing to which it has been subjected. These measures are really showing how well the testing process has been applied to the product in question, but are so intimately related to the product that they can almost be viewed as product measures.

Ensuring that both the process and the developing reliability-related characteristics of intermediate products are made visible in the ways described above helps to substantiate claims that the actual development process was of the standard necessary to the achievement of the reliability targets, and that the quality of intermediate products is consistent with the level of reliability required in the delivered system.

PRODUCT RELIABILITY

The final assessment stage is to measure the reliability of the final product, based on test results. This stage is carried out so as to provide further evidence to support the confidence already established by the previous four stages. It is almost never the case that a valid software reliability assessment can be carried out by addressing this stage alone, for reasons which are discussed below.

A host of statistical models exist for predicting the reliability of software, based upon test data [2]. All of these methods depend upon an assumption that the environment in which the software is tested is representative of the operational environment of concern (or that a relationship can be defined between these two environments, enabling a mapping of the reliability in the test environment to the reliability in use). To date, very few organisations have been willing to make the investment necessary to understand the operational environment well enough to enable this assumption to be validated - though some success has been reported from the IBM clean room development methodology [7]. Unless an attempt is made to replicate the real environment, any application of the statistical models for software reliability prediction will result in information of an essentially qualitative value - numerical values do not always mean that reliability has been quantified.

Even when the real world environment has been replicated, process assessment is necessary to ascertain that the testing process is of the necessary level of quality, and that the original specification was properly validated. So even in these circumstances, statistical measurement of the completed product cannot stand in isolation as the only assessment which is applied.

When the highest levels of reliability are sought, a more serious problem arises. The statistical models require amounts of data which cannot practically be obtained, in order to provide confidence in the measurement of the achieved level of reliability [3, 8]. In these circumstances, the statistical models can provide the reassurance required only if there are strong *a priori* reasons for believing that the necessary level of reliability has been achieved.

Despite these limitations, the exercise of measuring the achieved level of reliability should be carried out whenever reliability is an important requirement, either by applying a reliability growth model to data from testing, or by applying standard statistical techniques to a reliability test of the final product. In either case, the input data used should be statistically representative of data which will be seen in the environment of interest.

THE USE OF SOFTWARE METRICS

The success of the five stage assessment process outlined above depends upon the proper application of a software metrics programme. Properly assessing the achievability of targets is impossible without records of previous projects, and their achieved levels of reliability. Similarly, details of how previous projects were carried out is necessary to establishment of plan credibility. The assessment of development and product quality, and statistical measurement of reliability are in themselves concerned with the collection, analysis and interpretation of data.

The metrics programme then has two roles - to provide data about previous projects to aid in the assessment of achievability and plan credibility, and to extract information from the current project as necessary to the final three stages of the assessment process. The first role implies that relevant data about the current project should be retained, so as to enable the early assessment stages on future projects.

THE IMPACT OF EDUCATION AND TRAINING

An important implication of the above arguments is that software reliability assessment can be carried out only if the intrinsic intangibility of the software product and its development is overcome by making the process visible, and introducing a metrics programme. Unless special care is taken, the people involved in an activity which is being measured can feel very threatened, believing that the measurements may be used against them in a judgemental way.

The only way to overcome these fears is to provide education to the personnel involved. The management needs to understand how to obtain the required data in non-threatening ways, and that the data must be used only for specific product and process assessment purposes - never for personnel assessment. The development staff are also in most cases the providers of the data: if the data are to have integrity, the collectors must have a positive attitude, which can only be achieved if they know and understand why the data are being collected, and can see a benefit to themselves in providing data of high integrity. Appropriate education is the answer to winning the hearts and minds of these people, and training will also be needed in the specifics of the metrication and assessment processes.

The impact of the education and training will be reflected in a successful and convincing software reliability assessment activity.

THE ROLE OF MANAGEMENT

Software project management is concerned with the planning, monitoring and control of product development. Where reliability is a specifically expressed requirement of the software, then this so-called non-functional requirement should be planned, monitored and controlled along with all the other requirements. The role of management in these circumstances is then to recognise the need and implement the necessary mechanisms in the five stage assessment process outlined above.

There may be a feeling that the implied requirements for a metrics programme are onerous, except in the most extreme cases. In general, however, much of the data exists already if the development is well-managed, and it will be found that any additional data will be of value for reasons other than those associated directly with the reliability assessment need.

CONCLUSIONS

This paper has described a five stage software reliability assessment procedure, beginning with initial assessment of target feasibility, and culminating in measurement of the delivered level of reliability. In concept, the steps are simply those which need to be taken to manage the achievement of reliability, by enabling the responsible manager to monitor and control progress towards a target. They also provide the evidence necessary to support claims made to the customer or some independent assessor, concerning the reliability or safety of the product.

The five stage procedure described is an engineering approach to an engineering problem: confidence is built in parallel with the building of the product.

REFERENCES

1. O'Connor, P.D.T., Reliability Engineering, Heyden, 1981.

2. Dale, C.J., Software reliability models. In Software Reliability State of the Art Report, ed. A. Bendell and P. Mellor, Pergamon Infotech, 1986, pp. 31-44.

3. Miller, D.R., The role of statistical modeling and inference in software quality assurance. In Software Certification, ed. B. de Neumann, Elsevier Applied Science, 1989, pp. 135-52.

4. Fagan, M.E., Design and code inspections to reduce errors in program development. IBM System Journal, 1976, 3, 182-211.

5. Smith, D.J. and Wood, K.D., Engineering Quality Software, Elsevier Applied Science, 1987.

6. Hennell, M.A., Hedley, D. and Riddell, I.J., The LDRA testbeds: their roles and capabilities. In Proc. IEEE Soft Fair '83 Conference, Arlington, Virginia, July 1983.

7. Currit, P.A., Dyer, M. and Mills, H.D., Certifying the reliability of software. IEEE Transactions on Software Engineering, 1986, SE-12, 3-11.

8. Dale, C.J., Data requirements for software reliability prediction. In Software Reliability Achievement and Assessment, ed. B. Littlewood, Blackwell Scientific Publications, 1987.

SOME RESULTS OF THE ALVEY SOFTWARE RELIABILITY MODELLING PROJECT

C. McCOLLIN, D.W. WIGHTMAN, P. DIXON and N. DAVIES
Department of Mathematics, Statistics &
Operational Research,
Nottingham Polytechnic
Burton Street, Nottingham NG1 4BU

ABSTRACT

The Alvey software reliability modelling project is a multi-tasked project consisting of a collaborative team from UK industry and academia. Over the duration of the project, the membership has consisted of the National Centre of Systems Reliability (AEA Technology), British Aerospace, STC, Logica, Nottingham Polytechnic and City and Newcastle Universities. The objectives of the software reliability modelling project were to investigate a wide variety of methods, to judge the relative merits of each method, to effectively communicate the results of the research and to indicate the direction of future research. The project consisted of a number of tasks of which this paper describes areas in which Nottingham Polytechnic were task leaders; these are task 3 (statistical models with explanatory variables), task 4 (statistical models with different underlying assumptions) and task 9 (data collection and initial analysis).

INTRODUCTION

History

There has been growing concern in the software industry about unreliable software for many years and as a result there have been some initatives aimed at reducing the impact of the problem. Customers have imposed codes of practice on suppliers, lists of "approved" software have been specified and work has been done on better testing strategies, Up till now little coordinated research has been done nationally on modelling software reliability. The appearance in 1984 of the "Software reliability and metrics programme" document from the Alvey Directorate formed a natural focus for this work. A consortium was formed containing members from both academic and commercial backgrounds with the intention of conducting a research programme to improve the state of the art.

In July 1985, the Alvey Directorate placed a contract for the detailed study of software reliability modelling (SRM) with the aim of producing a plan for a National SRM Programme. The suggested course for the research was instilled into a set of project tasks. These tasks were as follows:-

Task 1:   Improving Current Statistical Models
Task 2:   Methods of Evaluating Statistical Models
Task 3:   Statistical Models with Explanatory Variables
Task 4:   Alternative Statistical Models
Task 5:   Functional Modelling
Task 6:   Models for Special Systems
6.1:   Models for VLSI Systems
6.2:   Models for Distributed Systems
6.3:   Concurrent/Real Time Systems
6.4:   Models for Fault Tolerant Systems
6.5:   Reusable Software Components
Task 7:   Cost Based Models
Task 8:   Testing and Reliability
Task 9:   Data Collection and Analysis

The project finished in June 1990 and more than sixty documents have been written during the project and a number of these are available to the public. The address for further details is given in the summary.

This paper describes some work carried out in each of the tasks 3, 4 and 9. The process of analysis of one of the datasets collected for the Alvey project is described with the problems of assessing software reliability for this dataset using the classical software reliability models e.g [1], [2], [3].

A description of other work carried out in task area 3 includes models which incorporate explanatory variables and extensions to published models. In task 4, the application of time series and multivariate techniques to software reliability is described.

WORK CARRIED OUT IN TASK AREA 9

In task area 9, considerable effort was invested in the creation of a software reliability database which is installed on a dedicated Vax computer at City University. Private computer links to the database are available to the other academic partners. It is envisaged that the usefulness of the database will continue beyond the duration of the project. The format of the data sets collected ranged from summaries of failure counts on networked systems, completed failure and repair reports on field data, software test and inspection information and cpu times to failure for individual computer installations.

The problems of collecting software failure data are highlighted in reference [4] and a description of a statistical analysis of one of the datasets appears in [5]. The main purposes of the data analysis exercise were to determine suitable models for software reliability estimation and to establish models which would incorporate explanatory factors found in analysis. The dataset was collected during the development phase of the project and the software was continually being operated and repaired after failure.

The analysis of Alvey dataset 3 and conclusions which led to further analysis were carried out in the following order:

Sorting, counting and merging file data to find any corrupt or missing data.

Plotting the number of failures per day against day.

Applying a parsimonious Box-Jenkins time series model to the data. This revealed a decreasing trend and a seasonal component.

An analysis using Proportional Hazards Modelling (PHM) was applied to the dataset to determine the effect of the seasonality. The analysis showed that the hazard rate was increasing on specific days of the week. Two observations were made during the analysis. The first was that a number of explanatory factors could not be fitted together in PHM. This was due to the factors being collinear.

The problem of multicollinearity of the covariates was investigated by applying multivariate techniques to the dataset and the results of this are described under the task 4 work heading. The second observation was that for this data set and a number of others execution time to failure was not available. The main software reliability models [1], [2], [3], use execution time as their time metric. It was found during the data collection exercise that companies do not usually collect execution time to failure of programs because:

It is a costly exercise to collect execution time to failure.

The customer only usually requires execution time if he wishes to estimate software reliability by using one of the available models.

The collection of execution time is not usually a requirement of general software guidelines or standards.

Based on these observations, further work on explanatory factors was carried out and this is described in the Task 3 section.

WORK CARRIED OUT IN TASK AREA 3
Review of models
A report [6] was written which reviewed the models and techniques which incorporate explanatory variables and can be adapted to software reliability modelling. Techniques and methodologies reviewed were Software Science, Information Theoretic approach, simple regression, multivariate analysis, proportional hazards modelling and generalized linear modelling.

Explanatory Variables highlighted in task 9:
Development of a modelling framework which incorporates these.
The following is a list of explanatory variables and their associated time metric. Each variable should be taken into account when deciding on a suitable software reliability model and its modelling assumptions.

Time metrics associated with the explanatory factors

CPU time
Execution time
Operating time
Calender time

Explanatory factors associated with CPU time

A1-Language
A2-Type of program
A3-Computational volume
  -Length of machine code/text
A4-Computational complexity
  -Nesting complexity
  -Number of calls to external modules
  -Number of conditional statements
  -Type of input/output
A5-Mathematical complexity
A6-Loading
A7-Programmer skill/experience

Explanatory factors associated with execution time

A1-A7
B1-CPU time
B2-Compiler status
B3-Parallel/serial processing
B4-Queueing
B5-Priority
B6-Available storage
B7-Systems availability
  -Peripherals

Explanatory factors associated with operating time

A1-A7
B1-B7
C1-Execution time
C2-Usage
  -Idle time
C3-Type of installation

Explanatory factors associated with calendar time

A1-A7
B1-B7
C1-C3
D1-Operating time
D2-Stoppages
  -Holidays
  -Strikes
  -Shutdown
D3-Seasonal variation
D4-Number of staff
D5-Project deadlines
D6-Data collection method
D7-Job priority

The analysis of the times between failures of the Alvey dataset 3 presented in [5] described a PHM formulation with days to failure of programs as the time metric and program type and program size as two of the explanatory variables.

Different program types and program sizes affect the cpu time directly and the calendar time only indirectly by the cpu time. For the results of [5] concerning the two explanatory variables and time metric, days between failure, it was known that the software was continually operating all the time for the duration of the project so that the calendar time between failures was the same as the cpu time between failures accumulated for the complete software package. However, it is not possible to relate the hazard function based on cpu times to program failure to the two explanatory variables unless certain assumptions are made about other explanatory variables, e.g. the usage of the individual programs.

As PHM uses the ranking of the failure times and not the failure times themselves, it can be shown that as long as the ranking of the days between failures remains the same for cpu time, execution time or operating time between failures, then the conclusions concerning the hazard for days between failures are valid for the other time metrics. For example, if execution time can be controlled so that it is always the same function of calendar time, e.g calendar time = a constant multiplied by execution time, then conclusions about the calendar time hazard function will apply to the execution time hazard function.

Extensions to models

The City University has contributed three reports on extensions to existing software reliability models. These cover task areas 3, 4 and 1 (improvement of current models). A Bayesian formulation of the Jelinski-Moranda software reliability model [7] reports that the model performance seems to be at least as good as some other models. In reference [8], a simulation study is reported which investigates if a general but simple adaptive procedure which improves the accuracy of predictions also increases the variability of the predictions. Reference [9] describes an extension to the "u-plot" (used for assessing predictive performance or for obtaining improved "adapted" or "recalibrated" predictors) to allow for discrete or mixed predictive distributions. Two further modifications of the u-plot are documented which improve the performance of recalibrated predictors.

A paper relating experience of applying a proportional hazards modelling formulation to data set 5 (data set number as defined in Task 9) has been written. The paper has still to be presented to and cleared by the data providers, [10]. Previous application of proportional hazards modelling has been based upon modulated renewal processes; where the explanatory variables modulate the underlying renewal process, [11], [12], [13] [14]. Recently, Lawless [15] has introduced model formulations which allow explanatory variables to be considered within a Poisson process. These proportional intensity Poisson process models allow the traditional non homogeneous poisson process software models to be combined with explanatory variables. A particular model formulation investigated at Nottingham Polytechnic is one which depending upon parameter values leads to either a modulated renewal process or a proportional intensity model.

Software has been developed at Nottingham Polytechnic for Poisson Proportional Intensity models with covariates and an unspecified baseline intensity. No data has been applied to this software as yet, however details of the approach are available from the authors.

Work has been carried out at Nottingham in expressing binomial type models and Poisson type models of exponential class (as classified by Musa et al [1], within a PHM framework. Details are available from the authors.

Font [16] derived a proportional hazards model with the Musa model as the hazard function. The following formulation of the Musa model within a PHM framework is useful as a goodness of fit test for the Musa model in that if the number of software failures is not a significant explanatory variable in the PHM formulation then the Musa model is not appropriate for the data analysis.

The Musa basic execution time model [1] takes the form

$$\lambda(t) = \lambda \exp(-\phi t)$$

where

t = total execution time

λ = initial failure intensity

$\lambda(t)$= failure intensity function

ϕ = "Constant hazard which characterises any individual failure"

The expected number of failures in time t is given by

$$\mu(t) = \int_0^t \lambda (w) dw$$

$\mu(t) = \lambda (1 - \exp(-\phi t)) / \phi$

From [1], the cumulative hazard function is

$H(t^1/t) = \mu (t + t^1) - \mu(t)$

Letting t = last failure time and hence t^1 = time since last failure.

Thus, $H(t^1/t) = -\lambda (\exp(-\phi(t+t^1)) - \exp(-\phi t))/\phi$

If we differentiate $H(t^1/t)$ with respect to t^1,

$d H(t^1/t)/dt^1 = \lambda \exp(-\phi t^1)\exp(-\phi t)$,   (1)

$[dH(t^1/t)/dt^1 = h(t^1)]$

Our PHM formulation is

$h(t^1,z) = h_0 (t^1) \exp(\beta z)$,   (2)

Where t^1:- time since last failure, z is an explanatory variable, β is a parameter of the model and $h_0 (t^1)$ is the baseline hazard.

Now comparing (1) and (2),

i) If $h_0 (t^1)$ from PHM $= \lambda \exp(-\phi t^1)$ and

ii) $\exp(\beta z) = \exp(-\phi t)$ with $\beta = -\phi$ and $z = t$

then the basic execution model is a sub model of PHM

WORK CARRIED OUT IN TASK AREA 4

Time series

Time series methods in reliability have been implemented by many authors [17], [18], [19]. The techniques have included the application of traditional linear ARMA models of Box and Jenkins [20]. Although 'discoveries' of trend and cyclical features have been made using these techniques, the whole area is rather unsatisfactory. Nottingham Polytechnic research concludes that invariably almost all the assumptions that are made in applying linear modelling techniques are violated by reliability, and in particular software reliability data. Violated assumptions are linearity, normality, constant parameters, change points and outliers.

Alternative, and more flexible model formulations are provided by the Dynamic Linear models and implementable using the BATS package, developed at Warwick University [21], [22]. Typically, time between failures or time to failures (TTF's) are described by an observation equation

$$TTF(i) = m(i) + r(i) + v(i)$$

where i is the failure number, $m(i)$ a level parameter that evolves with the failures, $r(i)$ is a set of possible covariates (failure dependent) and $v(i)$ is white noise. The extra flexibility is provided by allowing the evolving nature of $m(i)$ and $r(i)$ to be stochastic. The Bayesian (Kalman filter) recursion allows outliers to be handled/detected automatically, missing values, and user intervention with the model. Nottingham Polytechnic has used these techniques to model the MUSA data sets and Alvey data set 8. The approach also allows flexibility in traditional Weibull and hazard modelling. Some results of the above work have been presented to the Highlands local group of the RSS in March 1989.

Multivariate Techniques

The purpose of this section is to describe briefly attempts made to analyse software data by multivariate methods.

The data was designated Alvey dataset number 3 and contained 1198 observations on the following variables:

X_1 = program	X_6 = type of program
X_2 = program version	X_7 = first appearance
X_3 = programmer	X_8 = final appearance
X_4 = language	X_9 = number of faults
X_5 = size of program	X_{10} = time

Data screening and editing were necessary to overcome ideosyncracies and to render the data meaningful and suitable for analysis. The screening and editing were undertaken using MINITAB. Subsequent analysis was undertaken using MINITAB and GLIM.

Principal Components Analysis.

A commonly used multivariate technique is that of Principal Components Analysis, where p correlated variables are combined to obtain a new set of uncorrelated variables, called Principal Components.

The new variables are linear combinations of the original variables and are derived in decreasing order of importance so that PC(1) accounts for as much as possible of the variation in the original data. If the first few components account for most of the variation in the original data, the effective dimensionality of the problem is less than p.

Let $\underline{X}^T = [X_1, X_2, ..., X_p]$ be a p-dimensional random variable with variance-covariance matrix Σ and let

$$Y_j = a_{1j} X_1 + a_{2j} X_2 + ... + a_{pj} X_p = \underline{a}_j^T \underline{X} \qquad , \qquad (j = 1,2, ..., p)$$

where $\underline{a}_j^T = [a_{1j}, a_{2j}, ..., a_{pj}]$ such that $\underline{a}_j^T \underline{a}_j = \sum_{k=1}^{p} a_{kj}^2 = 1$ and $\underline{a}_i^T \underline{a}_j = 0$, $(i \neq j)$.

Y_1 is found by choosing a_1 so that Y_1 has the largest possible variance, Y_2 is found by choosing a_2 so that Y_2 has the next largest variance and is uncorrelated with Y_1; Y_3 is found by choosing a_3 so that Y_3 has the next largest variance and is uncorrelated with Y_1 and Y_2, ... and so on.

Thus obtained, $Y_1, Y_2, ..., Y_p$ are the Principal Components of $\underline{x}$ having variance equal to the eigenvalues of the sample variance - covariance matrix S $(= \hat{\Sigma})$; [23].

In the case of the software data it is wise to base PCA on the sample correlation matrix P rather than S, thus rendering the variables, which are heteroscedastic, equally important.

The Minitab results were:

(i) Examination of the correlation matrix P showed a sufficiency of non-zero elements to warrant the PCA worthwhile.

(ii) Eigenanalysis of P.

i	1	2	3	4	5	6
Eigenvalue λ_i	2.25	1.52	1.12	0.98	0.64	0.42 *
Proportion $\lambda_i/\Sigma\lambda_i$	0.32	0.22	0.16	0.14	0.09	0.06
Cumulative proportions	0.32	0.54	0.70	0.84	0.93	0.99

(*denotes that subsequent eigenvalues exist but account for only 1% of the variation).

Note that as many as five PC's are required before more than 90% of the variation in the data is explained. Ideally, it is desirable that the majority of the variation in the data should be explained by two or three components at the most. Unfortunately no such reduction of the effective dimensionality was obtained. Reduction to two or three components is useful in that 2D or 3D plots of component score might be examined for patterns or clusters and that attempts at reification might be made. However, it is of some interest that the effective dimension of the data reduces to about five, with this technique.

MINITAB also supplies the coefficients a_j^T from the eigenvectors corresponding to each eigenvalue.

Discriminant Analysis.

Multivariate discriminant analysis is a technique which allows the multivariate response for $\underline{X}_*^T = [X_1, X_2, ..., X_j, X_{j+2}, ..., X_p]$ to be attributed to known groups according to X_{j+1} provided X_{j+1} is a group indicator, via discriminating functions. The discriminating functions then may be used to assign further observations on $\underline{X}_*^T$, not so far identified on a X_{j+1}, to a X_{j+1}.

A feature of the software data is that, in a number of cases, the multivariate response has not been identified by programmer (X_3). It is of interest to use the data on cases where the programmer is known as a "learning set" for discriminating between programmers, thereby making it possible for cases in the "prediction set", with programmer unknown, to be identified with a programmer.

The procedure is to calculate

$$w_i = \underline{L}_i^T \underline{X}_* - 0.5 \underline{L}_i^T \overline{\underline{X}}_{*j} + \ln(\pi_i) \quad ; \qquad i = 1, 2, ..., j, j+2, ...m$$

where m is the number of distinct groups (programmers) indicated by X_{j+1}, $\overline{X}_{*i}$ is the mean vector for group i, S_* is the pooled within groups estimate of Σ_*, the variance-covariance matrix of $\underline{X}_*$ and $\underline{L}_i = \underline{S}_*^{-1} \overline{\underline{X}}_{*i}$, π_i is the prior probability that a case belongs to group i, and to allocate the individual to that group for which the w_i is the greatest, [23].

Minitab Results.

Unfortunately, the success rate for correctly identifying the multivariate response on $\underline{X}_*$ by known programmer in the training set was found to be low, with only 25.5% of cases correctly identified.

However, the success rate varied from programmer to programmer, ranging from no cases correctly identified to 88.9% of cases correctly identified. With a low overall success rate it is inappropriate to attempt to identify programmers for cases in the prediction set.

It is possible that the failure of the technique to achieve a reasonable success rate may be attributed to a violation of the theoretical assumptions of discriminant analysis, that the discriminating variables have a multivariate normal distribution and have equal variance - covariance matrices within groups, (programmers); [23]. The data under study, consisting mainly of variables having a discrete or categorical nature, do not conform to these requirements. Reference [24] gives a discussion on techniques of discrete discriminant analysis applied to data not conforming to the MV normal, homoscedastic groups pattern. Reference [25] gives a similar treatment.

Log - linear models.

It is possible to obtain from the software data multiway tables containing number of faults as response corresponding to variables such as X_4 = program language, X_5 = program size and X_6 = program type.

With such categorical data it is appropriate to fit log-linear models, beginning with the no-association model.

$$E(F_{ijk}) = N \, \pi_{i..} \, \pi_{.j.} \, \pi_{..k} \tag{1}$$

where F_{ijk} = number of faults in the cell of the multiway table corresponding to the ith language, j^{th} program size, k^{th} program type;

N = grand total of faults in the multiway table;

$\pi_{i..}$ = marginal probability in the ith category of X_4 (language) irrespective of X_5 and X_6 (size and type of program), $\pi_{.j.}$ = marginal probability in the jth category of X_5 (size) irrespective of X_4 and X_6 (language and type); $\pi_{..k}$ = marginal probability in the kth category of X_6 (type) irrespective of X_4 and X_5 (language and size).

Taking logarithms in (1)

$$\ln E(F_{ijk}) = \ln N + \ln \pi_{i..} + \ln \pi_{.j.} + \ln \pi_{..k} \tag{2}$$

With a little manipulation it is possible to write (2) in the form

$$\ln E(F_{ijk}) = u + u_{1(i)} + u_{2(j)} + u_{3(k)} \quad , \tag{3}$$

where the u's are functions of the theoretical marginal fault counts.

Now, (3) is reminiscent of a two-way ANOVA model, with no interaction. It is possible to fit (3) using GLIM, employing the deviance statistic equal to $-2\log(l_c/l_f)$ as the goodness-of-fit criterion, where
l_c = likelihood of the data under the current model and l_f = likelihood of the data under the fullest possible model, following the notation of [26].

Failure of the no-association model to fit the data encourages the inclusion of further model terms, firstly the two-way associations

$$u_{12(ij)}, \quad u_{13(ik)}, \quad u_{23(jk)}$$

corresponding to first-order interaction in ANOVA, and then, if necessary, the three-way association $u_{123(ijk)}$, corresponding to second-order interaction in ANOVA, (see [27]).

GLIM Results:

Model		Scaled deviance	change	residual df	change in df
A	(3)	89.75		16	
B	A+ SIZE.TYPE	77.78	11.97	11	5
C	B+ SIZE.LANG	26.94	50.84	10	1
D	C+ TYPE.LANG	0.44	26.50	5	5

Now, the scaled deviance (or change in scaled deviance) is approximately X^2 - distributed with the residual degrees of freedom (or change in degrees of freedom).

It can be concluded that

(a) there is a significant association between size and type of program,

(b) there is a significant association between size of program and language,

(c) there is a significant association between type of program and language,

(d) there is no significant three-way association, suggesting that
  (i)   the association between size of program and type of program is the same for all languages,
  (ii) the association between size of program and language is the same for all program types,
  (iii) the association between type of program and language is the same for all program sizes.

Resulting from (b), close examination of the model parameters suggests that a negative association between SIZE (2) and TYPE (5) is indicative of a tendency for a lower fault count with medium to large programs than with small programs of the type "INCLUDE FILE".

Also, resulting from (c), a negative association between TYPE (4) and LANG (2) suggests a tendency for a lower fault count with system operating language programs than with COBOL programs of the type "FIND CONTROL FILE".

Comments: The multivariate procedures described earlier in this section revealed relatively little. However this should not malign the power and usefulness of techniques such as PCA and discriminant analysis, and they should be used if appropriate on other examples of software data in attempts to reveal data structure.

Log-linear modelling, a useful example of which is discussed above, has a very positive usefulness in investigating data of the type considered and is recommended as an important tool in future work.

SUMMARY

Some results of the Alvey Software Reliability Modelling (SRM) project have been presented. A number of models have been generalised or extended to incorporate explanatory variables. Problems of software reliability have been highlighted and further work is required in this area. Multivariate techniques and time series analysis are shown to be applicable to software reliability data and a number of results have been presented. A number of documents have been written for the whole project including tasks 3, 4, and 9 and some of these are available to the public. The authors may be contacted initially for further details of their work in task areas 3, 4 and 9. For further details of other Alvey SRM work, the contact address is DTI/IED, Alvey SRM project, Kings Gate House, 66-74 Victoria Street, London SW1E 65W.

REFERENCES

1. Musa, J.D. and Okumoto, K., Application of Basic and Logarithmic Poisson Execution Time Models in Software Reliability Measurement. Software System Design Methods, Springer-Verlag Berlin Heidelberg 1986, Nato ASI series, Volume F22 pp275-298.

2. Littlewood, B. and Verrall, J.L., A Bayesian reliability growth model for computer software., Journal of the Royal Statistical Society., C (Applied Statistics), 22, 1973 pp 332-346.

3. Jelinski, Z. and Moranda, P.B., Software Reliability Research, in <u>Statistical Computer Performance Evaluation</u>, W. Freiberger New York Academic Press, 1972, pp 465-484.

4. Bendell, A., McCollin. C., Wightman. D.W., Linkman, S. and Carn, R. Software Reliability Data Collection - Problems and Possibilities , <u>Proceedings of the Sixth EureData Conference</u>, Siena, 1989.

5. McCollin, C., Bendell, A. and Wightman. D.W., Effects of Explanatory Factors on Software Reliability, <u>Reliability 1989</u>, Vol 2.

6. Wightman, D.W., Review of models/techniques which incorporate explanatory variables and may be applied to software failure data. Nottingham Polytechnic report (for Alvey SRM project).

7. Csenki, A., Bayesian Formulations of the Jelinski-Moranda Software Reliability Model. Unpublished City University report (For Alvey SRM project).

8. Brocklehurst, S., On the Effectiveness of Adaptive Software Reliability Modelling Unpublished City University report (for Alvey SRM project).

9. Wright, D.R., A Modified U-Plot applied to Failure Count Prediction., Unpublished City University report (for Alvey SRM project).

10. Wightman, D.W., McCollin. C. and Bendell. A., Proportional Hazards Modelling of an Alvey Software Reliability Data Set. awaiting publication.

11. Cox, D.R., The Statistical Analysis of Dependencies in Point Processes <u>Stochastic Point Processes</u> ed P.A.W. Lewis, Wiley, New York pp55-66.

12. Wightman, D.W., 1987. The Application of Proportional Hazards Modelling to Reliability Problems. Unpublished Ph.D thesis Trent Polytechnic.

13. Prentice, R.L, Williams, B.J and Peterson, A.V., On Regression Analysis of Multivariate failure time data. <u>Biometrika</u> Vol 68 No. 2, 1981 pp 373-379.

14. Anderson, P.K. and Gill, R.D., (1982) Cox's Regression model for counting processes a large sample study. <u>Annals of Statistics</u>, 10 pp 1100-1120.

15. Lawless, J.F. (1987) Regression Methods for Poisson Process Data. <u>Journal of the American Statistical Association</u>. Vol. 82, No. 399.

16. Font, V., Une approache de la fiabilite des logiciels: modeles classiques et modele lineaire generalise. Thesis L'Univerite Paul Sabatier de Toulouse, France 1985.

17. Singpurwalla, N.D., (1978). Time series analysis of failure data. <u>Proceedings Annual Reliability and Maintainability Symposium</u>., pp 107-112.

18. Singpurwalla, N.D. and Soyer, R. Assessing (software) reliability growth using a random coefficient autoregressive process and its ramifications., IEEE Transactions on Software Engineering 1985.

19. Walls, L.A. and Bendell, A., The structure and exploration of reliability field data; What to look for and how to analyse it. Proceedings of 5th National Reliability Conference 1985 pp5B/5/1-17.

20. Box, G.E.P and Jenkins, G.M., Time Series Analysis:Forecasting and Control Holden-Day, London, 1976.

21. Harrison, P.J. and Stevens, C.F., Bayesian Forecasting (with discussion). J.R. Statistic Soc., 1976, B38, pp205-247.

22. West, M. Harrison, J. and Pole, A., BATS: A User Guide. University of Warwick, 1988.

23. Chatfield, C. and Collins, A.V., (1980), Introduction to Multivariate Analysis, Chapman Hall.

24. Goldstein, M. and Dillon, W.R., Discrete Discriminant Analysis. Wiley, 1978.

25. Bishop, Y.M, Fienberg, S.E. and Holland. P.E., (1975) Discrete Multivariate Analysis, MIT Press.

26. Baker, R.J. and Nelder, J.A., (1978), GLIM System, Released by the Royal Statistical Society.

27. Everitt, B.S., (1977), The Analysis of Contingency Tables, Chapman Hall.

THE ACCEPTANCE OF SOFTWARE QUALITY AS A CONTRIBUTOR TO SAFETY

Dr Brian W Finnie and Iain H A Johnston
The Centre for Software Engineering Ltd
Bellwin Drive, Flixborough, Scunthorpe, South Humberside, DN15 8SN

ABSTRACT

Electronics and software are used increasingly in safety related
applications and the number of reported incidents in which
software is involved as a possible cause is growing. The emerging
standards for safety related systems stress the need for adequate
quality assurance during the design process. The nature of
software quality assurance and its relationship to safety is
discussed. Prevailing attitudes in industry are considered and it
is shown that there is a need for these to change. The range of
available guidance on the introduction and use of software quality
assurance is also increasing and the time is right for this to
find its way into every organisation writing software and not just
the larger corporations for whom it is now showing a return.

INTRODUCTION

Electronics are becoming increasingly common in both everyday
applications and in specialised applications in industry. These
Electronic Systems increasingly incorporate software in order to
increase flexibility and reduce costs. In many cases there is
provision for some means of modifying the program or data stored
in the electronics.

This trend has been in existence for some time and much
expertise has been built up in the design of both the hardware and
software components of such systems. Until recently there was
considerable reluctance on the part of designers to create systems
where safety depends on programmable electronics. Additional non-
programmable protection systems were installed which enabled
software to be excluded from any safety case.

Nevertheless programmable electronics have been used in safety related applications and this trend is continuing. Applications where safety does depend on the software are becoming more common. In this paper we intend to show not only that Quality Assurance is an important element in ensuring that such systems achieve an acceptable level of safety but also that though the necessary Quality Assurance is often not applied to such software little encouragement is needed to persuade software developers to adopt it on a wide scale.

THE PROBLEM WITH SOFTWARE

Software is a contrasting mixture of simplicity and complexity. It appears to be something simple to undertake and write, yet is used for solutions to the world's most complex problems. The resulting assemblage of computer instructions or code can be exceedingly complex, yet may be built by small and simple incremental steps. Behind this contrast lie the reasons for the problem with software. The apparent simplicity of the process has misled many into constructing software without the rigour and procedural control that is necessary for the success of such a complex end result and such as would be used in other branches of engineering.

The same simplicity leads to ease of change and this has further encouraged uncontrolled development on the grounds that change is less expensive than prior thought. The ease with which changes can be made alters the traditional ratio between implementation time and test time. A small change is assumed to require only a small amount of testing. The focus on overall impact is lost.

Ease of change also leads to uncontrolled change. It is easier to carry out the modifications immediately than it is to undertake the request and authorisation procedure. Such uncontrolled changes may not be adequately tested or even properly specified.

All these things undermine the integrity of software.

Where safety is involved another factor needs to be considered. The complexities of the underlying process in the plant may be beyond the understanding of the programmer. If the programmer works to incomplete requirements then he may introduce assumptions which take no account of consequences on the real plant.

SOFTWARE AS A CAUSE OF ACCIDENTS

The increasing use of programmable electronics in control
applications over the past 10 years has led to a steadily
increasing toll of accidents in which software has been implicated
as either a definite or possible cause. These incidents vary in
their characteristics, including their severity and the degree to
which they can be said to have been caused by software. An
excellent collected list of such incidents has been published by
the Association for Computing Machinery (ACM) [1] and a compendium
of these incidents with more detailed descriptions is to be
published in due course.

The software community is accustomed to stories of software
failures and speculation concerning the potential of software to
cause harm. The ACM list of over 500 incidents, mostly in the
last three to four years, puts the fears into perspective. The
fears are real and are justified, but the extent to which injury
has been caused to humans is as yet relatively small. It is
however important not to excuse software unjustifiably. The
extent to which problems with computer programs in applications
such as word processing are excused by blaming the user when
things go wrong is pointed out by Norman [2] and is recognised by
all who have experienced it. This tendency not to blame the
'computer' can be expected to distort the figures and raises
question marks against some incidents reported as being due to
human error.

An analysis of the ACM list shows that relatively few
incidents directly link software and loss of life or injury. Out
of over 500 incidents a total of 42 involve loss of life, a total
of 31 clearly involve software faults and only 1 involves both
loss of life and software faults. Of the 31 incidents involving
software faults 12 involve a potential for loss of life. This
might be considered a small proportion of software related
incidents, particularly since the list has been compiled by a
community with a particular interest in software. There are
however over 130 incidents which involve "flaws in system concept,
requirements, design, implementation, maintenance or system
evolution". These are typical sources of software error and a
question must rest on whether all or some of these incidents
deserve classification as software related. The same can be said
of the 100 incidents which involve "accidental misuse" or
"misinterpretation/confusion at a man-machine interface".

The experience of the Health and Safety Executive (HSE) is
noteworthy in respect of the extent to which software is a cause
of accidents. A figure of 17% has been quoted [3] as the
proportion of accidents attributable to software in a study of 34
industrial accidents.

282

EXAMPLES OF INCIDENTS

Incidents from the ACM list and elsewhere which have been reported as attributable to software.

a)   Production of 6000 bottles of cough medicine with 10 times the correct amount of one ingredient [4].

b)   Uncommanded inflight closure of throttles on Boeing 747-400s. [5].

c)   Failure of AT&T telephone networks [1] [6].

d)   Two deaths and 3 injuries caused by malfunctioning of linear accelerators during hospital treatment [7].

e)   Nuclear refuelling machine dropped at Bruce A power station in Canada [8].

f)   F-14 fighter went into uncontrollable spin [1].

g)   Flood gates opened in Norway during maintenance [1].

   Incidents from the ACM list and elsewhere which have been reported as associated with programmable systems but which have not been directly attributed to software:

h)   Boeing 737 autopilot changes height without warning [9].

i)   'Exxon Valdez' autopilot initially overrode crews efforts to change course [9]

j)   Fire service attendance at Hillsborough delayed because computer required full address [10].

k)   Workers arm crushed while cleaning metal forming machine [11].

l)   10 people killed by robots in various incidents in Japan [12].

WHAT IS SOFTWARE QUALITY ASSURANCE ?

We consider software Quality Assurance to be the process by which confidence is achieved that the software running in the system performs at all times in the manner intended by the user.

It is now well established that to achieve success in the control of any complex process the overall task should be broken into components and each stage considered separately [13].

Quality assurance requires that for each stage:

(1) The inputs are defined.

(2) There is a clear and agreed specification of what the stage is required to achieve.

(3) These requirements are implemented correctly.

(4) The correctness is confirmed by verification.

(5) The above steps are made visible for examination.

Software is no different from any other product or deliverable. It must satisfy the requirements of the user, it must be cost effective in doing this and it must be robust against adverse conditions and influences including misuse [14].

Software Quality Assurance is integrated with software development. It is not an overblanket to be used according to the prevailing conditions. As a result a Quality Assurance system for software must be reflected in the development plan. The Quality Assurance system should ensure [15] that:

(1) The project is divided into stages:

    Requirement definition

    Software specification

    Software design

    Software coding

    Software integration

(2) Following each stage the results are verified against the requirements for that stage.

(3) There are design reviews.

(4) There are codes of practice to ensure consistency and clarity of each stage. These codes should specify standards, style, methods and tools to be used during each stage.

(5) There is a fault/deficiency recording system and a configuration control system.

DISCUSSION

Our experience at CSE has been that there is an awareness of the
problem with software. Both users and developers are aware of the
concerns being expressed in the technical press and many users and
developers share those concerns.

Many believe from reports in the press that there is a clear
cut dividing line on one side of which the software is safety
critical and should be developed using all possible means to
ensure safety and one the other side of which no special methods
are necessary. Most see clearly that this all or nothing view
does not relate well to the applications with which they are
involved. This results in widespread uncertainty about whether
the software in any particular application is or is not safety
critical.

Many organisations who develop software have well
established Quality Assurance systems with procedures covering
design development and production, but which do not include
procedures to cover software.

Where software is covered it is often treated as hardware.
The EPROM which contains the software has a part number which
relates to the software version contained in the EPROM.
Frequently there is little or no <u>formal</u> documentation of
specifications, designs or changes to the software and no common
ground rules for the selection and use of software tools or the
structuring and commenting of source code. In many cases informal
procedures are in use with individual engineers keeping personal
records which vary in their completeness and may be unknown or
inaccessible to other engineers.

Where there are formal Quality Assurance procedures covering
software these vary considerably in their characteristics. There
is a tendency to make unjustifiable or at least unjustified
statements concerning the extent to which a change to a piece of
software can affect other parts of the software and therefore
about the extent to which regression testing is necessary. It is
difficult to determine whether this tendency is due to a lack of
appreciation of what is or is not possible or whether it is due to
a conscious or unconscious desire to avoid extensive testing for
commercial or other reasons.

Many large organisations are taking a lead on Quality Assurance and no less in software Quality Assurance. These organisations have established over time that Quality Assurance is a good investment and are taking the concepts well beyond basic quality systems. In software as well they have realised that the investment in good procedures and feedback of results at each stage has achieved measurable benefits in improved software.

Much of industry is made up of small companies, with a small team of software people, fully employed on producing what the company wants done. These organisations do not have the extra resources to provide software expertise in their quality department. However, Quality Assurance does not need a quality department since quality depends on operating departments doing the work correctly to clearly specified requirements. These small companies where much of the country's software work is done are not operating good software Quality Assurance systems. It is in these companies that we hear the argument that there is not enough time on this project to put in Quality Assurance , yet when the project goes wrong more and more effort is poured on the job to bail it out.

There are two contrasting management attitudes.

There are the converted, usually from large companies who have been able to break out of the traditional project mould where problems are sorted out as they arise with more and more effort. These people now swear by the application of Software Engineering principles and software Quality Assurance as the only means to achieve results on time and on cost.

Then there are the traditionalists, who are unwilling to commit the up front effort that Software Engineering dictates. This is built into the culture of the company and its accounting systems and will not be broken without strong support from the very top of the organisation.

In the parts of the organisation where the work is actually done we again find two attitudes.

First there is the school which believes that Quality Assurance is something which is measured by the thickness of the documentation. This is a conception which Quality Assurance auditors have often brought about through a severe insistence on evaluating a company's system through its documentation rather than by observing the results. This Quality Assurance by the inch attitude will prevail in many organisations unless a more creative approach to implementation and assessment can be found.

We also find many cases of technical staff saying that something must be done. That projects are suffering because good Quality Assurance practice and associated Software Engineering procedures need to be introduced, but there is no opportunity within existing project plans to bring this about. This is frequently linked to the second management attitude described above. The technical staff are using informal methods as their best effort to compensate for what they see as a company inadequacy.

In other fields of development and Total Quality the Japanese have shown the way on Quality Assurance and Total Quality. The same approach to software Quality Assurance avoids the introduction of new techniques and changes to the traditional development process but instead relies on careful and detailed measurement of results at each stage to identify and rectify problems with each part of the process. This perhaps relates to the large company experience that improvement in software quality is only possible over a long period of consistent application of the available methods.

SOURCES OF GUIDANCE

Software Quality Assurance is a developing discipline but is based soundly on the established mechanisms for Quality Assurance of other components. Software is uniquely easy to reproduce in quantity without introducing faults during 'manufacture'. Quality Assurance of software is therefore principally associated with the two aspects of

a)   Design and development

b)   Specification, selection and purchasing

There are a number of sources of guidance which address one or both of these aspects of software Quality Assurance.

BS5750 [16] is the standard document in the UK (as is ISO 9000 elsewhere) providing guidance on Quality Assurance. The document does not specifically address software and as such many of its requirements appear to be inappropriate to software. In particular there is an emphasis on the role of Quality Assurance during manufacture. The system of using Quality Assessment Schedules to help interpret BS5750 for different processes pays dividends for software. The QAS for the design and development of software provides excellent guidance on the provisions which are required for software Quality Assurance.

AQAP 13 [15] describes the software Quality Assurance requirements which suppliers to the NATO military organisations are required to fulfill. This document is addressed specifically at the Quality Assurance of software, other documents deal with Quality Assurance of other components. As such AQAP 13 provides a brief overview of the procedures and documentation which are necessary for the Quality Assurance of the specification, design and development of software. AQAP 13 is however an old document and is due for rewriting.

The STARTS guide [17] is directed at the procurement rather than the design and development of software. As such it provides much useful material dealing with the Quality Assurance mechanisms which are appropriate to the specification, selection and purchasing of software and to dealing with software sub contractors. Much helpful material is also included dealing with the Quality Assurance of the software design and development process and with the special requirements of safety critical or high integrity software projects.

RELATIONSHIP TO SOFTWARE SAFETY

The increasing use of software in safety related applications has led to the development of a number of standards dealing with the problem. Standards have existed for a number of years which relate to the use of computers in particular industries such as power generation and aviation [18] [19]. More such standards are appearing [20] [21].

The HSE guidelines [22] and the draft IEC standard [23] are more general and address the use of programmable electronics in a generic way which is intended to be applicable to any category of application. Both these documents regard the implementation of an acceptable level of Quality Assurance as the basis for the use of programmable electronics in any safety related application. Both choose the same standard (BS5750/ISO9000) as the base level Quality Assurance requirement. Software Quality Assurance is an essential element in the use of programmable systems in safety related applications.

CONCLUSIONS

The increasing use of programmable electronics in safety related systems has led to an increasing number of accidents in which software has been implicated. The contrasting characteristics of simplicity and complexity in software contribute to many occurrences of software failure. Such failures can cause accidents.

Software Quality Assurance is a means by which confidence can be achieved that software performs as intended. This confidence is necessary for safety related systems, as is recognised in the emerging standards. There is wide variation in industry in the extent to which software Quality Assurance is applied. There is, however, a general agreement that software Quality Assurance is desirable. This approval amongst engineers should be harnessed by management to ensure that enthusiasm is not lost and that high quality safe software is produced.

REFERENCES

1. Neuman, P.G., RISKS: Cumulative Index of Software Engineering Notes, ACM SIGSOFT SEN, Jan 1989 Vol 14, No 1.

2. Norman, D.A., The Psychology of Everyday Things, Basic Books Inc, 1988.

3. Report in Computer Weekly, 16th Nov 1989.

4. Report in Computer Weekly, 4th May 1989.

5. Report in Flight International, 11-17 April 1990.

6. Report in Financial Times, 19th Jan 1990.

7. Unattributed report, Legal Implications of Computerised Patient Care, Health Technology, May/June 1988, Vol 2, No 3, pp 86-95.

8. Report in Nuclear Engineering International, Mar 1990.

9. Report in Lloyds List International, 18 April 1989.

10. Report in Computer Weekly, 22nd June 1989.

11. Howard, J.M. and Askren, W.B., Paper published in Hazard Prevention, May/June 1988.

12. Unattributed report, Data disaster, SafetyNet, Oct/Nov/Dec 1989.

13. Melan, E.H., Process Management in Service and Administrative Operations, Quality Progress, June 1985.

14. Sullivan, L.P., The Seven Stages in Company-Wide Quality Control, Quality Progress, May 1987.

15. NATO, AQAP-13, NATO Software Quality Control System Requirements, NATO, August 1981.

16. BSI, BS5750, Quality Systems, British Standards Institution, 1987.

17. STARTS Purchasers Handbook, NCC Publications, May 1989.

18. IEC, IEC880, Software for Computers in the Safety Systems of Nuclear Power Stations, International Electrotechnical Commission, 1986.

19. RTCA, RTCA/DO-178A, Software Considerations in Airborne Systems and Equipment Certification, Radio Technical Commission for Aeronautics, 1985.

20. I.Gas.E., IGE/SR/15, Use of Programmable Electronic Systems in Safety Related applications in the Gas Industry, The Institution of Gas Engineers, 1989.

21. EEMUA, Publication 160, Safety Related Instrument Systems for the Process Industries (including Programmable Electronic Systems), The Engineering Equipment and Materials Users Association, 1989.

22. HSE, Programmable Electronic Systems in Safety Related Applications, Her Majesty's Stationery Office, 1987.

23. IEC, Sub Committe 65A, Working Group 9, Software for Computers in the Application of Industrial Safety-Related Systems, Proposal for a standard, International Electrotechnical Commission, 1989.

ENHANCED PARTIAL BETA FACTOR METHOD FOR QUANTIFYING DEPENDENT FAILURES

L P DAVIES
NNC Limited
Booths Hall, Chelford Road, Knutsford, Cheshire. WA16 8QZ, UK

ABSTRACT

The Enhanced Partial Beta Factor (PBF) method for quantifying dependent failures uses a checklist to guide the assessor through all the potential defences and provides a systematic and fully auditable approach to the quantification of dependent failures. It therefore reduces the degree of expert opinion required for a dependent failures assessment by breaking down the judgement of the effectiveness of defences against dependent failures into their component parts. The method is described and an example of its application by considering a simple standby diesel alternator system.

INTRODUCTION

For systems requiring a high reliability it is usual to introduce redundancy as a design feature. For such systems there is a potential for dependent failures to affect the assessed integrity of the system and in particular to affect the results of any analysis based on random (independent) failures. The effect of dependent failures is to degrade system reliability.

Probabilistic assessments generally reflect all engineered features of a system, including, for example, common services to redundant plant, and hence any such features which defeat the inherent redundancy are usually explicitly modelled. Dependent failure analysis treats faults outside the envelope of the design process that can lead to multiple failures in redundant systems. It therefore covers less tangible sources of multiple plant failures than any engineered features such as common services.

A number of methods exist for the quantification of dependent failures. The majority of these, however, rely strongly on expert judgement and therefore two assessors may calculate radically different probabilities or frequencies of dependent failures. To enhance the auditability of dependent failure quantification it is necessary to use a method which reduces the amount of expert judgement to an absolute minimum. This ensures that the likelihood of dependent failures reflects the features of the system rather than the subjective views of the assessor.

The Partial Beta Factor (PBF) method [1] and [2] has been developed by the Safety and Reliability Directorate (SRD) of UKAEA. This allows a systematic appraisal of the likelihood of dependent failures based on the effectiveness of a number of defences against them. Even so the method involves a sufficiently high degree of subjective judgement that there are problems with auditability of an assessment. The enhanced PBF method for quantifying dependent failures develops the PBF method by using a fully auditable method to quantify each of the individual factors required.

This paper describes the enhanced PBF method and presents a simple example of its use for a standby power system.

OVERVIEW OF DEPENDENT FAILURES

All projects undergo three phases up to the end of their operational life – design, construction and operation. Problems in any of these three stages have the potential to introduce dependent failures into a system which may affect its reliability. These three phases have been investigated by SRD and a classification scheme for potential causes of dependent failures identified. This is described in detail in [3].

The classification scheme presented in [3] develops the potential causes of dependent failures to a detailed level. For the purposes of this paper, however, the principal groupings are most pertinent and are as follows:

(1) functional deficiencies

(2) design realisation

(3) manufacture

(4) installation and commissioning

(5) maintenance and testing

(6) operation

(7) normal extremes

(8) internal and external hazards.

In order to counter any potential causes of dependent failures, defences against them can be built into the system again during each of the three phases of the project life cycle. Again, this has been considered by SRD and 20 defences against dependent failures have been identified [4].

Each of the defences against dependent failures is effective against one or more of the potential causes. A matrix can therefore be drawn up to identify, in qualitative terms, the effectiveness of each defence and this is presented in Table 1 (overleaf). It should be noted that Table 1 does not provide any measure of the effectiveness of each individual defence, but merely a statement of whether each defence may or may not provide a mitigating factor for each potential cause of dependent failures in turn.

TABLE 1
Defences against dependent failures

Project phase	Dependent failure	Design		Construction		Operation			
		Functional deficiencies	Design realisation	Manufacture	Installation and commissioning	Maintenance and testing	Operation	Normal extremes	Internal and external hazards
Design	Design control	✓	✓	✓	✓	✓	✓	✓	✓
	Design review	✓	✓	✓	✓	✓	✓	✓	✓
	Functional diversity	✓	✓	✓	✓	✓	✓	✓	✓
	Equipment diversity	✓	✓	✓	✓	✓	✓	✓	✓
	Fail safe design		✓			✓	✓	✓	✓
	Operational interfaces					✓	✓		
	Protection and segregation			✓	✓	✓	✓	✓	✓
	Redundancy and voting		✓			✓			
	Proven design and standards		✓						
	Derating and simplicity		✓					✓	
	Reliability assessment	✓	✓	✓	✓	✓	✓	✓	✓
Construction	Construction control			✓	✓				
	Construction standards		✓	✓	✓				
	Inspection			✓	✓				
	Testing and commissioning			✓	✓				
Operation	Operational control					✓	✓	✓	✓
	Reliability monitoring					✓	✓	✓	✓
	Maintenance					✓		✓	✓
	Proof test					✓			✓
	Operations						✓	✓	✓

It is noted that in addition to providing defences against dependent failures in their own phase of a project many of the defences are effective against potential causes during the other two phases. For example, design reviews provide a defence for faults during design, construction and operation.

Since by definition dependent failures are outside the design envelope, it is not possible to use the potential causes of dependent failures as a basis of assessment. Although clearly any design processes and the defences identified in Table 1 limit the effect of any dependent failures there is always some residue of unidentified causes. The defences against dependent failures, however, are assumed to be effective against the spectrum of causes including any unidentified causes. To quantify dependent failures, therefore, it is necessary to determine the adequacy of the defences afforded during design, construction and operation.

PARTIAL BETA FACTOR METHOD

The PBF method is described in [1] and [2]. For completeness this is outlined below by first reviewing the beta factor method and demonstrating that the PBF method provides the first stage of providing a systematic method of quantifying the beta factor.

The beta factor method assumes that if the failure rate of a single element of a redundant system is λ, then a proportion of this is due to dependent failures and hence the failure rate associated with dependent failures can be expressed as:

$$\lambda_{DF} = \beta\lambda \qquad (1)$$

Likewise for a system where the failure probability, P, is the relevant parameter, then the probability of dependent failures is given by:

$$P_{DF} = \beta P \qquad (2)$$

Considering failure probabilities, although the same arguments apply equally to failure rates, it can be concluded that the probability of an independent failure is given by:

$$P_I = (1 - \beta)P \qquad (3)$$

It is therefore clear that if the beta factor model is adhered to rigorously it is necessary to reduce the failure probability of independent failures to account for the inclusion of a dependent failure probability.

There is, however, some uncertainty in the calculation of beta factors and in general, the method of estimation of beta factors leads to a conservative estimate of the dependent failure probability. It is therefore inappropriate to reduce the independent failure probability as expressed in equation (3) as this leads to an optimistic estimate of the probability of independent failures and hence the following conservative approximation is made that:

$$P_I = P \qquad (4)$$

For systems where redundant items have differing failure rates or
probabilities (due to differences in design features or the provision of
partial diversity), the beta factor is applied to the smaller of the two to
give the failure rate or probability of dependent failure.

The above discussion relates to the beta factor method in general.
This method is used universally for quantifying dependent failures and
traditionally subjective judgement has been used to quantify beta factors.
For a fully auditable approach to quantification it is necessary to
consider each of the defences in Table 1 and to determine the effectiveness
of each in a systematic manner. The first stage of this is the partial
beta factor method.

The approach adopted in the PBF method is to assign a PBF, β_{pi}, for
each of M defences against dependent failures in the range:

$$1.0 > \beta_{pi} > \beta_{pim} \tag{5}$$

Where β_{pim} is the minimum value of the PBF which can be achieved with
the maximum level of effectiveness for each specific defence. The beta
factor is then calculated by taking the product of the PBFs:

$$\beta = \prod_{i=1}^{M} \beta_{pi} \tag{6}$$

Table 2 lists data for use with the PBF method. For convenience some
of the defences have been grouped together and therefore only 17 defences
appear in Table 2 rather than the 20 in Table 1.

TABLE 2
Minimum Partial Beta Factor (PBF)

Item	Defence	System generic	Minimum PBF (β_{pim})
1	Design control	✓	0.6
2	Design review	✓	0.8
3	Functional diversity		0.2
4	Equipment diversity		0.25
5	Operational interfaces	✓	0.8
6	Protection and segregation		0.8
7	Redundancy and voting		0.9
8	Proven design and standardisation		0.9
9	Derating and simplicity		0.9
10	Construction standards and control	✓	0.7
11	Testing and commissioning	✓	0.7
12	Inspection	✓	0.9
13	Operational control	✓	0.6
14	Reliability monitoring	✓	0.8

TABLE 2 (cont'd)

Item	Defence	System generic	Minimum PBF (β_{pim})
15	Maintenance	✓	0.7
16	Proof test		0.7
17	Operations	✓	0.8
		$\beta_{min} = \displaystyle\prod_{i=1}^{17} \beta_{pim} = 0.001$	

The product of each of the β_{pim} is the minimum achievable beta factor and has a value of 0.001. This value has been determined by SRD[2] as a practical limit by a survey of incidents in the US civil nuclear power and aircraft industries.

To determine the value of each β_{pi} it is necessary to select whereabouts in the range β_{pim} to 1 the beta factor should be based on the quantity of the defences afforded by the system. In the form above the PBF method still relies on a substantial degree of expert opinion to determine each PBF.

A number of defences in Table 2 are identified as system generic. This is defined in the next section.

ENHANCED PARTIAL BETA FACTOR METHOD

The PBF method provides a systematic framework for quantifying dependent failures by assessing the effectiveness of each defence. The choice of value for β_{pi}, however, still entails a high degree of subjective judgement. To reduce this an auditable method of determining β_{pi} has been developed.

In [2] an extended PBF method is described which uses the correlation between causes of dependent failures and defences against them to break down the partial beta factors further. A different approach is adopted here which concentrates solely on the effectiveness of the defences provided to determine the PBFs by a systematic and auditable appraisal of each defence in turn.

Further details on the features which are necessary for the maximum level of defence are included in [1]. As an example, Table 3 (overleaf) lists the detailed defensive measures for redundancy and voting and includes guidance on how many of the requirements can be complied with. Similar tables have been compiled for each of the defences in Table 2.

TABLE 3
Detailed defensive measures - redundancy and voting

Item	Defensive measure	Comment and guidance
1	The degree of redundancy and the voting logic of each sub-system must be related to both dangerous and safe reliable criteria, which have conflicting requirements	Only applicable to tripping systems
2	The redundancy and voting logic must include allowances for dependent failures	Only applicable to tripping systems
3	The consideration of dependent failures in the system design should consider the application of diverse or increased identical redundancy	Benefit should only be claimed if these have been specifically implemented as a result of dependent failure considerations
4	The redundancy and voting logic should allow the system and all its redundant components to be adequately proof tested	For in-service systems the design should allow testing without removing full system from service. For standby systems maximum benefit can be claimed if systems can be tested without preventing operation of full system if a demand arises
5	Where possible, redundant channels should not be bypassed for test and maintenance, which severely degrades the system. For example, a two from three system should become a one from two rather than two from two during channel maintenance or test	Strictly only applicable for protection systems. May be applied to other systems if multiple redundancy is present
6	The use of double m from n decision logic should be carefully considered because of the linking between guardlines, and therefore dependent possibilities, that are introduced	Only applicable to tripping systems
7	Separate input measuring equipment should be used for each guardline	Only applicable to tripping systems

A review of the detailed defences in Table 3 reveals that some of the entries are not applicable to all systems. As an example, item (1) relates solely to tripping systems and is not applicable to mechanical or electrical systems. The first stage of calculating each PBF is therefore to determine the applicability of each detailed defensive measure and assign each an applicability index A_j as follows:

(1) A_j = 0 if the defensive measure is not applicable to the potential CMF event

(2) A_j = 1 if the defensive measure is applicable to the potential CMF event.

This applicability index sorts the defensive measures and the sum of these is given by:

$$N = \sum_{i=1}^{n} A_j \qquad (7)$$

where n is the number of defensive measures identified in Table 3. N therefore reflects the total number of applicable defences for a particular system or dependent failure event.

For each point where A_j = 1 the degree of compliance with the defensive measure C_k is determined as follows:

(1) C_k = 0 if the system features do not comply with the defensive measure

(2) C_k = 1 if the system features do comply with the defensive measure

(3) C_k = 0.5 if there is some intermediate degree of compliance.

A total compliance C_T is then determined by summing over the defensive measures:

$$C_T = \sum_{k=1}^{n} C_k \qquad (8)$$

It is assumed that for each PBF full compliance with all of the applicable defensive measures allows the minimum PBF to be achieved and that each of the defensive measures is equally weighted. The PBF is thus given by:

$$\beta_{pi} = 1.0 - (1.0 - \beta_{pim}) \frac{C_T}{N} \qquad (9)$$

The method to determine the individual PBFs for all defences requires 158 detailed points to be addressed in order to quantify a single CMF event and for any single system there may be a number of dependent failures. It is therefore important that the more significant dependent failures should be identified in an auditable manner in order that the effort can be concentrated on these whilst the less significant dependent failures are given a less complete but nonetheless auditable treatment as outlined overleaf.

To perform this initial screening to determine the more significant dependent failures a conservative estimate of each beta factor can be calculated. It is noted that a number of the defences apply to a system as a whole, rather than individual potential dependent failures within the system. These are identified in Table 2 by being indicated as system generic.

A conservative estimate of the beta factor for any dependent failure event in the system can therefore be calculated by determining the PBFs associated with the system generic defences using the method described above and setting all of the others to 1.0. The resulting beta factor is then input to the full probabilistic analysis for each potential dependent failure and the system failure frequency or probability calculated. For any critical dependent failures the beta factor can be further refined and reduced by calculating the PBFs associated with the other defences.

Whilst it is not possible to fully eliminate subjective judgement from the quantification of dependent failures, the method above forces the assessor to consider, in a systematic manner, the defences afforded by a system. It therefore reduces the degree of expert opinion from that used in the PBF method. Furthermore, the assessor is forced to formally justify his assessment and an independent review can easily be performed to veriffy the validity of the assessment.

As with all probabilistic methods the absolute value of the results must be treated with caution. The enhanced PBF method does, however, identify the system strengths, or perhaps more importantly weaknesses, and hence identifies where improvements can be made.

EXAMPLE STUDY

To demonstrate the enhanced PBF method an example study is briefly summarised below.

The standby generating station, shown in simplified schematic form in Figure 1 (overleaf), is used to supply power to a number of essential loads in the event of a loss of external grid supplies. There are four diesel alternators arranged on two switchboards in two segregated engine halls. The essential loads are arranged on ring mains and any single diesel can supply the minimum level of essential load.

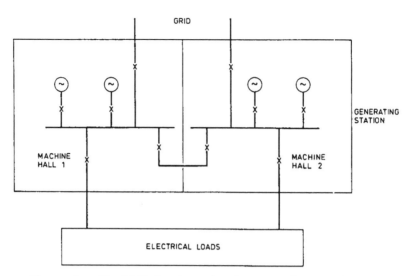

Figure 1. Simplified schematic of generating station.

Depending on circumstances the diesel alternators are required to run continuously for three hours or 24 hours. In the shorter case each diesel has sufficient stocks of consumable services stored local to the diesel for them to be independent of make-up. There is insufficient fuel oil stored locally for a diesel to run for 24 hours. Fuel oil make-up is therefore required and this comes from a common source for all four diesel engines.

The analysis assesses the probability of failure of all four diesel alternators by fault tree assessment and models all supporting services, although these are not described in this paper. Using standard failure data the probability of failure of a single diesel alternator is 0.039 for the three hour case and 0.102 for the 24 hour case.

The assessment has been carried out whilst the project is in the later stages of construction. The design is therefore fixed and only the operational strategy can be varied to reduce the likelihood of dependent failures. An initial assessment was performed assuming that the development of operating procedures does not involve specific consideration of dependent failures. A second sensitivity assessment assumed that specific defences against dependent failures are incorporated during the development of procedures and system operation.

The contribution of all 17 defences in Table 2 have been assessed and the calculation of the resulting beta factors for the diesel alternators are presented in Table 4 (overleaf). A table has been generated for each defensive measure to substantiate the calculation of each PBF. As an example, the calculation of N and C_T for redundancy and voting is presented in Table 5 (overleaf).

TABLE 4
Evaluation of β factor

Defence	Minimum partial beta	Initial study			Sensitivity		
		N	C_T	β_{pi}	N	C_T	β_{pi}
Design control	0.6	13	5.5	0.83	13	5.5	0.83
Design review	0.8	7	1.0	0.97	7	1.0	0.97
Functional diversity	0.2	8	0.0	1.00	8	0.0	1.00
Equipment diversity	0.25	8	0.0	1.00	8	0.0	1.00
Operational interfaces	0.8	6	2.0	0.93	6	2.0	0.93
Protection and segregation	0.8	10	5.5	0.89	10	5.5	0.89
Redundancy and voting	0.9	3	1.5	0.95	3	1.5	0.95
Proven design and standardisation	0.9	5	4.0	0.92	5	4.0	0.92
Derating and simplicity	0.9	5	4.0	0.92	5	4.0	0.92
Construction standards and control	0.7	3	1.0	0.90	3	1.0	0.90
Testing and commissioning	0.7	8	7.0	0.74	8	7.0	0.74
Inspection	0.9	5	4.0	0.92	5	4.0	0.92
Operational control	0.6	14	5.0	0.86	14	11.0	0.69
Reliability monitoring	0.8	8	0.0	1.00	8	8.0	0.80
Maintenance	0.7	10	2.0	0.94	10	10.0	0.70
Proof test	0.7	14	3.5	0.93	14	14.0	0.70
Operations	0.8	7	0.0	1.00	7	7.0	0.80
Assessed beta factor (β)		0.25			0.071		

TABLE 5
System compliance for redundancy and voting

Item	Defensive measure	Appraisal	A_j	C_k
1	The degree of redundancy and the voting logic of each sub-system must be related to both dangerous and safe reliable criteria, which have conflicting requirements	Not applicable – relevant only to systems incorporating redundancy/voting logic	0	–

TABLE 5 (cont'd)

Item	Defensive measure	Appraisal	A_j	C_k
2	The redundancy and voting logic must include allowances for dependent failures	Not applicable – relevant only to systems incorporating redundancy/voting logic	0	–
3	The consideration of dependent failures in the system design should consider the application of diverse or increased identical redundancy	Dependent failures were not fully considered during system design. Design was not reviewed for dependent failures. Good engineering practice applied only	1	0.0
4	The redundancy and voting logic should allow the system and all its redundant components to be adequately proof tested	Whilst proof testing of engine does not require it to be taken out of service (as it will be on local control and still available for use should a demand arise), limited benefit only is claimed for this defensive measure	1	0.5
5	Where possible, redundant channels should not be bypassed for test and maintenance, which severely degrades the system. For example, a two from three system should become a one from two during channel maintenance or test	All testing is undertaken in service, with administrative procedures written such that only one diesel is proof tested at a time	1	1.0
6	The use of double m from n decision logic should be carefully considered because of the linking between guardlines, and therefore dependent failure possibilities, that are introduced	Not applicable	0	–

TABLE 5 (cont'd)

Item	Defensive measure	Appraisal	A_j	C_k
7	Separate input measuring equipment should be used for each guardline	Not applicable	0	-

One particular design feature that is identified above is that the four engines are in two engine halls. The defence most affected by this is protection and segregation. From Table 4 the calculated values for this defence are $N \approx 10$, $C_T = 5.5$, which gives $\beta_{pi} = 0.89$ compared with a PBF of 0.8, and is consistent with the value which would be obtained for this sytem using a less auditable approach.

The overall results of the analysis are presented in Table 6.

TABLE 6
Overall results of the analysis

Description	Initial		Sensitivity	
	3 hours	24 hours	3 hours	24 hours
Diesel alternator failure probability	3.9×10^{-2}	1.0×10^{-1}	3.9×10^{-2}	1.0×10^{-1}
Beta factor	0.25	0.25	0.071	0.071
Diesel alternator dependent failure probability	9.6×10^{-3}	2.5×10^{-2}	2.8×10^{-3}	7.2×10^{-3}
System failure probability	1.1×10^{-2}	4.2×10^{-2}	3.9×10^{-3}	2.0×10^{-2}
Percentage contribution from diesel dependent failures	96%	59%	72%	36%

For the initial case the results are dominated by diesel dependent failure, although less so for the 24 hour case as a result of a number of failures associated with the common fuel oil supply.

Careful attention to providing defences against dependent failures during the development of operating procedures reduces the probability of dependent failures by a factor of about 3.5, and the system failure probability by a factor of 2 to 3.

CONCLUSIONS

The enhanced PBF method uses a systematic and fully auditable method to quantify dependent failures. It ensures that the assessor is compelled to consider the effectiveness of the detailed defences against dependent failures. In doing so, any areas of particular weakness can be readily identified so that, if appropriate, measures can be taken to reduce the impact of dependent failures.

The use of a checklist approach to quantify dependent failures and the selection of one of just three values for the degree of compliance with each detailed defence reduces the amount of subjective judgement employed in an assessment to the lowest practicable level.

An example study assessing the probability of failure of a diesel standby generating system is outlined and this demonstrates that preparation of operating procedures taking into account the potential impact of dependent failures can have a significant effect on the system reliability.

REFERENCES

1. Humphreys, P. and Johnston, B.D., SRD dependent failures procedures guide. SRD Report R418, March 1987.

2. Johnston, B.D., A structured approach for Dependent Failures Analysis (DFA). Reliability, 87, April 1987

3. Edwards, G.T. and Watson, I.A., A study of common mode failures. SRD Report R146, July 1979.

4. Bourne, A.J., Edwards, G.T., Hunns, D.M., Poulter, D.R. and Watson, I.A., Defences against common mode failure in redundancy system - A guide for management, designers and operators. SRD Report R196, January 1981

A BAYESIAN ANALYSIS OF POLLUTION INCIDENTS THROUGH TIME

Dr. NEVILLE DAVIES, Dr. JOHN C. NAYLOR and Dr. JOHN M. MARRIOTT
Dept. of Mathematics, Statistics, and Operational Research,
Nottingham Polytechnic, Burton Street, Nottingham.

ABSTRACT

Published data on oil spillage incidents are analysed using a variety of statistical techniques. Such data represents considerable past experience of such incidents in a form suitable for the application of Bayesian and time series analysis methods.

The data includes various measures of the severity of the spillage (eg quantity of oil) and of the response (eg cost). Questions about changes in these aspects during the last 20 years are considered. For example, is there evidence that we are "getting better" at dealing with, or at preventing, such incidents? Connected with this, we could ask 'What are the predictions for the future levels of incidence and costs?'

The techniques and style of analysis may find application in other incident data, for example nuclear and chemical, for which published figures may not be so readily available.

INTRODUCTION

This paper presents Statistical analyses of published data; the authors have had asccess to no more data than that readily available in a good library. This data is to be described, and some of the restrictions and features discussed.

Conventional modeling and Statistical techniques are inapplicable to this situation since they generally assume: i) large sample sizes, and ii) a limited class of possibly inappropriate models.

It will be shown in the next section that large samples of appropriate data are simply not readily available. In such circumstances great care must be exercised in defining the statistical models; routine assumptions of normality, independence and constancy of parameters may not be appropriate.

Three alternative Bayesian analyses are presented which avoid, or compensate for these assumptions in various ways. The first analysis presented is based on a non-normal model; next explicit dependence in the data is considered; and finally possible variation through time of the underlying model parameters is investigated.

The Bayesian approach also permits careful thought about the output from such analyses. There is some interest in model parameters, especially when these may have changed during the time period considered, but a major concern is for prediction of future events and data. The Bayesian methodology is especially suited to this need as will be demonstrated in these analyses to be presented.

THE DATA

The Advisory Committee on Oil Pollution of the Sea (ACOPS) has published various summary reports and annual reports, extracts from which are published by the Department of the Environment Digest of Environmental Protection and Water Statistics[1] (DEPWS). The data sets considered here have been extracted from these digest reports; the ACOPS reports being less readily available.

It is of some interest to see changes in what is reported in DEPWS. In the early 1970's quite detailed reports on individual (major) incidents were presented. Towards the end of the 1970's the use of summary tables increased, but considerable detail was still available. This included, for example, not only costs, but also details of methods used for clean up (9 categories) by region throughout Great Britain. Unfortunately this detail on costs and methods is not presented in the reports for the 1980's.

To examine the situation for changes through time or to attempt forecasts of future values, time series data on groups of variables for an extended period of time is desirable. However, this is not always readily available. For example, some sort of count on number of incidents, would seem to be a simple indicator of the severity of any perceived oil pollution problem. During the period of these digest reports the number of incidents have been reported variously for "all

---

[1] *formerly the Digest of Environmental Pollution and Water Statistics.*

reported spills", "spills over 100 gallons", and "spills requiring clean up". The regions for these counts has varied a little and the totals may be for GB or UK depending on the year. These all present serious restrictions on the usefulness of any Statistical analysis of these data. For these reasons relatively short sequences of values on consistently defined variables are available. For illustration three data series have been extracted from DEPWS and are detailed in Table 1 (SPILLS1) and Table 2 (SPILLS2 and Cost).

TABLE 1

Number of oil spill incidents reported

Year	SPILLS1
1973	447
1974	507
1975	497
1976	575
1977	636
1978	507
1979	558
1980	530
1981	541

Source: Digest of Environmental Pollution and Water Stastistics.   No. 5, 1982.

TABLE 2

Oil spill incidents requiring clean up

Year	SPILLS2	Cost (£000)
1978	365	966
1979	184	69
1980	181	113
1981	171	118
1982	170	66
1983	117	149
1984	159	68
1985	139	107
1986	126	134
1987	105	198

Source: Digests of Environmental Pollution and Water Statistics
or Digests of Environmental Protection and Water Statistics 1980 to 1988.

Simple plots of these series are given in Figures 1 and 2. Perhaps the dominant feature is the large initial value in both series in Figure 2. We might speculate that such a value relates to multiple pollution events, and hence multiple reports, for one large oil spill incident. It is difficult to ignore such a feature, but some analyses may omit this data point as having undue influence on the analysis and probably relating to a different model.

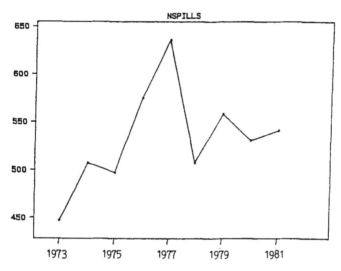

Figure 1. Annual number of Oil spills reported for Great Britain 1973-1981

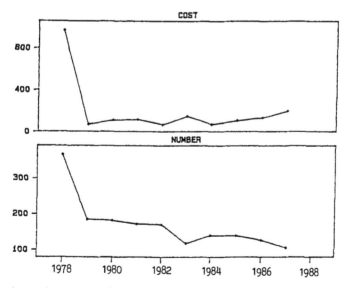

Figure 2. Annual number of Oil spills and cost of clean-up 1978-1987

A POISSON REGRESSION MODEL

A simple and natural model to consider for number of incident series such as SPILLS1 and SPILLS2 is a Poisson regression model. For this we suppose that:

(i) X_t = number of incidents in time period t, follows a Poisson distribution with mean λ_t;

(ii) λ_t in a simple function of t, eg linear.

Classical regression techniques assume normality and constant error variance. The normality assumption may well be approximately valid for large values of X_t (as we have here), but the variance of the Poisson is λ_t and so cannot be assumed constant. The variation in variance with current mean may go some way to modeling the "usually large" initial value in the SPILLS2 series.

We will accordingly adopt a Bayesian analysis in which probability distributions are used to represent knowledge about quantities of interest.

Suppose, initially, we are interested in the level and slope parameters of the regression of the model with $\lambda_t = \alpha + \beta t$ where t takes values 1,2, ... through the time period of the data. We might specify little or no prior knowledge about α and β with an improper prior density

$$p(\alpha,\beta) = 1.$$

The posterior density, $p(\alpha,\beta|x)$ to represent our knowledge after examining this sample data, is obtained via Bayes theorem in the form

$$p(\alpha,\beta|x) = \frac{\ell(x|\alpha,\beta)\, p(\alpha,\beta)}{\iint \ell(x|\alpha,\beta)\, p(\alpha,\beta)\, d\alpha d\beta}$$

Hence,

$$\ell(x|\alpha,\beta) = \prod_{t=1}^{n} (\alpha + \beta t)^{x_t} e^{-(\alpha+\beta t)}$$

is the likelihood function for this Poisson regression model.

The computations may be routinely performed using computer software described in, for example, Smith et. al. [1]. The posterior density may be represented by a contour plot such as that in Figure 3 for analysis of the SPILLS1 data set. This shows moderate dependence between slope and level which is related to the level parameter being defined for one end of the sample time period. The contours are seen to be elliptical suggesting normality in which case posterior moments provide a convenient compact summary; results in this form for both SPILLS1 and SPILLS2 are given in Table 3.

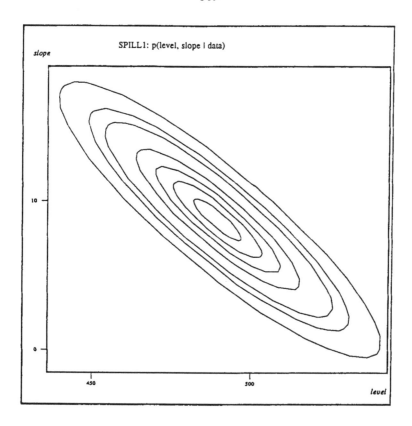

Figure 3. Contour Plot of $p(\alpha,\beta|x)$ for SPILLS1

TABLE 3
Posterior Moments for α and β

	SPILLS1		SPILLS2	
	α	β	α	β
Mean	489.6	8.74	261.6	-16.66
St.Dev.	16.8	3.06	9.4	1.32
Corr.	-0.89		-0.91	

The parameter α is in each case the level for the year before the data set began; it has little practical significance. In comparing results for these two data sets only inference about slopes (β) is of practical interest. We may obtain the marginal posterior density for β as

$$p(\beta|x) - \int p(\alpha,\beta|x)d\alpha$$

and plots of these for both SPILLS1 and SPILLS2 are given in Figure 4.

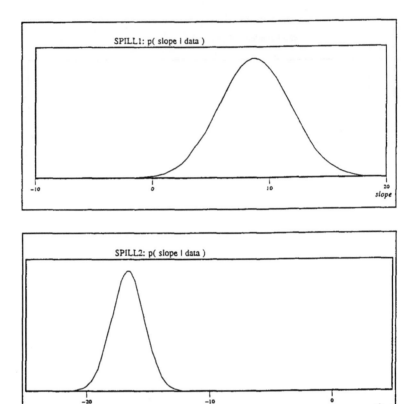

Figure 4. Marginal Posterior densities for slope (β)

Examining these plots and tables there is evidence of a positive slope for SPILLS1 and a negative slope for SPILLS2. This apparent decrease in number during the time period for SPILLS2 should be treated with caution; the results being influenced by the "unusually large" first observation and dependent on the criterion for "requiring clean up" which may itself change with time.

Frequently the major practical questions must be about values of future data, eg given $x_1, \ldots, x_n$ what might we infer about X_{n+1}? If we knew the values of α and β we could report its distribution; Poisson in this case

$$p(x_{n+1} | \alpha, \beta) = \frac{(\alpha + \beta(n+1))^{x_{n+1}} \, e^{-(\alpha + \beta(n+1))}}{x_{n+1}!}$$

and one approach would be to report this with estimates of the parameters α and β - a so-called estimative procedure. However, using Bayesian methods we may obtain directly the predictive density:

$$p(x_{n+1}|x) = \int\int p(x_{n+1}|\alpha,\beta)\ p(\alpha,\beta|x)\ d\alpha d\beta$$

which represents our knowledge about future data conditioned only on the data (and modeling assumptions), but not on particular values of, often artitrary, parameters. A predictive density for the number of spills requiring clean up in 1988 based on SPILLS2 data is given in Figure 5 . The mean of this distribution provides an optimal point estimate of X_{n+1}.

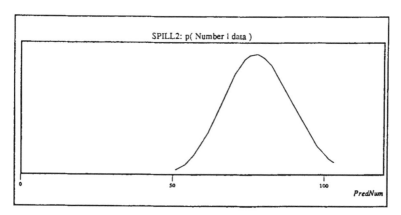

Figure 5. Predictive posterior density of number of spills requiring clean-up 1988

Of course, these results depend on the use of the simple linear model which can apply, at best, only for relatively short periods of time. More realistic models, incorporating asymptotic limits on growth or decay, may be considered using the same basic methodology, but would involve more parameters. Space does not allow a proper treatment of such models here.

ARMA MODELS FOR PREDICTION

Insofar as these data have been collected through time, it is natural to ask whether the family of simple autoregressive, moving average models, *ARMA(p,q)*, of the type discussed in Box and Jenkins [2] could be used to provide forecasts of number of oil spills or costs of cleaning operations. A Bayesian implementation of these models has been provided by Monahan [3] and extended by Marriott [4] and examples of its implementation are provided, for example, by Marriott and Tremayne [5].

The starting point for such an analysis is to consider whether the data exhibits any trend, which is certainly the case for the data considered here, and then to remove any such trend by computing the differences of successive observations of the data. These differenced series are then examined for structure by fitting a small family of low order *ARMA* models to each series and computing the posterior probability of each of the models to decide which of the models provides the most likely explanation for a given set of data.

The family of models considered here is those *ARMA(p,q)* models ‹for which $p + q \leq 2$, this restricted choice is necessitated by the small sample sizes.

Two prior specifications for these models are also considered; the first is one that expresses no prior knowledge as to which is a preferred model and the second is a prior specification that expresses a firm belief in as simple, or parsimonious, a model as possible. A more detailed discussion of this approach is provided by Marriott in Tremayne [5] and the references therein.

The results of this stage of the analysis can be seen in Tables 4(i) and (ii). From Table 4(i), for which no prior knowledge was assumed, it is clear that for the cost data it would not be possible to select any one model structure in preference to another, the data provides no help, whereas for the other two series there is clearly a preferred model, white noise for SPILLS2 and *ARMA*(0,1) for SPILLS1, although in both cases there are at least two other models that appear reasonably likely.

When a strong prior belief is expressed in favour of low order models, Table 4(ii) shows a different picture emerging. For SPILLS2 there is a clear suggestion that the best explanation of the behaviour is white noise, which is also the first preference for cost. However, for cost there are also two closely competing models, namely *ARMA*(1,0) and *ARMA*(0,1). These same three models are singled out for SPILLS1, but here two models, white noise and *ARMA*(0,1) were singled out as jointly most likely.

It is this latter situation that highlights the problem of obtaining forecasts when there is no over-whelming evidence in favour of a single model. If a single model had emerged an optimal Bayesian prediction would have been provided by the predictive mean from that model. However, in the situation described here Marriott [4] has shown that, when prediction is the main consideration, an optimal Bayesian forecast is obtained by taking a weighted average of the predictive means from each model using the posterior probabilities as the weights.

Table 4

Posterior Model Probabilities for Different data

(i) No prior knowledge

ARMA

Model		SPILLS1	SPILLS2	COST
p	*q*			
0	0	0.1529	0.2794	0.1840
1	0	0.1713	0.1845	0.1595
0	1	0.3122	0.1280	0.1974
2	0	0.0648	0.0986	0.1127
1	1	0.1746	0.1272	0.1683
0	2	0.1242	0.1824	0.1782

(ii) Strong prior belief in low order models.

ARMA

Model		SPILLS1	SPILLS2	COST
p	*q*			
0	0	0.3788	0.5957	0.4606
1	0	0.1693	0.1569	0.1593
0	1	0.3085	0.1088	0.1971
2	0	0.0255	0.0335	0.0449
1	1	0.0689	0.0432	0.0671
0	2	0.0490	0.0619	0.0710

Table 5 illustrates this for the case of the cost data for both prior choices, the column total providing us with the prediction for the differences in each case, this number being added to the 1981 figure of 541 spills to provide the prediction for 1982. As can be seen the prior beliefs about the models make little practical difference in this case.

In the interests of brevity no plots of posterior distributions for the parameters or predictive densities have been presented in this section, although it is a simple matter to obtain them, as has been illustrated in the previous section. It is also straightforward to produce predictions for 2, 3, 4 and further steps ahead, although clearly with only 8 data points available for the differenced data, such predictions would be over-ambitious.

Table 5.
One step ahead Predictive mean calculations for differential SPILLS1

ARMA Model p	q	Predictive weighted means mean	no prior knowledge	low order performance
0	0	0	0	0
1	0	-3.2008	-0.5483	-0.4419
0	1	-3.3452	-1.0444	-1.0320
2	0	2.5952	0.1682	0.0662
1	1	-3.4889	-0.6092	-0.2404
0	2	0.7503	0.0932	0.0368

predicted means -1.9405   -1.6113

BAYESIAN TIME SERIES MODELS

A simple form of the dynamic linear models developed by West and Harrison [7] relates an observation equation for Y_i (that could be cost of oil spillage clean-up) by

$$Y_i = \mu_i + \beta_i r_i + v_i \tag{1}$$

where i indexes the year, μ_i is a level, r_i is a regression variable (that could be the number of oil spillages observed in year i) and v_i is zero mean normally distributed white noise. Here β_i corresponds to a time dependent regression coefficient. The level of cost of oil spillage, μ_i, and regression coefficient β_i are allowed to evolve according to the parameter equations

$$\mu_i = \mu_{i-1} + \omega_{1i}$$
$$\beta_i = \beta_{i-1} + \omega_{2i} \tag{2}$$

where ω_{1i} and ω_{2i} are independent zero mean normally distributed white noise variables. The Bayesian learning approach involves parameter estimates being updated at each time point i in a forward recursive way that accounts for each new data point, but also allows the potential to intervene in the model if the user wishes. For example, the modeller could override a current estimate of cost level if he knew there was a sudden escalation in oil spillage clean up costs due to, for example, an increase in raw chemical costs.

The application of the model (1) can be implemented using the Bayesian Analysis of Time Series (BATS) software of West, Harrison and Pole [8]. Much of the output from this package is graphical.

We first apply a model of the form (1) to SPILLS1. Since we have no regression variables, Y_i must represent the observed number of oil spillages in a year and we shall use only historical values to model and predict the future. In the level term, μ_i we included a possible linear growth factor since, as can be seen from Figure 1, there appears to be two local time trends in the data. Using initial prior estimates of level and growth of 450 and 0, respectively, we applied the BATS software program [8].

Figures 6 and 7 are the time trajectories of the year to year estimates of level and growth of the number of oil spills in Great Britain from 1973 to 1981. As can be seen from Figure 7 the rate of growth of oil spills is decreasing over the last years of the 1970's and the early eighties. We would like to point out at this stage that a conventional trend analysis over time would have yielded a single time-constant growth factor over the whole period. The Bayesian time series approach <u>builds</u> <u>in</u> a time dependency in the growth term, and this is reflected by the clear picture of decreasing growth in Figure 7.

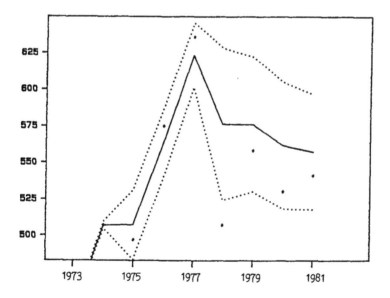

Figure 6. Year on year estimates of level of oil spills in Great Britain 1973-1981
(Dotted lines indicate 90% limits for estimates)

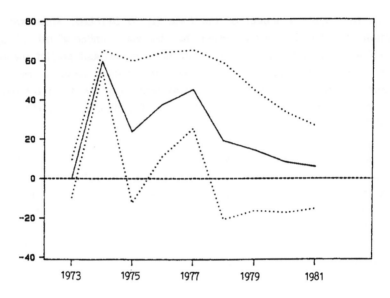

Figure 7. Year on year estimates of rate of change of cost per oil spill in
Great Britain 1973-81
(Dotted lines Indicate 90% limits for estimates)

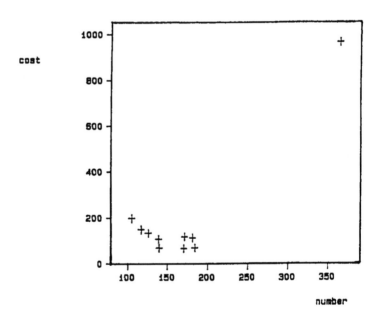

Figure 8. Scatter plot of cost of clean-up and number of oil spills 1979-1988

Our second data set SPILLS2 and associated annual cost of clean-up are plotted in Figure 2. The figures for 1978 are clearly rogue points compared with the rest. This is made clear by examining Figure 8, the scatter plot of cost of clean-up and number of oil spills for 1978 to 1987. As we are interested in predicting cost of clearing up oil spills and 1978 is such a clear outlier compared with the rest of the data, we decided to proceed with the analysis, ignoring the first year's figures.

We applied model (1) with Y_i now being cost of clearing the oil spills. We set μ_i to be a steady (no trend) level and r_i the corresponding number of spillages in year i. We set our prior cost level to be 60 with standard deviation of 40 and assumed the prior coefficient of the associated number of spillages to be zero with standard deviation of 2. In Figures 9 and 10 we plot the year to year estimates of the cost level and the rate of change of cost per oil spill.

We feel that the interpretation possible from Figure 10 is most important. Even though, over the last few years the absolute annual cost of clearing a decreasing number of oil spills has increased (Figure 2), the year on year estimate of the rate of change of cost per oil spills appears to be dropping. From this we conclude that the amount spent per spill may be decreasing. In particular a step change occurs between the year 1983 and 1985, where, of course, there may have been a change in policy towards spending on oil spills.

To predict future costs of clearing oil spills using the current model, we would need the number of oil spills in future years. The final posterior estimate of level of cost of oil spill for 1987 is 364 with standard error 62 and the posterior estimate of rate of change of cost per oil spill is –1.6 with standard deviation 0.6.

The Bayesian approach allows the user to change these values, if current information suggests the model parameters should change. For example, if there were to be 170 oil spills in 1988, with the above posterior estimates of level remaining the same, but the rate of change of cost now being unity, (with the same standard deviation of 0.6), we would predict a cost of approximately £480,000.

318

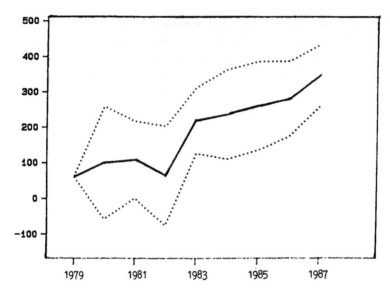

Figure 9. Year on year estimate of level of cost of cleaning up oil spills

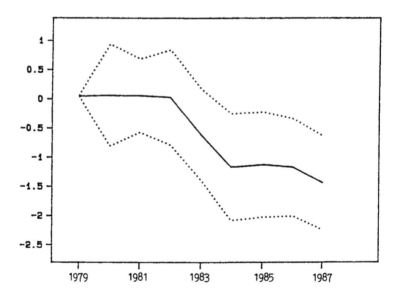

Figure 10. Year on year estimate of rate of change of cost per oil spill

CONCLUSIONS

We have shown that a Bayesian approach to modelling pollution incidents through time has great potential from a number of viewpoints. The approach can handle the limited amount of data available in the literature, whereas conventional modelling procedures cannot. Perhaps more importantly, other relevant information can be made part of the modelling exercise. If necessary, changes in policy such as increased expenditure rates or changes in costs of raw materials can be incorporated into the models so that prediction of future costs can be made to adapt to these sudden changes.

REFERENCES

[1] Smith, A.F.M., Skene, A.M., Shaw, J.E.H., and Naylor, J.C. (1987). Progress With Numerical and Graphical Methods for Practical Bayesian Statistics. *The Statistician*, **36**, pp 75-82.

[2] Marriott, J.M. and Tremayne, A.R. (1988). Alternative statistical approaches to time series modelling for forecasting purposes. *The Statistician*, **37**, pp 187-197.

[3] Marriott, J.M. (1988). Reparameterisation for Bayesian Inference in *ARMA* Time Series. *Bayesian Statistics 3*, pp 701-704. J.M.Bernardo, M.H.DeGroot, D.V.Lindley and A.F.M.Smith (Eds.) UUP.

[4] Box, G.E.P. and Jenkins, G.M. (1976). *Time Series Analysis: Forecasting and Control* (2nd Edn.) Holden-Dry.

[5] Monahan, J.F. (1983). Fully Bayesian analysis of *ARMA* time series models. *J.Econometrics*, **21**, pp 307-331.

[6] Marriott, J.M. (1989). Bayesian Numerical and Approximation Techniques for *ARMA* Time Series. PhD Thesis, Department of Mathematics, University of Nottingham.

[7] West, M. and Harrison, P.J. *Bayesian Forecasting and Dynamic Models*. Springer Series in Statistics. Springer-Verleg, New York 1989.

[8] West, M., Harrison, P.J. and Pole, A. (1987) BATS: Bayesian Analysis of Time Series. *The Professional Statistician*, **6**, pp 43-46.

AN EXPERT SYSTEM FOR INTEGRATING THE PHYSICAL MODELLING AND RELIABILITY ASSESSMENT OF ENGINEERING SYSTEMS

G A Richards
Z A Gralewski
R M Consultants
Suite 7 Hitching Court
Abingdon Business Park
Abingdon, Oxon
OX14 1RA

ABSTRACT

This paper describes an expert system having the potential to reduce the man-effort required to perform safety and relibility assessment by providing a common interface for the physical modelling and reliability assessment of engineering systems. The approach taken is to describe an engineering system as a network of connected nodes, the physical and reliability characteristics of each node being described by the analysts and stored in expert system knowledge bases. The physical state of the system under specified boundary conditions can then be determined and reliability analyses performed. The present paper describes the physical modelling of hydraulic/pneumatic systems and the reliability analyses of these systems using FMEA and FTA techniques. The approach however is also suited to other types of systems and other reliability analysis techniques including HAZOP/HAZAN.

INTRODUCTION

At present, the design of engineering systems proceeds typically with minimal interaction between designers, safety analysts and reliability analysts, particularly at the working level. Safety and reliability analysts are sometimes not involved until the latter stages of design perhaps due to the cost and man-effort associated with disseminating information, answering queries and updating the safety assessments. Toward the end of design, financial penalties associated with proposed design changes can be prohibitive in spite of potentially significant engineering advantages. The use of a common software interface for the physical modelling and reliability assessment of systems would have a number of advantages, including:

- communications would be more direct since all groups would have access to a common information base, resulting in reduced man-effort requirements for disseminating information.

- safety and reliability assessments could be conducted more efficiently since most of the required information would be in a readily retrievable form.

- quality assurance documentation would be easier to compile since a common set of information would be used by every group of analysts.

The work on which this paper is based was the development of a practical prototype software tool which would be flexible, relatively easy to develop and which could be written and tested on a PC. With this in mind it was decided to develop the software using a commercially available Expert System shell (Leonardo) on an IBM type PC. The prototype Expert System was restricted to hydraulic and pneumatic type engineering systems and to FMEA and FTA reliability analyses. Theoretical aspects of the work were considered after, and in the light of, lessons learned during development of the prototype system but are excluded from the scope of the present paper.

GENERAL CONSIDERATIONS

To facilitate both physical modelling and reliability analysis within one system, a suitable set of programs is necessary which will minimise the data input. After some consideration of different modelling options a network approach has been adopted for representation of physical phenomena, based on computer codes developed in-house. However, the approach is not restricted to the use of these codes.

A network is the representation of the system as a number of simple "transfer links" connected to each other via discrete "boundary conditions" (nodes). For example, in fluid flow through pipe systems, the "transfer link" is a pipe section through which under flow conditions a pressure drop is associated only with frictional effects. The "boundary conditions" are defined as any sources of pressure discontinuity, e.g. pumps, valves, bends, etc. The subdivision of the network into links and nodes allows both physical and reliability modelling to be carried out on the basis of a coherent representation of the system.

For heat transfer problems the "transfer link" represents the heat flow passage through a given material by means of conduction, convection and radiation as appropriate. The boundary conditions are then any heat sources and sinks.

An advantage of such an approach is that any system under investigation can be modelled and analysed rapidly. In addition systems can be modified with ease by changing the boundary conditions or altering part of the network, without a major change to the main network. Both steady state and transient problems can be solved using this approach.

From the reliability point of view there is little difference between links and nodes. In each case the failure modes of the components representing the links and nodes are of interest to the reliability analyst. The connectivity provided by the link/node representation allows the total system to be sub-divided into component groups according to function. This grouping can be conducted at various levels in order to build up a failure model of the overall system.

A functional representation of the Expert System is shown in Figure 1. The analyst interacts with the Expert System through six input/output modules which provide a user friendly interface. The top three modules shown in Figure 1 allow information about the system to be collected and stored. The bottom three modules allow boundary conditions to be specified and provide the interface for the analyst to perform physical modelling and reliability analysis. The Expert System can be interfaced with other software and in Figure 1 three software interfaces are shown: a link between the reliability data module and a reliability data base; a link between the physical models module and software for the analysis of physical problems (e.g. heat transfer, fluid flow) and a link between the reliability analysis module and reliability software (e.g. FMEA, FTA). Other software interfaces are possible but only the above were implemented for the prototype Expert System.

The system description module has the facility for entering a basic description of the components comprising the engineering system and for defining the connections between components. As described above, typically, each component and section of pipework in the system is identified as a separate node eg. valve, pump, heat exchanger. For the example hydraulic system shown in Figure 2 connections between nodes are as follows: 1-2, 1-6, 1-10, 2-3, 6-7, 10-11 etc. This indicates that node 1 is connected to three nodes (2,6,10) ie. it is a multi-junction. Associated with each node is the component type along with any other information required by the analyst e.g. component sub-type, manufacturer, materials. Ideally, the basic system description would be performed through a graphical display. However, for the prototype Expert System nodes are defined by the analyst entering text in accordance with prompts from the module.

The physical data module allows the analyst to enter physical characteristics of each node required for physical modelling of the system. The physical links and characteristics of the system of interest to the analyst depend upon the node type and physical problems to be solved. Standard lists of parameters of interest can be defined on the basis of component type and the analyst prompted for these values. As for all the modules the analyst is not

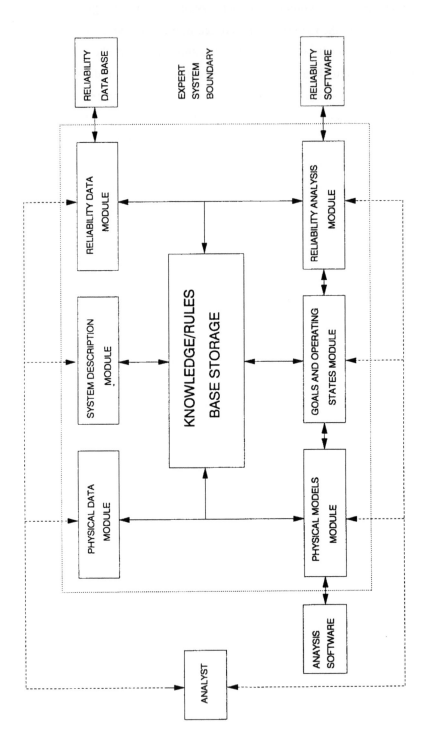

FIGURE 1 - FUNCTIONAL REPRESENTATION OF THE EXPERT SYSTEM

325

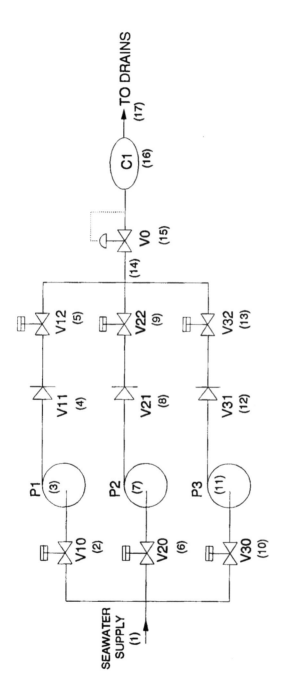

FIGURE 2 - HYDRAULIC SYSTEM EXAMPLE

restricted to the standard list of parameters and can add, ignore or delete parameters as necessary.

Reliability data for each node is collected by the reliability data module. For the prototype Expert System a simple component failure data base was constructed and interfaced with the module, but the reliability data module could be interfaced with more comprehensive data bases. The analyst specifies a standard set of failure modes for each component type (e.g. pumps: failure whilst running, failure to start on demand) and a standard set of reliability parameters of interest for each failure mode (e.g. failure whilst running: failure rate, repair time) and the relevant reliability data is taken from the data base and stored in the Expert System. Failure modes, reliability parameters and data assigned to each node can be examined by the analyst through the reliability data module and modified as necessary.

RELIABILITY ANALYSIS

The first step toward constructing fault trees and FMEA tables using the Expert System reliability module is to define the component failure modes of interest for the engineering system being analysed. This is achieved by defining for each node type the 'operational states' of interest, the failure modes of interest and any 'prerequisites' associated with a failure mode. A standard set of values can be defined for each node type and automatically assigned to each system node. The values assigned to each node can be modified by the analyst as required. An example of a standard set of values is shown in Table 1. The operational states shown in Table 1 are those having associated failure causes of interest, although success states could be shown as well. The list of states shown is not comprehensive and could be expanded to include, for example, external leakage. Similarly, the failure modes shown in Table 1 are not comprehensive and could be modified to suit the assessment. The column entitled 'prerequisites' defines the node state required in order for the failure mode to be valid, e.g. a valve can only fail to close if the valve is initially open. The prerequisites form the basis of the construction of rules

327

NODE TYPE	OPERATING STATE	FAILURE MODE	PREREQUISITES
ISOLATING VALVE	PATH OPEN	- FAILURE TO CLOSE	OPEN
		- INTERNAL LEAK	CLOSED
	PATH CLOSED	- FAILURE TO OPEN	CLOSED
		- PLUGGED	OPEN
FLOW CONTROL VALVE	HIGH OUTPUT	- FAILURE TO MODULATE(OPEN)	IN USE
	LOW OUTPUT	- FAILURE TO MODULATE(CLOSED)	IN USE
PUMP	NO OUTPUT	- FAILURE TO START	STANDBY
	NO OUTPUT	- FAILURE WHILST RUNNING	RUNNING
COOLER	NO OUTPUT	- BLOCKAGE	IN USE
CHECK VALVE	NO OUTPUT	- FAILURE TO OPEN	CLOSED
	REVERSE PATH OPEN	- FAILURE TO CLOSE	OPEN
PIPE	-	-	-
MULTI-JUNCTION	-	-	-

TABLE 1 - STANDARD SET OF RELIABLITY PARAMETERS FOR EXAMPLE NODES
PARAMETERS OF INTEREST FOR NODES

which are used by the Expert System to determine whether a particular failure mode should be included during fault tree construction.

Once the engineering system has been fully described and the reliability parameters of interest defined it is possible to automatically construct an FMEA table with minimal input from the analyst using the physical models module and some description of system operation. However, in order to simplify the prototype Expert System the above facility was not included. It was decided instead to base the construction of FMEA tables on the information presented in Table 1, the reliability analyst being required to manually input text associated with an 'effects' column. A facility was provided whereby the reliability analyst could raise a flag in situations where the effects of a particular failure mode were not known and some physical modelling was required. A physical analyst could then interrogate the Expert System for all questions concerning failure mode effects requiring resolution and perform the necessary analysis through the physical models module.

It is possible to determine combinations of failures having particular result by 'back-tracking' through the physical models module. Although potentially useful for checking failure logic, this approach is computing-intensive, and because it is difficult to generate well constructed, easily checked, fault trees using this approach it was not implemented in the prototype Expert System. Instead it was decided to describe the information required to construct a fault tree in piecemeal fashion by defining groups of nodes ie. component trains, sub-systems and systems. For each node group, information similar to that shown in Table 1 is collected: nodes included in the group; description; node group operating states of interest; node operating states causing the group operational state. When this information has been compiled for all defined groups of nodes then a request for construction of a fault tree can be made. In order to construct a fault tree the initial states of each node and a definition of the top event must be given as requested by the goals and operating states module. The Expert System then constructs fault tree input files for evaluation by an in-house fault tree graphics and analysis program. The method by which the Expert System compiles the fault tree information should be made clear by the following example.

EXAMPLE

Construction of a fault tree for the example shown in Figure 2 is described below. The system is a once-through cooling system comprising three 50% pumping trains supplying a single cooler through a flow control valve. Two pumping trains are normally in operation, the third train being on standby with the pump shutdown and outlet valve closed. Instrumentation is not shown on Figure 2 and was not modelled in the fault tree described below. On indication of low flow in any pumping train the standby pump is signalled to start and the associated outlet valve signalled to open. No operator actions are credited and routine change of pump duty is assumed to be perfect. Component leakage was not considered in this instance.

Nodes were identified for the example system as shown in Figure 2. Pipework was excluded from the reliability analysis since component leakage was not considered but must be included in the physical model to provide the transfer links. The standard reliability parameters shown in Table 1 were assigned to each node. Node groups were defined as follows: 2 through 5 (pumping train 1); 6 through 9 (pumping train 2); 10 through 13 (pumping train 3); 1 through 14 (pumping sub-system); 1 through 17 (cooling system). For each node group, operating states of interest were defined and applicable node and/or group states identified as shown on Table 2. Note that combinations of states can be identified as in the case of reverse flow through a pumping train where 'reverse path open' (ie. check valve fails to close) AND 'no output' (ie. pump fails) must occur to cause the node group to be in the 'reverse flow' state.

In addition to node and node group information it is necessary to define the initial state of the system and the system response to failures, both of which are described above, through the goals and operating states module. It is now possible to construct a fault tree. The analyst is asked to enter the 'goal' of the analysis ie. define the top event of the fault tree. In this case the only event of interest for the overall system is low flow and so the analyst identifies the node group as 'cooling system' and the state as 'low flow' (it is possible to indentify more than one node group and operating state e.g. if low

330

NODE GROUP	DESCRIPTION	OPERATING STATE	NODE/GROUP STATE
a) 2 - 5	PUMPING TRAIN 1	LOW FLOW REVERSE FLOW	- NO OUTPUT - PATH CLOSED - REVERSE PATH OPEN AND NO OUTPUT
b) 6 - 9	PUMPING TRAIN 2	AS ABOVE	
c) 10 - 13	PUMPING TRAIN 3	AS ABOVE	
d) 1 - 14	PUMPING SUB-SYSTEM	LOW FLOW	- 2 OF : a,b,c LOW FLOW - 1 OF : a,b,c LOW FLOW AND 1 OF : a,b,c REVERSE FLOW
e) 1 - 17	COOLING SYSTEM	LOW FLOW	- LOW FLOW d - LOW OUTPUT 15 - NO OUTPUT 16

TABLE 2 - RELIABILITY PARAMETERS FOR EXAMPLE NODE GROUPS

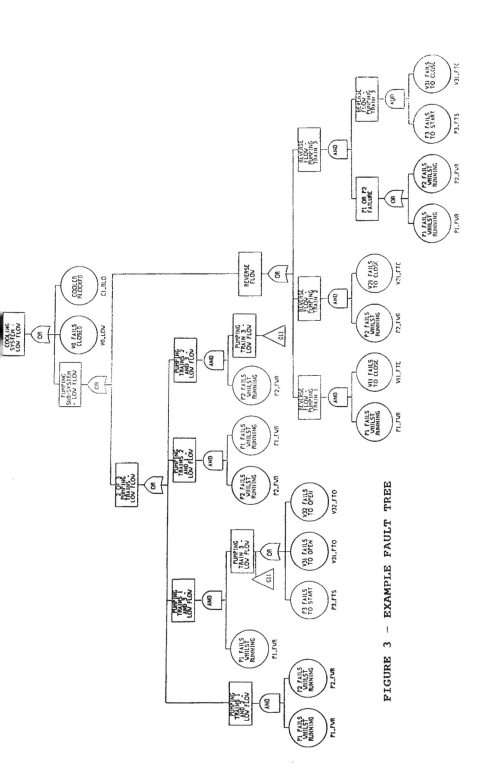

FIGURE 3 – EXAMPLE FAULT TREE

pressure was also of interest then the state would be entered as 'low flow OR low pressure'). The Expert System finds three node/group states as per Table 2: low flow group d (pumping sub-system); low output node 15 (flow control valve); no output node 16 (cooler). Upon looking up entries for the above the Expert System finds two group states of interest for the pumping system and only one failure mode of interest for each of the flow control valve and cooler. Having determined that the prerequisites for the flow control valve and cooler failures are satisfied the Expert System further assesses the branch 'low flow - pumping sub-system'. Analysing node group state '2 of: a,b,c low flow' the Expert System, starting with node a (pumping train 1), assesses the states 'no output' and 'path closed' for each node in group a. Given the initial conditions and the prerequisites, only one failure mode is identified: pump 1 fails whilst running. Having failed group a, this process is repeated for group b (pumping train 2) and group c (pumping train 3) in turn to obtain the failure combinations for: group a AND (group b OR group c) failure. The above is repeated for group b AND (group a OR group c) failures etc. Having assessed '2 of: a,b,c low flow' and having performed a similar assessment of '1 of a,b,c low flow AND 1 of a,b,c reverse flow' the fault tree is constructed by a reliability analysis sub-module. The fault tree compiled for the example system is shown as Figure 3.

FURTHER WORK

Further areas of work include: expanding the approach to include electrical supply and control failures; identification of those areas of the Expert System where provision of greater intelligence would significantly enhance its usefulness; development of interfaces to other physical models; and investigation of the applicability of the basic approach presented in this paper to other types of reliability assessment.

HARIS - A PC-BASED DATA SYSTEM TO SUPPORT RISK AND RELIABILITY STUDIES

J P Stead
Senior Consultant
R M Consultants Ltd
Genesis Centre, Birchwood Science Park, Risley, Warrington, WA3 7BH.

ABSTRACT

HARIS - Hazards And Reliability Information System - has been developed to meet the need for an integrated data system to support risk and reliability studies. This paper describes the philosophy behind the design and implementation of the original system, enhancements brought about through feedback from users, and outlines the advanced features currently available to users of the latest version of the system. Some other databases available to the safety, risk and reliability professional are briefly reviewed, and compared and contrasted with the HARIS system. The viewpoint throughout is that of a data user not necessarily skilled in programming but with an appreciation of the problems facing the risk and reliability engineer.

INTRODUCTION

As consulting scientists and engineers to the petrochemical, nuclear, offshore and more recently the defence and aerospace industries, R M Consultants have long been aware of the requirement for an in-house data system to support projects in these areas. Such systems must be capable of storing and manipulating a variety of data types. They must also be convenient to expand with new data as further projects are completed, thus adding to the experience already in the database.

HARIS has been developed to meet the need for an integrated data system to support safety, risk and reliability studies. The system has been required to possess the seemingly conflicting characteristics of being comprehensive with sophisticated data handling facilities, yet manageable, being easy to operate from an occasional user's point of view.

THE DRIVING FORCE

The concept of HARIS was developed in the early 1980's. The company's Managing Director had previously been manager of a large reliability data bank in the nuclear industry and had become convinced of the need for a simple, personal-computer based system to support safety, risk and reliability studies at the individual engineer level. It was believed that this would be a great improvement on the paper-based or large mainframe-resident databases prevalent at the time, which tended to be both slow and cumbersome.

At this time, the Commodore PET microcomputer was released onto the UK market. The PET, affectionately referred to as "the grandfather of the modern PC" immediately fulfilled the requirement for a self-contained machine with disk-drive on which the prototype HARIS system could be developed. Although primitive by today's standards, it provided the missing link to enable database power to be economically brought to the engineer's desk top.

Discussions with potential users of the system were held to establish the data formats most likely to be required by risk and reliability engineers. From these studies, the four main areas of abstracts, incidents, reliability and maintainability were identified and each was given a datasheet format as illustrated in figures 9 to 12 respectively (see end of text). These formats have changed very little since their inception. The abstract database was seen as the hub of the system, each new reference being given a unique identification number which was to be quoted on all other sheets containing data derived from it. Thus traceability of all data was ensured; the fundamental requirement. The relationship between different data types within HARIS is illustrated in figure 1.

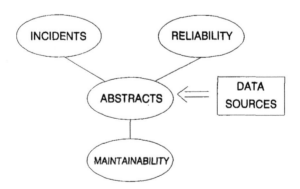

Figure 1 Relationship Between data types in HARIS

DATA

All organisations, groups and individuals collect, process and distribute data as part of their daily activities. Data is handled generally in two forms; raw and processed. Raw data is unprocessed, unstructured and often incomplete. Processed data is structured and "complete", in so far as the gaps have been filled by the structuring process. Safety and reliability data can take either raw or processed forms.

All data is used to improve the quality of organisational output. Theoretically larger databases contain more experience and should therefore improve the chance that the product will be of high quality. However, in reality large databases become unwieldy and eventually unusable unless they have a clearly defined structure and an established set of rules for adding new information. Experience has demonstrated that it is far better to clearly define the database structure at the outset and fit new data into it than to wait until a large quantity of amorphous data exists which will then require a huge restructuring effort. In the latter case a "point of no return" is normally reached after which it is impossible to justify the time required and the raw data is simply archived. This is an unfortunate waste of resources. In a risk assessment environment quality data is essential as actual failure experience is needed on which to base assessments. Given that good quality structured data exists within the organisation then it is also important to ensure that people are aware of it.

DEVELOPMENT OF THE SYSTEM

From the very beginning it was considered essential that the engineer needing data would find the HARIS system simple and convenient to operate. It was clear that only if this basic criterion was satisfied could the system be expected to have any success. Although technically very capable, many risk and reliability engineers are slightly apprehensive of computers and it was therefore necessary to ensure that no previous experience of computing would be required other than a simple 10 minute "driving lesson". To this end it was decided that the selection of system facilities would be menu-driven by single alpha-numeric keystrokes, each assigned to an option beginning with that letter; for example, "D" to display and "P" to print. In addition there would be a few cases where sections of displays larger than one single screen needed to be accessed using other familiar keys such as the space bar. The desktop computer would be assumed to be connected to a basic dot-matrix printer capable of reproducing datasheets on A4 sheets of continuous stationery. Personal computers at this time were not generally fitted with hard-disk drives, such devices being expensive luxuries reserved for mainframe machines, and so a single floppy disk-drive was assumed. This had the advantage that the user only needed to insert one floppy disk containing program and data, and did not need to consider the "path" command which is very important on the more sophisticated personal computers in use today.

Working to these basic requirements, a suite of 4 prototype programs were created using BASIC as detailed in [1]. Each of these programs enabled the user to perform a single pass text or keyword search on one field within a corresponding datasheet and listed out the matched sheet numbers and their top-level information either on the screen, on paper or on both. The programs did not allow for multiple-pass searches as will be described later in this paper, but nevertheless gave engineers direct access to data which could be viewed, printed and incorporated into assessments and reports. A typical main menu is illustrated in figure 2.

```
HARIS ABSTRACT PROGRAM
23 of 325 sheets generated so far
select the required function


        generate a new sheet ..........................................g
        modify an existing sheet ....................................m
        display a sheet ...................................................d
        print a sheet .......................................................p
        list the sheets .....................................................l
        search through the sheets ...............................s
        second level menu .............................................2
        ........or finish ......................................................f
```

Figure 2 Early HARIS Main Menu

Over the following years HARIS was transported from the PET microcomputer to a DEC PDP-11 lookalike machine and eventually to an IBM Personal Computer having a hard disk enabling the data to reside permanently on the machine. The system was constantly improved and extended and in the course of this work was re-coded using FORTRAN for a while and then back again into a more advanced version of BASIC which is used today. Throughout, the task became a trade-off between the competing priorities of making full use of ever more sophisticated machine capabilities and keeping the system within the understanding of an occasional computer user. A special version of the system using a "virtual RAM disk" was tried for a while when hard-disk search times were still slow, but this was eventually abandoned since it had to be coded specifically for the machine in use taking into account its memory arrangement. Hard-disk performance on modern machines is considered satisfactory for this system; 1000 datasheets can be searched in most configurations in around 25 seconds (25ms per sheet). These HARIS developments took place gradually over a period of years. Then in 1987 a major re-appraisal of the system was made and plans were prepared for HARIS Version 5.

HARIS VERSION 5

It became apparent in the middle of 1987 that the continuing program of modifications, which had been successfully used up to that point to improve the system, were no longer sufficient to maintain pace with hardware developments and the increasingly sophisticated demands of the users. It was clear that a significant re-design was required. A list of required improvements to facilities was prepared as follows:

1)   Scanning of sheet display in both upward and downward directions.
2)   Use of colour to highlight important areas of menus.
3)   Ability to perform multiple-level search passes via the incorporation of a variable search "pool" known as the Current Working Set (CWS) which could be expanded, restricted or reset according to the wishes of the user before any display needed to be invoked.
4)   Ability to exit from the system temporarily to operating system level and return without loss of the CWS memory.
5)   Use of insert/typeover and delete functions to enhance line editing,

especially in the remarks fields.

6) Access to help screens at strategic points in the user interface to allow examples of acceptable responses or input to be displayed.

7) Ability to change datasets or disks without leaving the system.

In addition to program re-design it was agreed that some work was required on the abstracts database to consolidate its position at the hub of the system:

8) Fast and efficient flagging of those abstract datasheets which were confidential as opposed to public domain information.

9) Double checking of all abstract sheets to ensure anonymisation where necessary.

A timescale and work programme to implement the above facilities and features was agreed and resources were made available. The result was HARIS Version 5 (V5). A new user-manual to incorporate all of the above features was also prepared. Figure 3 illustrates the main menu screen presented to the user by HARIS V5.

```
┌──────────────────────────────────────────────────────┐
│              HARIS ABSTRACT PROGRAM                     │
│   DATA SET TITLE: RMC ABSTRACTS DATABASE 1990 PART A   │
│      data set range 1 to 1000    last sheet created is 530 │
│   currently 530 data sheets in the current working set (CWS) │
├──────────────────────────────────────────────────────┤
│ ════════════════ MAIN MENU ════════════════            │
│   generate a new sheet ........................... G    │
│   modify an existing sheet ....................... M    │
│   display a sheet ................................ D    │
│   print sheets in the CWS ........................ P    │
│   list sheets in the CWS ......................... L    │
│   select a subset of sheets from the CWS ......... S    │
│   append a subset of sheets to the CWS............ A    │
│   reset the CWS to the full data set.............. R    │
│   backout to the previous CWS .................... B    │
│   use a different data set or disk................ U    │
│   second level menu .............................. 2    │
│   finish.......................................... F    │
├──────────────────────────────────────────────────────┤
│     select the required function = >                    │
└──────────────────────────────────────────────────────┘
```

Figure 3   HARIS V5 Main Menu

The most significant difference from its predecessors is the incorporation of a CWS status into the top of the display, together with a group of CWS-related options within the menu; namely S (Select), A (Append), R (Reset) and B (Backout). Since the CWS commands provide the new system with most of its enhanced searching capability it would be appropriate to consider them in more detail.

The Current Working Set (CWS)

On first entry into a dataset, the CWS contains pointers to all sheets and therefore the size of the CWS will be exactly equal to the size of the dataset as illustrated in figure 4.

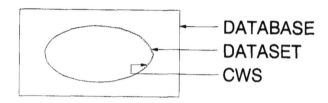

Figure 4   Relationship between Database, Dataset and CWS

Following a first search pass using the SELECT option the CWS will contain only pointers to those sheets which satisfy the search criterion specified by the user. It will therefore have reduced in size as illustrated in figure 5.

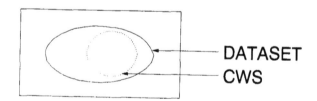

Figure 5   Reduced size of CWS after searching

It is important to note that further use of the SELECT option will only consider those sheets in the new CWS ie those which have already satisfied the previous search conditions. Conversely, use of the APPEND option will consider only those sheets which are outside the CWS but still within the dataset as illustrated in figure 6. Datasheets located by APPEND are added to the existing CWS if they are not already present in it; duplication does not occur. Therefore the size of the CWS may increase.

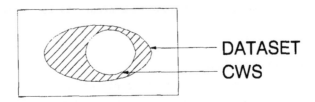

Figure 6   Datasheets available for searching using APPEND

In general, the size of the CWS is decreased by SELECT and increased by APPEND.   The BACKOUT option allows the user to take one step backwards from either an APPEND- or SELECT- produced CWS, replacing the latest CWS with that just prior to the search.   This is particularly useful following entry of a mis-spelled phrase, or when the CWS is unnecessarily cut down by a search operation.   The RESET option simply makes the size of the CWS equal to that of the dataset, in preparation for a new search strategy.   (This is carried out automatically if the user changes datasets with the USE command).

The Second Level Menu

The HARIS V5 main menu contains an option to move into the second level menu covering a variety of further options which are less often required.   Figure 7 illustrates the HARIS V5 second level menu.

```
╔══════════ HARIS 2ND LEVEL MENU ══════════╗

    alter the data set title.............................   T
    sort the sheets in the CWS ...............................   S
    copy the CWS to another data set ...................   C
    execute a DOS command .............................   E
    return temporarily to DOS .............................   D
    return to main menu .........................................   R


╚═══════════════════════════════════════════╝

    select the required function = >
```

Figure 7   HARIS V5 Second Level Menu

341

SORT enables the user to arrange the contents of the CWS into
alphabetical or date order etc. for listing purposes. It is not possible to
save the contents of the CWS in this special order since doing so would
scramble the numerical sequence carried by the datasheets. However it is
useful for visually scanning a comparatively large CWS for a required
datasheet. COPY will save the contents of the CWS to an alternative disk or
directory, so that they may be separately accessed as a project-specific
dataset at some later time. EXIT and DOS return the user to MS-DOS operating
system level to execute, respectively, a single command or a sequence of
commands. The important feature of these options is that the CWS is retained
on re-entry into the system, unlike FINISH which is final.

THE DATABASE

Database Structure
A data system may be considered to consist of a database plus database
management software (DBMS). In order for the DBMS to operate effectively, the
database must be properly structured so that information can be quickly
located.

HARIS has a multi-level structure. At the most fundamental level there
are fixed length fields in pre-determined positions in each of the four types
of datasheet. The positions of these fields ensure that when data is drawn
from a dataset for display, each item appears in its correct place on the
computer screen. At a higher level there is a hierarchical structure of
dataset directories or disks as illustrated in figure 8 below:

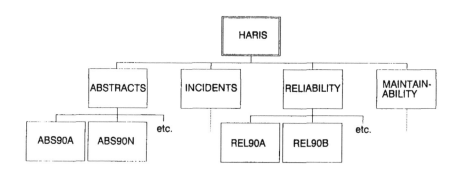

FIGURE 8 Hierarchical Structure of the HARIS Database

Furthermore there are some pseudo-relational links between the four datasheet types which enable connections to be made for traceability and cross-referencing. These links are illustrated in figure 1.

Database Content
HARIS contains data gathered from sources used by the company over a period of approximately 10 years. The main areas of interest are the petrochemical, nuclear and offshore oil and gas industries. However, more recently the company has become involved in work for the aerospace and defence industries and this is now beginning to be reflected in the database content. Tables 1 and 2 give a guide to the total size of the HARIS database in May 1990 and to the coverage of certain subject areas as denoted by keywords.

TABLE 1
Contents of HARIS Abstracts, Incidents and Reliability Databases

Data Type	Sheets	Offshore	Nuclear	Chemical	Industrial	Oil/Gas
Abstracts	530	245	132	117	-	-
Incidents	2578	-	52	1074	-	1270
Reliability	1676	521	671	-	587	-

TABLE 2
Contents of the HARIS Maintainability Database

Data Type	Sheets	Instrument	Electrical	Mechanical
Maintainability	364	126	131	284

It can be deduced from tables 1 and 2 that the total size of the database in May 1990 was 5148 sheets.

```
-------------------------------------------------------------------------------
     'HARIS' ABSTRACT DATA SHEET
-------------------------------------------------------------------------------
               DATA SET TITLE : RMC ABSTRACTS DATABASE 1
-------------------------------------------------------------------------------
```

```
    SHEET NUMBER:  a 299                    BASE DSN:

    LOCATION       RMC DB S/SC2/            SEE ALSO:

    TITLE:         CHERNOBYL - BEFORE AND AFTER

    AUTHOR:        JOHN DEVINE

    PUBLISHER:     I MECH E

    YEAR:          1986

    CLASS:         textbook REPORT   data      standard
```

MAIN KEYWORDS
==============

```
storms          SAFETY           SYSTEMS
lightning       reliability      EQUIPMENT
missiles        availability     components
fires           maintainability  computers
explosions      consequences     software
toxic releases  testing          human factors

electrical      probability      NUCLEAR
instrument      statistics       chemical
electronics     weibull          offshore
mechanical      corrosion        military
process plant   fatigue          aerospace
transport       creep            power generation
```

    Note: Additional keywords can be included in REMARKS.

REMARKS
=======

INFORMATION SHEET NO.1 COVERING BIBLIOGRAPHY OF REPORTS CONCERNED WITH

DESIGN AND SAFETY AT CHERNOBYL POWER STATION AND THE IMPACT OF THE

ACCIDENT ON SAFETY IN THE NUCLEAR POWER INDUSTRY

Figure 9   HARIS Abstract Datasheet

```
-----------------------------------------------------------------------
      'HARIS' INCIDENT DATA SHEET
-----------------------------------------------------------------------
                DATA SET TITLE : RMC INCIDENT DATABASE 1
-----------------------------------------------------------------------


   SHEET NUMBER: i 37              BASE DSN:     inc001

   ABSTRACT: 301                   SEE ALSO:

   DATE: 740601                    USED FOR:

   TOWN: FLIXBOROUGH

   COUNTRY: UK

   INCIDENT TYPE:  bleve   UCVE   cve   fireball   FLASHFIRE   poolfire

   FATALITIES: 28

   INJURIES: 89

   AREA DAMAGED: TO 8 MILES

   COST: $48M-$180M

   CHEMICAL INVOLVED: CYCLOHEXANE

   EQUIPMENT INVOLVED: BYPASS ASSEMBLY


INCIDENT KEYWORDS
=================

      MATERIAL        TYPE     REACTION      RELEASE        SIZE
   ---------------------------------------------------------------
           GAS      LEAKAGE   FLAMMABLE        spill       small
        LIQUID         FIRE       toxic     V.CLOUD      medium
   particulate    EXPLOSION radioactive      aerosol       LARGE


ENVIRONMENT KEYWORDS
====================

      INDUSTRY    LOCATION INSTALLATION    ACTIVITY  CONTAINMENT
   --------------------------------------------------------------
       nuclear       LAND   GENERATION  MANUFACTURE    P.VESSEL
      CHEMICAL        sea distribution   processing        tank
       oil/gas        air      storage    transport    pipeline


REMARKS
=======

A BYPASS ASSEMBLY FAILED RELEASING CYCLOHEXANE WHICH IGNITED. 433,000
GALLONS OF FLAMMABLE LIQUIDS WERE INVOLVED IN FIRES BEGUN WHEN A HUGE
VAPOUR CLOUD EXPLODED. PRESSURE WAVES, FIRES AND THE EXPLOSION CAUSED
THE DESTRUCTION OF THE PLANT AND SURROUNDING BUILDINGS.
```

Figure 10   HARIS Incident Datasheet

```
----------------------------------------------------------------------
    'HARIS' RELIABILITY DATA SHEET
----------------------------------------------------------------------
                DATA SET TITLE : RMC RELIABILITY DATABASE 1
----------------------------------------------------------------------
```

SHEET NUMBER: r 462   BASE DSN: mesyl

ABSTRACT: A259   SEE ALSO:

CLASS: VALVE   TYPE: DHSV

USED FOR: GAS   FAILURE RATE: 0.015 /DEMAND

FAILURE MODES   %   EFFECT ON SYSTEM
==============   =   =================

   ALL   100

SAMPLE STATISTICS
=================

     POPULATION: 1651 TESTS
     NUMBER OF FAILURES: 26
     CALENDAR TIME INTERVAL: 2.5 YEARS
     AV. OPERATING TIME/ITEM: -
     TEST INTERVAL: 1 MONTH

DESIGN KEYWORDS
===============

     DISCIPLINE   SIZE   MOTION   SPEED   COMPLEXITY
     ----
     instrument   small   static   slow   simple
     electrical   MEDIUM   rotating   high   complex
     MECHANICAL   large   recip   v.high   V.COMPLEX

OPERATIONAL KEYWORDS
====================

                                                     FAILURE
     ENVIRONMENT   FUNCTION   STRESS DISTRIBUTION MECHANISM
     ----
       nuclear   passive   low   unimodal   RANDOM
      industrial   process   normal   bimodal wearout
       OFFSHORE   SAFETY   HIGH   multimodal   burnin

REMARKS
=======

FLAPPER TYPE. COMPARISON GIVEN OF BALL & FLAPPER FAILURE RATES FOR THE

SAME WELLS.

Figure 11 HARIS Reliability Datasheet

```
------------------------------------------------------------------------
     'HARIS' MAINTAINABILITY DATA SHEET
------------------------------------------------------------------------
              DATA SET TITLE : RMC MAINTAINABILITY DATABASE 1
------------------------------------------------------------------------
```

SHEET NUMBER: m 102   BASE DSN: mmes1

ABSTRACT: A130/ORE   SEE ALSO: 451/80/3

CLASS: DRYER   TYPE: DESICCANT

USED FOR: AIR

MEAN TIME TO RESTORE
====================

```
    TOTAL
    mode 1    INCIPIENT         8 MAN HOURS
    mode 2    INSTAL NEWDRIER  38 MAN HOURS
              CALENDAR TIME 0.0628MH
```

TEST AND MAINTENANCE INFORMATION
================================

FREQUENCY   ACTION
MONTH   CHECK PROTECTION CIRCUITS

YEAR   ELECTRICAL & ALARM TESTS.

OTHER   2 YEARS FULL INSPECTION.

REPAIR KEYWORDS
===============

DISCIPLINE	SKILLS	SITE	SPARES COSTS	TEST EQUIPMENT
instrumental	operator	PLANT	MODERATE	bite
ELECTRICAL	TECHNICIAN	w/s	high	mobile
MECHANICAL	specialist	manuf.	v.high	special

LOGISTIC KEYWORDS
=================

COMPLEXITY	VISIBILITY	ACCESS-IBILITY	ENVIRONMENT	SPECIAL TOOLS
simple	good	good	good	SIMPLE
COMPLEX	poor	poor	POOR	complex
v.complex	v.poor	v.poor	v.poor	v.complex

REMARKS
=======

ABSORBER TYPE AUTO REGENERATED TWIN TOWER AIR DRYERS. NO IN-SERVICE
FAULTS.

Figure 12 HARIS Maintainability Datasheet

Methods of Use

No strict rules are enforced on how HARIS must be used. Indeed, given its portability and ease of ease, users are encouraged to experiment in their own time both to investigate the extent of the system and to develop an operating style with which they feel comfortable. However, many people in the Company have adopted the following approach when embarking upon a new client project:

1) Write down a number of key phrases or words which describe the equipment and work to be carried out, or known previous work in the area.

2) Perform both high level (title, author, keywords) and low level searches (remarks) of the abstract database to identify relevant documents and data sources. APPEND can be usefully used at this stage to gather together the results of both types of search.

3) Perform a high level search (equipment, chemical, keywords) of the incident database to identify actual accidents which may have the potential to occur in the system being studied. If extensive coverage of an incident exists then add the appropriate abstract to the CWS created at step 2.

4) Perform a high level search (equipment class, type, keywords) of the reliability database to identify failure data for equipment of interest. Add appropriate abstracts to the CWS created at step 2.

5) Perform a high level search (equipment class, type, keywords) of the maintainability database to identify repair data for equipment of interest. Add appropriate abstracts to the CWS created at step 2.

6) Search both reliability and maintainability databases using the abstract numbers originally generated by step 2 to uncover any additional data not matched in steps 4 and 5.

7) If appropriate, restrict the Abstracts CWS to public domain documentation only.

8)   Print out the complete list of abstracts in the latest CWS and collect together all necessary documents and sources from the library.

OTHER DATA SYSTEMS

Databases of interest to the safety and reliability engineer fall conveniently into three categories; bibliographic, accident and failure rate. Bibliographic databases are the most prevalent and widely-used in all science and engineering disciplines. Accident (or incident) databases are mainly used by engineers involved in risk or safety fields, and tend therefore to be more specialised.   Component failure rate (reliability) databases are highly specialised and used only in the safety, risk and reliability domain as input to fault trees, to predict outages and to plan maintenance programs.

Bibliographic Databases
Bibliographic databases are simply lists of public domain document abstracts covering all the major conferences, journals and reports throughout the world. They may be searched using software which is accessed via remote terminals connected to dedicated telephone lines. Such databases are usually very large and have therefore little scope for desktop containment or portability. However they are wide-ranging in scope and are consequently of interest to other users outside the risk, reliability and safety professions.   Examples are the DIALOG, Compendex, Inspec, Ismec and Scisearch databases.

Accident Databases
Accident Databases consist of details of actual incidents which have occurred in industrial processes, transport and storage.   Dates and locations are given, although these are sometimes less than specific in order to protect commercial interests.   Equipment, chemicals and fatality and injury figures are normally provided.   An indication of incident size may be deduced from these casualties, or alternatively damage area or repair cost data may be provided. As the numbers of accidents worldwide over a relatively short time period is large, such databases tend also to be extensive and as a result access may only be available via remote interrogation or application direct to the database operating organisation. With the advent of high capacity disk drives and CD-ROM technology, accident databases will become more accessible.

Examples are the AEA Technology MHIDAS, TNO FACTS and HARIS INCIDENT databases.

Failure Rate Databases

Component failure rate data is expensive and time-consuming to produce. Companies which have gathered such data tend therefore to restrict its use to within their own organisation. The number of public access databases of this kind is very few, although there is considerable amount of failure data buried in the open literature. In order to properly define failure rate, a measure of the loads, temperatures, vibration and many other factors will need to be made on the failed component. This information is not easily collected and adds to the task, resulting in the highly specialised nature of these databases. Although failure rate databases need not be large to be useful, they are expensive in PC software terms. Examples are the RAC NPRD-3 Non-electronic Parts Reliability database [2], and HARIS RELIABILITY database.

SUMMARY

It is the author's belief that in order to improve safety and reliability in the coming decade there must be a marked improvement in the methods used to store and retrieve experience, and in the application of that experience to new equipment and systems. This paper has described the realisation, development and use of one system which the author believes will help to fulfil that need.

It has shown how the need for an in-house data system has been translated into a marketable product which is now used by a number of organisations both in the UK and abroad. The most important system facilities have been described with particular reference to the ease of operation required by novice or occasional users. Other databases of interest to the safety and reliability professional have been briefly reviewed and compared with the system described. The paper should serve as a guide to those wishing to design or develop an information system in this or a related field. Most importantly it has demonstrated how experience from accidents and failures in hazardous processes may be stored in an organised fashion and thus used to guard against re-occurrences in the future.

The author wishes to express thanks to the management of R M Consultants Ltd for permission to publish this paper.

REFERENCES

1. Moss, T.R. and Piotrowicz, V., HARIS - An information system for reliability and risk studies, R M Consultants Limited, 1985.

2. Rossi, M.J., Nonelectronic Parts Reliability Data Floppy Disk NPRD-3, Reliability Analysis Center, RADC/RAC, Griffiss AFB, New York, January 1986.

RELIABILITY ANALYSIS OF THE INDUSTRIAL ELECTRICITY SUPPLY
PART I: OUTLINE OF THE PROBLEM.

M.H.J. BOLLEN AND P. MASSEE
Eindhoven University of technology, Faculty of Electrical Engineering
P.O. Box 513, 5600 MB Eindhoven, The Netherlands.

ABSTRACT

This paper concerns the industry-owned part of the electric power supply to an industrial load, in short: the industrial electricity supply. In the reliability assessment of the industrial electricity supply we can distinguish four main areas: finding a suitable description for the behaviour of network components; calculating the (stochastic) system behaviour from the known component behaviour; describing the (deterministic) system behaviour during a disturbance; and linking the behaviour of the protection to the system reliability.
After a discussion on the reliability problem of the electricity supply the four main problem areas are discussed in detail.

INTRODUCTION

The high interruption costs associated with modern industrial plants calls for a highly reliable electricity supply. The general design problem can be described as follows: given a connection to the public supply grid and one or more industrial loads (both with their own demands), find the best electrical network to supply energy to the loads. The "best" simply means the cheapest: i.e. the total costs during the economic life-time of the network should be as low as possible. The total costs consist of a deterministic term: the building and exploitation costs that increase with increasing reliability, and a stochastic term: the interruption costs that decrease with increasing reliability. This optimization should take place within the boundary conditions set by the public supply, by the industrial load, by the security and by rules and regulations.

The most important difference between the industrial and the public electricity supply is the economic value of the reliability. The owner of the industrial electricity supply is also the consumer of the electric energy. The building and exploitation costs thus come from the same purse as the interruption costs. This makes the above reliability optimization possible. In the public grid only a small part of the interruption costs come at the expense of the owner, making such an optimization of little value.

Some technical differences between an industrial electricity grid and the average public grid are:
- the large concentration of load in the industrial grid, which leads to short connections;
- the highly predictable time behaviour of the industrial load;
- the large fraction of an industrial load which consists of electrical machines;
- an industrial grid mainly consists of underground cables whereas the majority of public grids (especially the ones being studied) consists of overhead lines.

All this calls for a reliability assessment directed especially towards the industrial electricity supply. Most of the techniques that will be used and developed can be used for the analysis of public distribution grids as well, and, to a lesser degree, also for high-voltage grids.

A reliability analysis has been performed quite often for extended high-voltage grids, where is has been determined to what extent the power stations can generate the energy needed by the users and to what extent the high-voltage grid is able to transport this energy [e.g. 1,2,3]. Much less studies have been performed into the reliability of public distribution grids [e.g. 4]. Only a few studies have been performed into the reliability of the industrial electricity supply [5,6,7,8].

THE RELIABILITY PROBLEM

Figure 1 shows the different aspects related to the reliability of the industrial electricity supply. During the reliability assessment of an (industrial) electricity supply all aspects of Figure 1 should be taken into account.

Network state. The probability of the occurence of a disturbance and the probability that this disturbance leads to an interruption are related to the network state: the electrical loading; the physical loading and the availability.

The **electrical loading** is the amount of energy being transported from the public supply grid to the industrial load. As an example we distinguish between no-feed (there is no connection to the public grid), no-load (there is a connection to the public grid yet no transport of energy to the load), loaded (there is transport to the load) and the transitions between loaded and not loaded (switching on and switching off of the load). An industrial load often shows an intermittent behaviour (periods of high and low power demands alternate in a regular pattern). In that case a distinction can be made between no-load, half-load, full-load and their transitions.

The **physical loading** indicates all external phenomena that might influence the chance of failure of the electricity supply. One can think about weather influences (rain, storm, lightning, very dry periods), mechanical activity (digging in the ground) and chemical influence (Two extreme examples: desert sand that sticks to insulators in electrical installations in Oman; unknown chemical compounds that affect the electrical parts of a waste combustion furnace).

The **availability** indicates which components are available for the transport of electrical energy. If one or more components are not available, the remaining components will, in general, be subjected to a higher loading. The probability of interruption given a disturbance is determined by the availability too. At each moment the availability is determined by the maintenance that is performed at that moment: preventive maintenance as well as repair (corrective maintenance).

Threats. Everything that threatens the interruption-free operation of the electricity supply is described by the term "threats". A distinction can be made between ageing (an increase of the momentary failure rate with increasing age), transient overload (a short overvoltage or overcurrent due to a switching operation, a lightning stroke or a short circuit), temporary overload (a long-lasting overcurrent or overvoltage e.g. due to one cable being out of operation), and damage (e.g. of the insulation of a cable). The appearence of these threats is influenced by many other events and aspects. Figure 1 shows some of them.

The degree of ageing is influenced by the (normal) electrical loading, the frequency and severity of transient and temporary overloads, the physical loading and by quantity and quality of the preventive maintenance. Quite general it can be stated that increasing the loading will lead to an increased ageing. No such general relation is available for preventive maintenance, although preventive maintenance is intended to decrease the failure rate.

Transient overload occurs or may occur due to a short circuit, due to transitions in the normal electrical loading (from no-load to half-load to full-load and back), due to an intervention of the protection (this can lead to overvoltages but it reduces the severity of overcurrents), due to a component being taken into operation after repair or after preventive maintenance, and due to a component being taken out of operation for preventive maintenance. Transient overvoltages can further enter the industrial network from the public supply grid.

Temporary overload will generally not occur during normal operation. The design will be such that the highest normal loading is not an overload. Due to extreme weather (a long-lasting period of drougth) or chemical influence (the before mentioned Arabian desert sand; or Dutch sea water) the permitted loading can go down. The dry weather in combination with a high ambient temperature prevents a cable from loosing its heat. As the temperature should not exceed a certain upper level, the maximum permitted current shrinks. In a similar way the maximum permitted voltage shrinks due to the influence of the sea water.

A temporary overload that will probably appear more often is related to the availability. In case one or more network components are not available the others will suffer a higher loading. Whether this is considered an overload depends on the dimensioning of the components and the ("arbitrary") boundary between no-overload and overload.

A temporary overload can also be due to a change in voltage of the public grid. An increase in voltage might cause a temporary overvoltage, a decrease an overcurrent (the electrical power demand of the industrial load will be nearly constant).

Damage can occur due to physical loading, but also during repair or preventive maintenance. Preventive maintenance can also repair the damage.

Faults. Further on we will distinguish between disturbances and interruptions (sometimes referred to as interruption of electricity supply or interruption of plant operation). A disturbance is any deviation from the normal state. A disturbance can be "on purpose" (e.g. taking a component out of operation for preventive maintenance) or "by accident" (a short-circuit, an overload, or an incorrect intervention by a protective device).

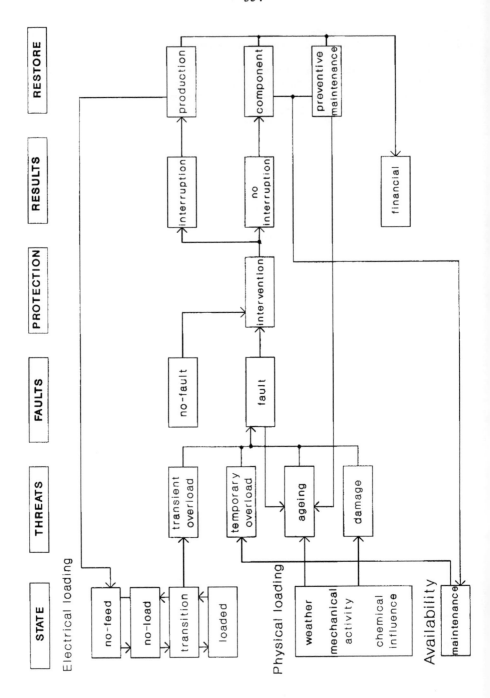

Figure 1: Visual summary of the different aspects of the reliability of the electricity supply.

The disturbances can be divided into faults and non-faults. A fault is a situation that is so dangerous to the components and/or to the electricity supply somewhere else, that we are willing to disconnect one or more components, even if this causes an interruption of the electricity supply. The best-known example is the short circuit. The currents appearing during a short circuit are such high that components would suffer considerable damage if no intervention would occur. As normal electricity supply is impossible during a short circuit, no-intervention would lead to an interruption for shure. The result of an intervention will not be worst than no-intervention.

The weighing is more difficult in case of an overload. The electricity supply will not be (directly) influenced by the overload, the component can however be damaged, leading to a (future) short circuit.

All disturbances that are not considered as a fault will be referred to as non-faults.

Protection. The appearance of a fault calls for an intervention by the protection (in fact, we have defined a fault as a situation for which we want the protection to intervene). A protective device might also intervene in case there is no fault, or it may not intervene in case of a fault. Both cases will be reffered to as failure of the protection. The function of the protection is completely different from that of the primary equipment (the current-carrying parts). We therefore consider the fail and restore behaviour of the protection seperately from that of the primary equipment. The influence of the protection on the reliability of the system is generally considered to be one of the most difficult aspects of the reliability assessment of the (industrial) electricity supply.

Results and restore. The protection's intervention (due to a fault or not) can lead to an interruption of the electricity supply or not. In both situations the failed component should be repaired. In case of an interruption the industrial production has te be restored too.

With modern industrial plants the financial consequences of the restore of the production will be much higher than those of the repair of a component. This makes the introduction of parallel connections attractive. If sufficient redundancy is created, preventive maintenance can be performed. A healthy component will be taken out of operation in order to perform all kind of work to reduce the component's failure rate. The costs of preventive maintenance and the temporary reduction of the availability should be compensated by the reduction of the failure rate.

From the above we came to a few areas of interest:
- setting-up a data base containing all kind of data from actual electricity networks;
- developing computer software to calculate the reliability of a given system;
- describing the system's behaviour during a disturbance;
- modelling the influence of the protection;
- economic aspects of reliability and design of the electricity supply;
- developing methods for the optimal design of an industrial electricity supply;
- finding the relations between ageing, loading and preventive maintenance.

The first four aspects will be discussed in detail in the forthcoming chapters.

THE DATA BASE

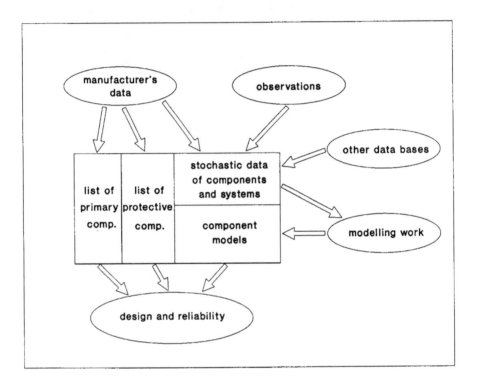

Figure 2: A schematic view of the reliability data base. The box
denotes the data base with its different functions. The
ellipses show the origin of the inputs and the direction of
the outputs.

One of the problems in the reliability assessment is to obtain field data
about the actual fail and restore behaviour of power system components. One
of the methods to obtain these data is the observation of actual power
systems. An alternative is to perform testing on a set of identical
devices. A disadvantage of the first method is the limited number of
observations (an electric power system is designed in such a way that a
failure will be a rare event). A disadvantage of the second method is the
lack of data on ageing.

We decided to start building a kind of data base containing information
on the actual fail and restore behaviour of network components. The
contents of the data base might be obtained from a number of sources:
 - manufacturer's data;
 - results of laboratory testing;
 - theoretical consideration of the ageing process;
 - study of the causes of past interruptions;
 - analysis of past experience with large industrial electricity networks;
 - data from similar data bases.
This data base will grow slowly and can be used by future reliability
assessments of the (industrial or public) electricity supply.

Apart from these statistical data the data base will contain some non-statistical data of importance to reliability assessment and network design. The non-statistical data mainly consists of possible choices during the design stage.

The final data base has to contain the following information:
- a list of commercially available primary equipment;
- a list of commercially available protective equipment;
- stochastic models for the fail and restore behaviour of primary equipment as well as protective equipment;
- a list of possible disturbances in the power system;
- deterministic models for the electric behaviour of primary equipment as well as protective equipment;
- a large amount of observations of large and small failures in (industrial) electricity networks.

The latter information will be the growing part that can be used to develop and check new models for the stochastical behaviour of electricity networks and network components. Figure 2 summarizes the different functions of the data base.

SOFTWARE FOR RELIABILITY ANALYSIS

In this area of interest the industrial electricity supply is given in the form of a reliability system consisting of components. The behaviour of the components is known, the task is to determine relevant reliability parameters of the system. New techniques are developed for this and will be implemented in a computer program. In the future this program will be a part of a computer-aided design package for the industrial electricity supply.

To what extent the model used can be considered as realistic, is not a matter of primary concern within this area of interest. We therefore have to distinguish between network components and model components. The electricity supply consists of (network) components like power transformers, underground cables, cuircuit breakers, and protective devices. A model component is just a simplyfied description of the behaviour of one or more network components.

There will often be a one to one relation between a network component and a model component, but certainly not always. For example it might be suitable to model a power transformer as the (reliability) series connection of two components: one that fails due to an incorrect switching action (by man or by machine); one that fails due to a short-circuit within the transformer.

For some other condition it might be more suitable to model the (electric) series connection of a circuit breaker, a power transformer and an underground cable as one component. Figures 3 and 4 give an example of the transition from an electricity network to a reliability model.

We distinguish between component events called "disturbances" and system events called "interruptions". A disturbance is the failure of one or more components. A disturbance leads to a cascade of events (sometimes a cascade of one event) that will or will not lead to an interruption. New disturbances can be part of the cascade.

An interruption is a state in which the electricity supply is not able to supply sufficient power to the industrial load. The definition of "sufficient" will be different from case to case.

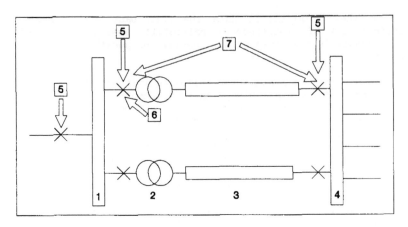

Figure 3: An example of an industrial electricity supply. It consists of a medium-voltage substation (1), two parallel connections formed by a power transformer (2) and an underground cable (3), a low-voltage substation(4) and a number of low-voltage connections to the industrial load. The crosses denote circuit breakers. The squares denote protective devices; 5 is an overcurrent relay that generates a tripping signal in case the current exceeds a certain threshold; 6 is a Bucholtz relay that detects overpressure in the transformer's tank; 7 is a differential relay that observes the difference between incoming and outgoing currents.

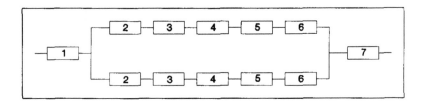

Figure 4: The reliability model of the industrial electricity supply of Figure 3; 1 represents the (correct or incorrect) opening of the circuit breaker between the public supply and the medium-voltage substation; 2 a short circuit in the transformer; 3 an incorrect action of the Bucholtz relay; 4 a cable fault; 5 an incorrect ·action of the overcurrent relay or of the differential relay; 6 preventive maintenance; 7 a fault in the low-voltage substation as well as all phenomena causing the outage of both connections.

The occurance of a disturbance is a completely stochastic process, it can be tackled by standard reliability techniques like Monte-Carlo simulation [9] or markov's method [10]. In case of a detailed model Monte-Carlo simulation is prefarble for this. The link between a disturbance and an interruption is a more or less deterministic process where electric network theory is of more use that the theory of probabilities. This can be combined more easily with a Markov model than with a Monte-Carlo simulation.

The compagnion paper [11] describes two examples of reliability analysis of the industrial electricity supply by using Monte Carlo simulation.

SYSTEM BEHAVIOUR DURING A DISTURBANCE

In order to find out whether a disturbance will lead to an interruption we have to study the phenomena that take place because of a disturbance (e.g. the failure of a generator unit or a short circuit in an underground cable).

1. In the electricity network transient currents and voltages will appear due to the transition from the steady state before the disturbance to the new steady state.
2. One or more protective devices might intervene to prevent further damage or because of a failure of the proetction. The protective device can react on the transient phenomenon or on the new steady state.
3. A new steady state will appear, eventually followed by an intervention, and a transition to another steady state, etc.
4. Other network components might show a fast ageing or might even fail due to a transient phenomenon or due to an overload during the new steady state.

According to the characteristic time scale one can distinguish between three (families of) transient phenomena: travelling waves; electromagnetic oscillations and electromechanic oscillations.

Travelling waves are the messengers that bring electrical inforamtion from one place in the network to the other. This takes place with almost the speed of light (100,000 - 300,000 km/s). Because of the limited dimensions of the industrial network their time scale is not of any importance for the current research.

Electromagnetic oscillations are caused by the disturbance of the equilibrium between electric and magnetic energy in the network. The period of these oscillations is of the order of 10 to 100 microseconds; the oscillations are damped with a time constant between 1 and 10 milliseconds. Some of these phenomena can influence the behaviour of the protective devices. Due to the importance of the protection for the reliability, these "potentially-dangerous" situations have to be identified.

Apart from the electromagnetic energy, there is mechanical (kinetic) energy present in the rotating parts of generators and motors. Electromechanic oscillations occur when the equilibrium between electromagnetic and mechanical energy gets disturbed. The corresponding time scale is of the order of seconds. For the industrial electricity supply this will be the most important transient phenomenon. The large number of motors (sometimes some generators too) will cause strong electromechanic oscillations; the long time scale enables an influence on the protection's behaviour.

To predict the behaviour of the protective devices and other network components during the disturbance, voltages and currents in the network have to be determined. This calls for the use of network models and for the

development of new models where necessary. The models should be suitable for incorporation in the reliability software.
Models are needed for the following situations:
- the calculation of currents during a short circuit; these currents determine the operation of the short circuit protection and the accellerated ageing of network components;
- the calculation of voltages in the network during and after a disturbance to determine the operation of undervoltage protection;
- the calculation of load flow after a disturbance, to determine the operation of overload protection and the accellerated ageing of network components.
The main problem in all these models are the electromechanical oscillations in systems consisting of many motors.

PROTECTION

The first aim of the protection is to intervene in the electricity grid in case of a fault. To prevent further damage a part of the grid (containing the faulted component) will be disconnected. This will reduce the availability and thus increase the probability of interruption.

The second aim of the protection is therefore to disconnect no more components than neccesary, even in case of a failure of the protection.

To fulfil both tasks as far as possible the electricity grid is devided into "zones": a zone is the smallest part of the network that can be disconnected from the rest of the network.

In case of a fault within a certain zone, the primary protection needs to disconnect this zone. In case of a failure of the primary protection the fault needs to be disconnected by a local backup (disconnecting only the faulted zone) or by a remote backup (diconnecting non-faulted zones too).

These three tasks (primary protection, local backup and remote backup) should be co-ordinated in such a way that each fault will be disconnected (even when the primary protection fails) and no more zones will be disconnected than neccessary. This co-ordination of protection tasks for a given network will be called the "protective strategy". The protective strategy describes the "ideal" behaviour of the protection. Each deviation from this ideal will be considered as failure of the protection. Figure 5 shows an example of a protective strategy.

To fulfil the ideal of the protective strategy, protective devices are needed. The implementation of the protective strategy with protective devices and their settings is called the "protective concept". As protective devices are not perfect (like all other equipment) the ideals of the protective strategy cannot be fulfilled completely.

We consider a protective device to consist of four functional parts (cf. Figure 6):
the measurement element that measures one or more physical quantities (in most cases current or voltage, although temperature and pressure are used too) and translates them to suitable quantities;
the detection element that compares the output of the measurement element with a certain setting acoording to the detection criteria;
the tripping element that gives an order to intervene (the tripping signal) when the detection criteria are met;
the disconnecting element that intervenes in the network on receiving a signal from the tripping element.

361

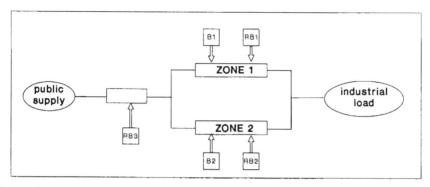

Figure 5: A protective strategy for a parallel connection. The different tasks of the protection have to be co-ordinated such that the following requirements are fulfilled:
in case of a fault in zone 1, B1 should react immediately, RB1 should react within a certain time, and B2, RB2 and RB3 should not react (within the clearing time corresponding to RB1);
in case of a fault in zone 2, B2 should react immediately, RB2 should react within a certain time, and B1, RB1 and RB3 should not react (within the clearing time corresponding to RB2);
in case a fault in zone 1 is not cleared by R1, RB1 should react as fast as possible;
in case a fault in zone 2 is not cleared by R2, RB2 should react as fast as possible;
in case a fault in zone 1 is not cleared by R1 or RB1, RB3 should react as fast as possible;
in case a fault in zone 2 is not cleared by R2 or RB2, RB3 should react as fast as possible.

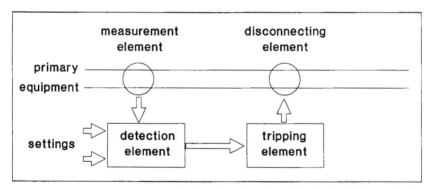

Figure 6: The four functional parts of a protective device.

In the failure of the protection we will distinguish a number of situations.

Failure of the protective concept: these are the fail situations that are known during the design stage. One of the demands of the protective strategy might be impossible to fulfil or only against too high costs. The designer then chooses for a non-ideal solution.

Failure of the model: these are the fail situations that were already present in the design but where not known because of the limitations of the models used. The designer can anticipate failures of the concept but she cannot foresee failures of the model. The only thing he can do is incorporate certain probabilities for failure of the model. These probabilities can be estimated from past performances. We can further distinguish between failure of the model of the measurement element, of the detection element, of the tripping element, of the disconnecting element and failure of the network model.

Failure of the setting: the protective concept includes setting of the different parameters of the detection element. However actual settings will not be exactly equal to the ones choosen by the designer. Mistakes can be made during the installation, or the setting can change as time passes. An incorrect decision due to such an erronous setting will be called failure of the setting.

Failure of a device: apart from all these failures it might happen that a protective device simple failes; i.e. it disconnects its protective zone without reason or it does not react at all to any fault. We make a subdivision in failure of the measurement element, failure of the detection element, failure of the tripping element and failure of the disconnecting element.

If we take into consideration that each fail situation can imply an incorrect disconnection (incorrect trip) as well as a fail-to-disconnect (fail-to-trip), a total of 22 fault situations can be distinguished. Figure 7 summarizes them.

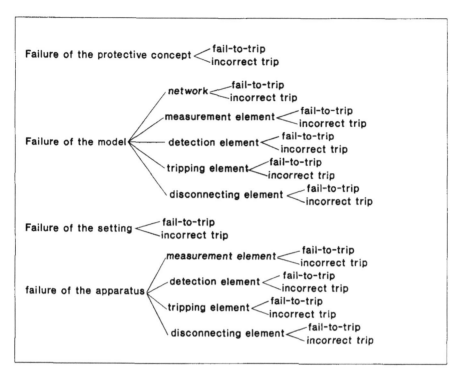

Figure 7: An overview of the different failure modes of the protection.

CONCLUSIONS

For the design of the optimum industrial electricity supply techniques have to be developed to determine the reliability of the (industrial) electricity supply. Our main areas of interest at this moment are:
- setting up a data base containing lists of available components as well as models for their fail and restore behaviour. This data base can be used during the design as well as during the reliability analysis.
- developing computer software to calculate the reliability of a given system, the current emphasis is laid on the Monte Carlo simulation.
- devoloping numerical techniques to describe the electrical behaviour of an industrial power system during a disturbance. Emphasis must be laid on the behaviour of many-motor systems.
- modelling the behaviour of the protection and its relation to the reliability of the supply.
- analysing existing techniques for the design of electric power systems and developing new techniques where neccessary.

REFERENCES

1. Billinton, R. and Allan, R.N., Reliability elaluation of power systems, Pitman, New York, 1984, Chapter 2-6.
2. Theil, G., Zuverlässigkeitsabschätschung von Energieverteiliungsnetzen met berücksichtigung der Netzlast, Habilitationsschrift Technische Universität Wien, 1985.
3. Edwin, K.W., Neue Planungsmethoden, Energiewirtschaftliche Tagesfragen, 1987, 37, 874-879.
4. Billinton, R. and Allan, R.N., Reliability evaluation of power systems, Pitman, New York, 1984, Chapter 7-9.
5. Dickinson, W.H., Economic evaluation of industrial power system reliability, AIEE Transactions Pt.II, 1957, 76, 264-272.
6. Fransen, H., De enkelvoudige reserve in de elektriciteitsvoorziening van een chemische industry, Elektrotechniek, 1989, 67, 719-722.
7. Reliability of electrical utility supplies to industrial plants, in Record of 1975 Industrial and Commercial Power Systems Technical Conference, IEEE, New York, 1975, pp.131-133.
8. Kloeppel, F.W. and Illing, F., Knowledge-based strategy to improve the co-ordination of the reliability in industrial electrical energy systems, in Proc. 10th Int. Conf. on Electricity Distribution, Institution of Electrical Engineers, London, 1989, pp.406-410.
9. Massee, P. and Bollen, M.H.J., Including voltage solidity in the reliability analysis of industrial electricity supply, 25th Universities Power Engineering Conference, Aberdeen, 12-14 September 1990.
10. Kochs, H.-D., Zuverlässigkeit elektrotechnischer Anlagen, Springer, Berlin, 1984, S.72-109, 296-312.
11. Massee, P. and Bollen, M.H.J., Reliability analysis of the industrial electricity supply, part II: two examples, this Symposium.

RELIABILITY ANALYSIS OF THE INDUSTRIAL ELECTRICITY SUPPLY
PART II: TWO EXAMPLES

P. MASSEE and M.H.J. BOLLEN
Eindhoven University of Technology
Faculty Electrical Engineering
P.O. Box 513, 5600 MB Eindhoven, The Netherlands

ABSTRACT

In this paper the reliability of the industrial electricity supply is determined by means of Monte-Carlo simulation using two different approaches. In the simplest approach the reliability structure of the network and the components is set up and the system function describes if there is a connection from source to load. This can be easily done by means of a Boolean expression but this approach does not allow to include specific electrotechnical problems. An example, which is considered in the paper, is the reaction of the load to short duration voltage dips. It is known from the literature [1] that the undervoltage protection can disconnect the load after the actual voltage dip has already been terminated. A criterion for this problem of voltage solidity is taken from the literature [1] and is used in simplified form in the paper. For this the voltage at the load after the failure of a component has to be calculated. In this second approach the calculated voltage is compared with the voltage solidity criterion from which it is concluded if the failure leads to plant interruption. The computerprogram needs further development since the voltage solidity criterion has to be refined and since short circuits have to be taken into account.

INTRODUCTION

The high interruption costs associated with modern industrial plants, calls for a highly reliable electricity supply. In order to weigh installation costs against interruption costs a quantification of the reliability is needed. One of the methods to obtain reliability parameters like expected number of interruptions and duration of interruption, is a Monte-Carlo simulation. Therefore the power system is divided into components for which the fail and restore behaviour is known. The statistical behaviour of each component is simulated, as well as the relation between the state of the components and the state of the complete system. Observations on this simulated system will give estimates of the reliability parameters. All this is contained within a computer program.

The term voltage solidity is new in this context and needs explanation. A short circuit somewhere in the electricity supply to an industrial plant causes the voltage at the plant bus to become zero, or at least far below its nominal value. After the protection has disconnected the fault, the voltage returns to a value close to nominal. We consider here the situation of a parallel system where one of the branches gets faulted and the other feeds the plant after disconnection of the short circuit. So, due to the fault, the plant experiences a voltage dip at its bus. The duration of the dip is determined by the speed of the protective devices (and automatic transfer systems) and by the reactive power that the (large number of) motors demand during their re-acceleration. If the voltage dip is too wide, the undervoltage protection of the motors will intervene and the industrial process will be disrupted, in that case the plants voltage solidity is not sufficient. Knowing the reaction time of the protective equipment one can determine voltage-solidity requirements that the plant has to fulfil [1]. From the other side, if the voltage solidity of the plant is known one can identify the events that will cause interruption of plant operation.

The reliability analysis of industrial electricity supply will be extensively studied by the Division Electrical Energy Systems. The long term goal of the study is the development of an optimum design strategy of an industrial electricity grid in which protective equipment, circuit breakers etc. are included with their reliability. The computer program must weigh the cost of increasing the reliability against the expected cost of interruption of the production process, as discussed in detail in part I.

A SIMPLIFIED INDUSTRIAL NETWORK

The industrial electricity networks that are considered in this paper,

are shown in the figures 1, 2 and 3. They are highly simplified versions
of the actual situation in order to limit the required computer time for
the Monte-Carlo simulation. They do, however, contain the essential
features of the actual system. The figures 1 and 2 show situations where
the industrial generation of electricity is unimportant. In the
situation shown in figure 3 the industrial generation of electricity is
essential. The figures 1 and 2 show the situation for the two DSM plants
now and in the near future. The reliability of this plant will be
evaluated in the two situations by means of the Boolean system function.
For the situation shown in figure 3 the problem of voltage solidity will
be taken into account which implies that a more complicated system
function is required. In the analysis of the systems shown in the
figures 1 to 3 the reliability of the protective equipment has not been
considered in detail. The failure and repair process of cables,
transformers and substations has been simulated. Besides the failure and
repair of transformers the behaviour of the Bucholtz relais is also
taken into account. These relais that have to protect the transformers,
fail so frequently that this has to be taken into account for a
realistic analysis. Further, preventive maintenance of the parallel
connections is included but in such a way that the maintenance can not
immediately lead to plant interruption (i.e. it will be postponed when
the other connection is out of operation). In this part of the work the
objective will be to investigate if the change in configuration from
figure 1 to figure 2 will improve the reliability of plant B.

Load 1 and load 2 in figure 3 are part of one chemical plant and
plant interruption occurs when either load 1 or load 2 is switched off
according to the voltage solidity requirement. The aim of this part of
the work is to analyze if the plant reliability is influenced by the fact
if the switch S (the switch between the 11 kV buses in figure 3) will be
normally in the open or in the closed condition. Here the computer
program only simulates the failure and restore process and the
maintenance of cables, generators and transformers.

POSSIBLE METHODS OF ANALYSIS

Three methods are available for the determination of the reliability of

368

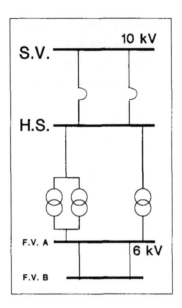

Figure 1. Electricity supply for plant B in the present situation showing the main distribution station (SV), the local distribution station (HS) and the plant buses (FV).

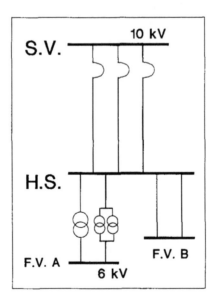

Figure 2. Electricity supply for plant B in the future situation showing the main distribution station (SV), the local distribution station (HS) and the plant buses (FV).

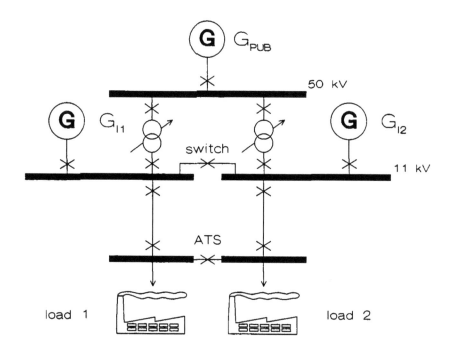

Figure 3. The simplified industrial network; G_{I1} and G_{I2} are industrial
generators, G_{PUB} represents the public grid.

an (industrial) electricity supply i.e. the network method, the Markov-
method and the Monte-Carlo simulation. The first two methods are
analytical and thus can be applied only when certain restrictions are
fulfilled. The network method can be used when, among other things, the
conditions of monotony are obeyed. These conditions demand that taking a
component out of operation should never make the system more reliable,
while taking it in operation should never make the system less reliable.
This is often not the case in electricity networks where taking a
component out of operation can end an overload or short circuit and
taking a component in operation can cause transient instability or
increase the chance of interruption due to short circuits (i.e. decrease
the voltage solidity).

Additional complications appear when the disturbance behaviour is
included. This might be incorporated by introducing "stochastical
dependencies", but this will be at the expense of the simplicity and

clarity of the method. A simplified form of the Markov-method has been used extensively already in determining the reliability of high voltage transmission grids [4, 5, 6]. The Markov-method has the disadvantage, however that the repair process has to obey the negative-exponential distribution (see equation (1)). This generally does not lead to a correct description of the repair process but abolishing this restriction is not easily done within the Markov-method. In Monte-Carlo simulation restrictions on the repair process and on the number of component states are not necessary. The disadvantage of simulation is the required computer time which can increase strongly for large systems with high reliability. Because of the important, fundamental advantages the Monte-Carlo simulation will be further used in this study.

THE MONTE-CARLO SIMULATION-MODEL

In simulating the system behaviour times are generated stochastically at which the components go out of or into operation and it has to be concluded if this implies interruption of plant operation or not. Two problems have to be implemented in the computer program for the simulation:
- generating the stochastical times that the components go out of or into operation, and
- determining the run length in order to reach the predetermined accuracy.

The starting point for generating the stochastical times has to be the knowledge of the distribution functions which are followed by the failure, maintenance and repair process. For simplicity the negative-exponential distribution is chosen for the failure and repair processes according to the cumulative distribution function

$$F(t) = 1-e^{-\lambda t} \tag{1}$$

where λ is constant. The reciprocal of the parameter λ can be called the mean time to failure (MTTF) respectively the mean time to repair (MTTR). For the preventive maintenance process of the system in figure 3 a mean time between maintenance (MTBM) of 5 years is taken and a mean time to maintain (MTTM) of 24 hours for a transformer and of 720 hours for a generator. Before a stochastic time can be determined a random number has to be generated. For the algorithm of the random generator, which is contained in many software packages, the reader is referred to the literature (e.g. [2]). After a random number r that is uniformly distributed on the interval (0,1), has been drawn the stochastic time is obtained by transformation to equation (1) according to the procedure

$$r = 1 - e^{-\lambda\tau} \quad \text{or} \quad \tau = -\frac{1}{\lambda} \ln (1 - r) \tag{2}$$

The computer program that has been created for the simulation, proceeds as follows. At time t = 0 all components are in operation, then random numbers are created and stochastic failure times derived for all components. These times are stored in a sorted row ("event table"). At the moment of the first event δ_1 a certain component fails and a repair time δ^* is generated. The time $\delta_1 + \delta^*$ is inserted at the proper place in the row and the program moves on to the next time in the row. This can either be the repair of the considered component or the failure of another component. The program stores the times that the system changes its state and the duration of interrupted plant operation. The stochastic changes of the state of the components are illustrated in figure 4 for a very simple system of two parallel components. For this simple system it is very easy to determine if the system continues normal operation or not. This is indicated by means of the system function Φ which can be written as a Boolean expression. The variable x[k] in figure 4 indicates if component k is in operation (x[k]=0) or has failed (x[k]=1). Similarly Φ indicates if the system is in operation (Φ=0) or has failed (Φ=1).

For the determination of the system function for the figures 1 and 2 the reliability structures which can be derived from this are shown in the figures 5 and 6. Besides the number of the component its type is indicated by means of a character. The maintenance of a connection is taken into account as an additional series component. It is clear from these figures that the reliability structure is not identical with the physical structure. The reason is that the transformers (numbers 11 and 15 in figure 5) have insufficient capacity to transport the total required power when the other transformer has failed. Thus the failure of one transformer in this connection will lead to interruption of the entire connection. After the reliability structure has been drawn it is

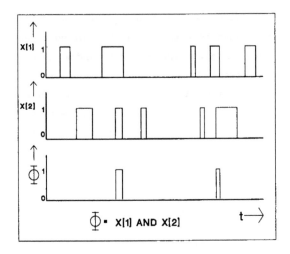

Figure 4. Illustration of a Monte-Carlo simulation for a simple system of two parallel components; the system function Φ has been written as a Boolean expression.

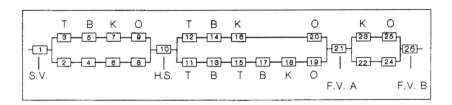

Figure 5. Reliability structure for plant B in the present situation.

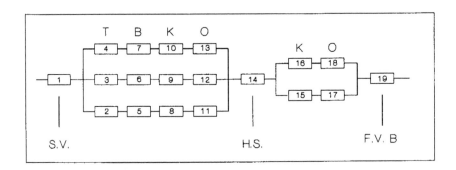

Figure 6. Reliability structure for plant B in the future situation. The characters in the figures 5 and 6 indicate the component: T = transformer, B = Bucholtz relais, K = cable, O = maintenance, SV = main distribution station, HS = local distribution station and FV = plant bus.

easy to determine by means of a Boolean expression if there is a connection from left to right. Taking figure 5 as an example the system function which is used in the simulation, is written as:

$$
\begin{aligned}
\Phi = \ &x[\ 1]\ OR\ (\ (x[\ 2]\ OR\ x[\ 4]\ OR\ x[\ 6]\ OR\ x[\ 8]) \\
&\qquad AND\ (x[\ 3]\ OR\ x[\ 5]\ OR\ x[\ 7]\ OR\ x[\ 9])\ ) \\
&OR\ x[10] \\
&OR\ (\ (x[11]\ OR\ x[13]\ OR\ x[15]\ OR\ x[17]\ OR\ x[18]\ OR\ x[19]) \\
&\qquad\qquad AND\ (x[12]\ OR\ x[14]\ OR\ x[16]\ OR\ x[20])\ ) \\
&OR\ x[21]\ OR\ (\ (x[22]\ OR\ x[24]) \\
&\qquad AND\ (x[23]\ OR\ x[25])\ ) \\
&OR\ x[26].
\end{aligned} \tag{3}
$$

When voltage solidity has to be taken into account this Boolean (or network method)approach is not possible. The system function for this situation is discussed in the next paragraph.

The next problem to be solved is the determination of the run length such that the predetermined accuracy is reached. For this the central limit theorem has been used. In our simulation we are interested in the expected time between system failures. In general the time between system failures (TSF) is a stochastic parameter, which follows a distribution with expected value μ and standard deviation σ. By storing the TSF values and analyzing them we can obtain estimates m and s of the true values μ and σ. For this the central limit theorem is used which can be expressed as follows:

"When n mutually independent samples are drawn from a population with expectation μ and standard deviation σ then the distribution of the calculated average m is approximately normally distributed with expectation μ and standard deviation $\sigma/\sqrt{n}$."
The central limit theorem can thus be written as [2]:

$$
\lim_{n \to \infty} P(|m-\mu| < t\, \tfrac{s}{\sqrt{n}}) = \frac{1}{\sqrt{2\pi}} \int_{-t}^{t} e^{-0.5\ u^2}\, du \tag{4}
$$

P(x < y) indicates a chance, n is the number of observations and the integral in the right hand side of equation (4) can be evaluated easily by using a table for the standard normal distribution, leading to

$$t = 1, \quad P(|m-\mu| < \frac{s}{\sqrt{n}}) \approx 0.68 \qquad (5)$$

$$t = 2, \quad P(|m-\mu| < 2\frac{s}{\sqrt{n}}) \approx 0.95 \qquad (6)$$

$$t = 3, \quad P(|m-\mu| < 3\frac{s}{\sqrt{n}}) \approx 0.997. \qquad (7)$$

In our study we have used the value $t = 2$. The relation (6) expresses that the expected value of TSF will lie with 95 percent statistical certainty in the interval $(m - 2s/\sqrt{n}, \ m + 2s/\sqrt{n})$.

Choosing a relative error ε then the required number of observations n follows from (4) according to

$$\frac{2\ s}{m\sqrt{n}} \leq \varepsilon \quad \text{or} \quad n \geq \frac{4\ s^2}{m^2 \varepsilon^2} \qquad (8)$$

This relation shows that the required value n may change during the simulation as the values of m and s become more accurate.

VOLTAGE SOLIDITY AND THE SYSTEM FUNCTION

The term voltage solidity has been explained in the introduction and a consequence of this effect has been mentioned. Line a in figure 7 could have been obtained when the voltage at a certain plant was monitored during the time that a failure (shortage) occured at a certain point in the network. In fact line a will have a different shape when the failure occurs at a different point in the grid. From several measurements a line could be constructed which represents 90 percent of the cases that may occur and which can then be used as a criterion for voltage solidity. Line a in figure 7 should be interpreted in this sense implying that normal plant operation continues as long as the voltage distribution after a failure stays above line a. As a first step for including voltage solidity in the analysis line a is approximated by line b in figure 7. This implies that four voltage values have to be determined when a

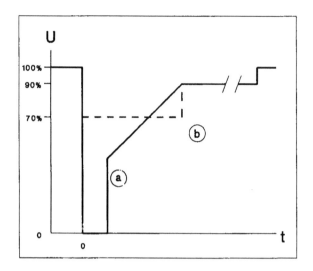

Figure 7. The voltage-solidity requirement; curve (a) is found in the
literature ([1]), curve (b) is the approximation used in the simulation.

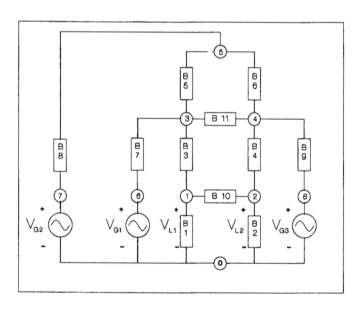

Figure 8. Single phase representation of the simplified industrial network in figure 3.

certain component fails during the simulation. In order to apply the voltage solidity criterion the voltages both at load 1 and at load 2 must be known and both under the condition that the automatic transfer system (ATS) is open or closed.

Line b in figure 7 indicates that normal plant operation continues after a failure when the voltages at load 1 and load 2 are larger than $0.7 \, V_n$ (V_n = nominal voltage) during the time that the ATS is open and larger than $0.9 \, V_n$ after the ATS has been closed. In order to determine the voltages at the loads after the failure of different components the network shown in figure 3 is translated into an impedance network. If the generator is substituted by a voltage source and an internal impedance this leads to the single phase representation shown in figure 8. The relation between figure 3 and figure 8 is easily seen from table 1 which also shows the reliability parameters that have been used in the simulation.

TABLE 1

Reliability parameters of the simplified network and meaning of components in figure 8.

Branch	MTTF [yr]	MTTR [hrs]	Component
B1	∞	N.A.	Load 1
B2	∞	N.A.	Load 2
B3	30	48	Cable
B4	30	48	Cable
B5	10	48	50/11 kV transformer
B6	10	48	50/11 kV transformer
B7	1	12	Generator G_{I1}
B8	5	2	Generator G_{PUB}
B9	1	12	Generator G_{I2}
B10	∞	N.A.	Automatic Transfer System
B11	∞	N.A.	Switch S (between 11 kV buses)

In order to determine the required voltages at the loads the Kirchhoff equations have to be solved leading to relations that can be written in the form

$$\bar{V} = \bar{\bar{Z}} \cdot \bar{I} \tag{9}$$

In this relation the vectors $\bar{V}$ and $\bar{I}$ contain the complex values of the voltages and the currents between the modes and the matrix $\bar{\bar{Z}}$ contains the impedances in complex form. It should be clear that all transient phenomena in the network are neglected when the relation (9) is used. The elements of the bus-impedance-matrix $\bar{\bar{Z}}$ are calculated directly by means of the procedure formulated in [3]. Compared to the procedure using the bus-admittance-matrix this has the advantage that matrix inversion can be avoided. Besides that the chosen procedure provides a simple introduction (or removal) of additional branches which is used many times to simulate the effects of a failure. For all possible situations the voltages at the loads are calculated before the Monte-Carlo simulation starts and are stored in a large table. Thus when a failure occurs during the simulation the resulting voltage dips are easily found in this table and can be simply compared with the criterion of line b in figure 7. In this way the conclusion if plant interruption is caused by this failure can be drawn very fast so that the program execution time can be limited.

SIMULATION RESULTS

The reliability of both DSM plants has been determined both for the present and for the future configuration which are shown in the figures 1 and 2. The reliability parameters used for the simulation have been investigated for the prevailing working conditions and are shown in table 2. The results of the simulations are shown in table 3. Here λ is the failure rate of the system up to the plant and the mean time to system failure (MTSF) equals $1/\lambda$. Further the average time of plant interruption H and the number of hours per year U that the plant cannot operate (U = λH) are given in table 3. It can be concluded from the table that the transition from the present to the new situation has no consequence for the reliability of plant A. The difference in the two situations for plant A is the transition from double to (2 out of 3) connections. Due to the redundancy the influence of the connections on system reliability is small. Table 3 shows that the transition to the new situation leads to a slight improvement of reliability for plant B. This transition implies that one plant bus is missing from the reliability structure of plant B (compare figures 5 and 6). We can conclude from this that the series elements in these figures determine largely the reliability. This

TABLE 2
Reliability parameters used in the simulation of the systems in the figures 1 and 2.

Component	MTTF [yr]	MTTR [hrs]
Cable	50	24
Transformer	300	168
Reactance coil	300	168
Bucholtz relais	100	4
Maintenance	2	8
Main distribution station (SV)	250	72
Local distribution station (HS)	250	72
Plant bus (FV)	100	48

TABLE 3
Simulation results for plant B in the situations of the figures 1 and 2.
Case-01: Present situation, Case-05: Future situation.

Case	MTSF [yr]	λ [1/yr]	r [hr]	U [hr/yr]
B-01	35	2.8E-2	54	1.5
B-05	54	1.9E-2	58	1.1
A-01	53	1.8E-2	58	1.1
A-05	52	1.9E-2	58	1.1

TABLE 4
Simulation results for plant B, comparing single and double connections.
Case-01: Present situation (double connections), Case-03: Present situation but with single connection.

Case	MTSF [yr]	λ [1/yr]	r [hr]	U [hr/yr]
B-01	35	2.8E-2	54	1.5
B-03	6	1.6E-1	34	5.6

conclusion is also illustrated by the value of H in table 3 ranging from 54 to 58 hours when we realize that failure of connection will lead to system interruption of no more than 8 hours (the maintenance time of the parallel connection). Apparently the value of H in table 3 is a weighed average of the repair time of a main distribution station, a local distribution station and a plant bus. These observations are further substantiated by determining the reliability of plant B for the present situation (fig. 1) in the case of a single connection. A comparison of this situation with the present situation with double connections can be obtained from table 4. From this it appears that the mean time between failures decreases with almost a factor 6 and the mean repair time with a factor 1.6. In the case B-03 maintenance has not been taken into account since this would lead immediately to plant interruption. In this situation maintenance has to be done during a production stop of the plant. It should be noted that a decision on changing the electricity grid can only be taken after evaluating the additional investment costs against the decreased costs of interruption of plant operation. This subject will not be further elaborated at this moment.

The results of one simulation run for the system of figure 3 are shown in the tables 5 and 6. Table 5 shows all component failures that occurred. Table 6 shows only those events that have led to interruption of plant operation and is split up into the situations that switch S is normally open and normally closed. In these tables the state of the components is indicated in the form of a binary number, where 0 or 1 indicates that a component is in or out of operation, respectively. It can be seen from the tables 5 and 6 that 20300 component failures had to be simulated in order to obtain 488 plant interruptions (switch S open). It is further seen from table 6 that the difference between the situation with switch S open or switch S closed is formed by two multiple outage events (numbers 404 and 456). It appears that the number of times that events 404 and 456 occur is much smaller than the total number of events leading to plant interruption. Therefore the normal situation in which switch S is operated, has only a small influence on the reliability of the system. It should be noted that the normal operation state of switch S will probably have a larger influence on the system reliability when short circuits are included also. Therefore this will be the next extension of the computer program.

380

TABLE 5
Simulation results for the system in figure 3; distribution of all
component failures.

Event	Event (BIN)	Mn.Time	Occ.	0	25	50	75	100
385	110000001	47.61h	221					
386	110000010	51.53h	224					
388	110000100	32.30h	1876					
389	110000101	6.03h	1					
392	110001000	32.43h	1842					
393	110001001	17.18h	1					
394	110001010	28.30h	1					
396	110001100	17.62h	3					
400	110010000	16.88h	7357					
401	110010001	0.88h	1					
404	110010100	5.93h	6					
408	110011000	11.60h	8					
416	110100000	1.94h	1435					
420	110100100	2.19h	2					
432	110110000	0.60h	3					
448	111000000	17.53h	7262					
449	111000001	3.27h	1					
452	111000100	8.64h	19					
456	111001000	5.71h	8					
464	111010000	8.57h	24					
480	111100000	3.25h	2					

TABLE 6
Simulation results for the system in figure 3; distribution of events
leading to plant interruption for switch S open (above) and switch S
closed (below).

Event	Event (BIN)	Mn.Time	Occ.	0	25	50	75	100
385	110000001	47.61h	221					
386	110000010	51.53h	224					
404	110010100	5.93h	6					
432	110110000	0.60h	3					
456	111001000	5.71h	8					
464	111010000	8.57h	24					
480	111100000	3.25h	2					

Event	Event (BIN)	Mn.Time	Occ.	0	25	50	75	100
129	010000001	47.61h	221					
130	010000010	51.53h	224					
176	010110000	0.60h	3					
208	011010000	8.57h	24					
224	011100000	3.25h	2					

CONCLUSIONS

In order to get experience with Monte-Carlo simulation we first used a Boolean expression for Φ. As a second step we included a simplified version of voltage solidity. We observed the following:

- From the simulation using the Boolean system function it can be concluded that the series elements in the reliability structure are the weakest links. This may be no longer true, however, when voltage solidity is taken into account,
- The reliability of plant B increases slightly after the transition from the present to the future electricity supply. The explanation is that one plant bus is missing from the power supply to plant B in the new situation,
- The voltage-solidity criterion used in the simulation should approach the actual criterion better, and
- Short circuits should be included as the next step in the computer program.

ACKNOWLEDGEMENT

The authors acknowledge the contributions to the work by R.W.E. Cerfonteijn and R.E. van de Vijver within the framework of their MSc study at the Eindhoven University of Technology.

REFERENCES

1. Spijkers, P.J.J.M.E., "Voltage solidity of electrical installations", Elektrotechniek, Vol. 67, No. 8, pp. 715-718, 1989 (in Dutch).

2. Verein Deutscher Ingenieure, "VDI handbook technical reliability: Monte-Carlo-Simulation", VDI-Richtlinien, VDI 4008 Blatt 6, VDI-Verlag, Düsseldorf, West-Germany, 1985 (in German).

3. Heydt, G.T., "Computer analysis methods for power systems", MacMillan Publishing Company, New York, 1986.

4. Wellßow, W., "A contribution to the calculation of reliability in network-planning", Dissertation, Darmstadt, 1986 (in German).

5. Koglin, H.J., "Models and methods", Energiewirtschaftliche Tagesfragen, Jg. 37, S. 886-890, 1987 (in German).

6. Roos, E., "The program ZUBER", Energiewirtschaftliche Tagesfragen, Jg. 37, S. 892-897, 1987 (in German).

382

PROBABILISTIC SENSITIVITY ANALYSIS OF MATERIALS FOR STRUCTURAL FATIGUE AND FRACTURE

WANGWEN ZHAO, Research Assistant
and
MICHAEL J BAKER, Reader in Structural Reliability
Department of Civil Engineering
Imperial College of Science, Technology and Medicine
London, SW7 2BU, U.K.

ABSTRACT

The properties of materials in structural components are never known with certainty and can be assumed to vary according to some distribution law depending on manufacturing conditions, the degree of quality control and the extent of non-destructive testing. This paper describes how modern methods of structural reliability analysis can be used to assess the sensitivity of various modes of structural failure to the material properties that govern their behaviour. In particular, a study has been made of the fracture of cracked steel components under static loading and of the failure of other components by fatigue-fracture when subjected to dynamic loads. It has been found that the material sensitivity varies depending on the type of failure mode, for both load cases. This means that it is difficult to assign a fixed set of partial safety factors to the design variables which will be universally applicable.

INTRODUCTION

For structures or structural components whose failure would result in severe consequences, there is generally a need to achieve a balance between safety and economy. In particular, there is a need to identify those design variables which have a significant influence on the risk of failure so that these can be controlled in an appropriate manner. In general, however, it is rarely economic to test every piece of material in a structure and in many cases this is impossible - for example, where the only possible test is a destructive one. As a result, material properties, as well as loads, are uncertain and this needs to be taken account in the design and assessment of structures.

Traditionally, the safety of structures has been achieved by the use of safety factors in combination with conservative estimates of structural behaviour. For example, in fatigue it is common to use both a conservative S-N curve and a safety factor of 2 or more on the required design life. The basic problem with this approach is that it leads to unquantified levels of safety and generally uneconomic design.

In the last decade, the traditional methods of safety assessment have started to be replaced by modern methods of probabilistic reliability analysis. These methods can be used both to determine suitable partial safety factors for use in deterministic design procedures and to assess the sensitivity of the failure probability to the various design quantities. Knowledge of these sensitivity factors can be used in various ways: to improve control procedures for the important quantities, to decide where data should be collected and in the calculation of safety factors.

The variation of material properties in fatigue and fracture problems has been a source of interest to researchers both in deterministic and in probabilistic studies. The earliest application of reliability methods was in the calculation of partial safety factors for use in structural codes. Recently some attempts have been made to determine the partial safety factors for design against fracture by using results from reliability methods.

This paper describes the results of a study of the sensitivity factors on material variables for two structural failure modes where material parameters are important in governing the risk of failure - namely, fatigue and fracture. The reliability of cracked components under both static and dynamic loads has been studied using first-order reliability methods (FORM)[1].

A brief review of structural reliability methodology is given in the next section of this paper, and in the two following sections these methods are applied to fatigue and fracture problems.

STRUCTURAL RELIABILITY METHODOLOGY

The Need For Probabilistic Methods In Safety Assessment

In the 1980's, there have been considerable achievements in structural engineering: in the design and construction of offshore platforms, nuclear pressure vessels, large bridges, etc. Increasing knowledge of structural mechanics and material characteristics has enabled engineers to design structures with greater confidence. However the 1980's have also seen a series of disasters, e.g. the Chernobyl nuclear accident, the Piper Alpha offshore platform disaster, structural damage brought about by various natural causes including earthquakes, floods, hurricanes, etc. In many areas there are increasing demands for higher safety standards.

Safety of the living space has always been a prime consideration for human kind. The understanding of structural mechanics has advanced civil engineering from the trial and error process of former times to design and construction based on mathematical models of structural behaviour. However, most of these models relate to idealised situations and are usually simplified, so that in engineering design practice safety factors must be used to ensure sufficient safety. Traditionally these safety factors have been based on engineering experience, data and even on human intuition, and differ from one type of structure to another.

Structural failure occurs as a coincidence of various events which are probabilistic in nature. Structural reliability methodology can be used to calculate the failure probability by treating failure as an event arising from an unfavourable combination of random variables at some time during service. The methodology aims to model the real situation and has therefore emerged as a rational way of safety assessment. There is now a reasonably common agreement on the philosophy behind the use of probabilistic methods in decision making, safety assessment and structural design.

Overview Of Structural Reliability Methods

Methods of structural reliability analysis require the physical uncertainty in each of the basic variables (e.g. loads, material properties and geometry) to be modelled by a random variable or random process. These uncertainties, together with model uncertainties arising from inadequacies in the mathematical modelling of the structural behaviour and the statistical uncertainties (i.e. parameter uncertainties) arising from limited statistical data must be included in the analysis [1].

For problems in which time does not need to be included explicitly, the probability of failure can be expressed as:

$$P_f = \bigcup_{j=1}^{m} \int_{-\infty}^{\infty} \int_{-\infty}^{\infty} .... \int_{-\infty}^{\infty} f_{X_1 X_2 .... X_n}(x_1, x_2, ..., x_n) dx_1 dx_2 ... dx_n \qquad (1)$$
$$M_j \leq 0$$

where $X, X_2, ..., X_n$ are the n basic design variables

where joint probability density function $f_{X_1 X_2 .... X_n}(x_1, x_2, ..., x_n)$ and where M_j is the safety margin for failure mode j given by

$$M_j = g_j(X_1, X_2, ..., X_n) \qquad j = 1, 2, ..., m \qquad (2)$$

where $g_j(.)$ is the deterministic failure function for mode j, with $g_j(.) > 0$ denoting safety.

The evaluation of Eq.(1) is rarely straightforward and can become difficult if either the number of potential failure modes, m, for the system or the number of basic variables, n, is large. For typical structures, analytical solutions to Eq. (1) do not exist and it is necessary to resort to one or more numerical procedures (see [1], [2]). Direct numerical integration of Eq.(1) is rarely feasible since $g_j(.)$ are generally non-linear functions of the basic variables. A further problem is that in practice the joint density function may not be well defined.

For the class of problems described above essentially only two types of solution strategy have been developed, each with a number of sub-options, as follows:

1)   Simulation

- crude Monte Carlo methods
- simulation with importance sampling
- directional simulation

- the above, with exclusion of part of the safe region.

2)   Level 2 methods

- first order reliability methods (FORM)
- second order reliability methods (SORM)

together with procedures for determining the probability of the union of the various failure modes.

For complex structural problems involving, for example, the progressive collapse of large structural systems it has recently been found [3] that the optimal solution strategy may often involve a combination of these techniques.

In the study reported in this paper, only relatively simple structural configuerartions have been studied - namely, the failure by fatigue and/or fracture of an isolated structural component containing an initial defect. The complexity lies in modelling the fatigue and fracture processes in a probabilistic context. The reliability calculations have been carried out using FORM (Level 2 methods) and hence these procedures are briefly described here.

Outline of First Order Reliability Methods

The principle of FORM is relatively straightforword and has been well documented (see [1], [2]) but the methods have not been used to any great extent outside the area of structural engineering. For the benefit of readers not familiar with the subject a brief description is given here.

For any given failure mode, the probability of failure is *not* obtained by integrating over the failure domain (c.f. Eq. (1)) but is approximated through the use of a reliability index β which is related to the failure probability P_f by

$$P_f \approx \Phi(-\beta) \quad \text{or} \quad \beta \approx -\Phi^{-1}(P_f) \tag{3}$$

where $\Phi(.)$ is the standard normal distribution function.

First, the set of n basic variable $\mathbf{X}$ is transformed to the space of un-correlated standard normal variables $\mathbf{Z}$. For example, when the basic variables are independent and normally distributed the following transformation can be used.

$$Z_i = \frac{X_i - \mu_{x_i}}{\sigma_{x_i}} \qquad (i = 1,2,....,n) \tag{4}$$

where μ_{x_i} and σ_{x_i} are the mean value and the standard deviation of

random variable X_i.

Note that the Z_i's now have zero mean and unit standard deviation. For dependent basic variables and for non-normal distribution functions other transformations are available [1] [2]. With this transformation the failure surface in the basic variable space is mapped to a failure surface, $g'(\mathbf{Z}) = 0$, in the $\mathbf{Z}$-space.

The reliability index β is then defined as the shortest distance from the origin to the failure surface in the transformed space. This concept was first proposed by Hasofer and Lind [4]. It should be noted that although the reliability index is defined in the space of standard normal variables, the concept is completely general and no restriction is placed on the form of the basic random variables used in modelling the particular reliability problem.

The point on the failure surface which is closest to the origin, normally called the 'design point', can be determined by any suitable algorithm. One approach [1] is to determine the coordinates of the design point $(z_1^*, z_2^*, ..., z_n^*)$ by an iterative solution of the following n equations.

$$z_i^* = b_i \frac{\sum_{i=1}^{n} b_i z_i^* - g'(z_i^*)}{\sum_{i=1}^{n} (b_i)^2} \qquad i=1,2,....,n \qquad (5)$$

where

$$b_i = \frac{\partial g'}{\partial z_i}$$

The reliability index is then given by

$$\beta = \left[\sum_{i=1}^{n} (z_i^*)^2 \right]^{\frac{1}{2}} \qquad (6)$$

Using Eq. (3), the actual failure probability is than approximated by effectively replacing the actual failure surface in the transformed variable space by a tangent hyperplane at the design point. Hence this method is called a 'first order reliability method' (FORM).

Reliability Sensitivity Factors

A powerful feature of any Level 2 method is the direct calculation of the sensitivity factors α_i which are a direct measure of the sensitivity of the reliability index to the various basic variables. In the Z-space,

$$z^* = \beta a$$

where **a** is the normal unit vector and

$$\sum_{i=1}^{n} \alpha_i^2 = 1$$

$$\alpha_i = \frac{\partial \beta}{\partial z_i} = \frac{z_i^*}{\beta} \qquad i = 1,2,....,n \qquad (7)$$

The values of α_i are the direction cosines of the line joining the origin to the design point in the Z-space. Furthermore, the α factors may be used in the direct calculation of partial safety factors on the basic variables to achieve a selected target reliability in deterministic design procedures[1]. However, these sensitivity factors must be used with care since they may vary considerably over the range of possible design solutions. The aim of this paper has been to study these sensitivity factors for the two cases of fracture and fatigue/fracture failure.

MATERIAL SENSITIVITY FACTORS FOR CRACKED COMPONENTS UNDER TENSILE STATIC LOADING

Using a direct integration method and a simplified joint pdf, Gates[5] has obtained solutions to Eq. (1) for a range of input values and has shown the inappropriateness of a single safety factor when using the R6 approach. In the following section an examination of the sensitivity factors for the individual variables has been carried out using first order reliability methods. This approach could be used in the selection of partial safety factors for deterministic design procedures, if sufficient parametric studies were to be carried out.

Failure Functions

For a cracked structural component under tensile static loading, two failure modes are possible: fracture and plastic collapse. The R6 method from CEGB[6] has combined these two failure modes into one failure assessment diagram as shown in Fig. 1. The assessment procedure includes the two extremes of brittle fracture and plastic collapse and an intermediate condition of elastic-plastic fracture. In the R6 method stresses have been divided into two classes 1) those contributing to plastic collapse, and 2) those not contributing to plastic collapse, e.g. residual stress.

Using the concepts, defined earlier, the safety margin for R6(Rev. 3) [6] can be written as follows:

$$M = (1 - 0.14 L_r^2) [0.3 + 0.7 \exp(-0.65 L_r^6)] - K_r \qquad \text{for } L_r < L_r^{max} \qquad (8)$$

$$M = L_r^{max} - L_r \qquad \text{for } L_r \geq L_r^{max} \qquad (9)$$

where

$$L_r = \frac{\text{total applied load contributing to plastic collapse}}{\text{plastic yield load for the flawed structure}}$$

$$L_r^{max} = \frac{\text{flow stress}}{\text{yield stress}}$$

and

$$K_r = K_r^p + K_r^s$$

$$K_r^p = \frac{K_I^p}{K_{mat}}$$

$$K_r^s = \frac{K_I^s}{K_{mat}} + \rho$$

where $K_I^p = Y(a) \, \sigma^p \sqrt{\pi a}$

$$K_I^s = Y(a)\,\sigma^s\,\sqrt{\pi\,a}$$

$Y(a)$ is the geometry factor
a is the crack size
σ^P is the stress contributing to plastic collapse
σ^S is the stress not contributing to plastic collapse
r is the plasticity correction factor,   see [6]
K_{mat} is the material fracture toughness, e.g. K_{Ic} for brittle fracture.

The figure below shows the R6 FAD with $L_r^{max}=1.15$

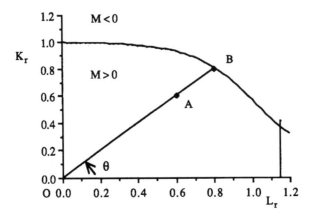

Fig.1 R6 Failure Assessment Diagram (FAD)

In the FAD, any point A can be defined in polar coordinates (d,θ) where

$$d = \frac{OA}{OB} \qquad \text{and}$$

$$\theta = \arctan\left(\frac{K_r(A)}{L_r(A)}\right)$$

Examples

In the example considered, a semi-elliptical surface crack is assumed to be found in a finite plate forming part of a pressure vessel. The geometry factor is obtained from the Newman-Raju formulae [7]. The plastic yield load is assumed to be equal to the area of the ligament section multiplied by the yield stress.

The variables considered are: σ_y, K_{mat}(or K_{Ic}), a and σ^P. Using the data obtained by Gates[8], σ_y is taken to be lognormally distributed with COV=0.07 and K_{Ic} is normally distributed with COV=0.1. The presence of a crack is assumed to be known so that the uncertainty in crack size is due to errors of sizing. Crack size is thus assumed to be normally distributed with COV=0.04. Further details are given in [9].

A parametric study has been carried out to examine the variations in the sensitivity factors α_i when the mean values of the input parameters (i.e. σ_y, K_{mat}(or K_{Ic}), a and σ^P) take a range of different values.   Three different sets of *mean values*   have been used corresponding to points (K_r, S_r) lying in different regions of the FAD and resulting in fracture by linear-elastic fracture, elastic-plastic fracture, and plastic collapse. This has been achieved by selecting the mean values of the input parameters so that each set of corresponding   (K_r, S_r) values lies on an approximate straight line in the FAD defined by a fixed value of the parameter θ. The data points are   shown in Fig. 2   which approximately correspond to $\theta = \pi 3/8$, $\theta = \pi/4$ and $\theta = \pi/8$.

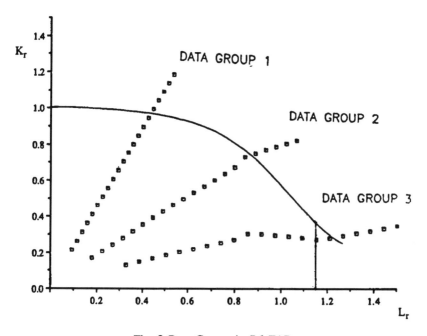

Fig. 2 Data Groups in R6 FAD

In this parametric study the values of the parameter d have been changed by increasing the mean value of σ^P but taking a constant COV of 0.1. In each case, the residual stress σ^s has been taken to be deterministic (i.e. COV = 0) with a mean value equal to σ^P. The value of d should not be confused with the load factor in R6 method, which only takes account of the variation in the applied stress.

The absolute values of the sensitivity factors (α_i values) obtained from the reliability analysis using different mean points, are shown in Figs. 3-5 with d ranges from 0.1 to 1.1 (it should be noted that in FORM, resistance variables have negative α values as a result of the design point being in the lower tails of the resistance distributions).

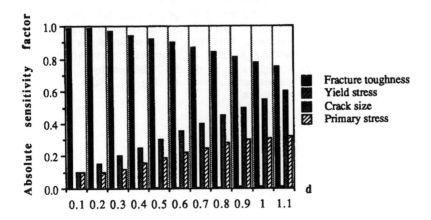

Fig.3 Sensitivity Factors For Data Group 1 (Linear-Elastic Fracture Region)

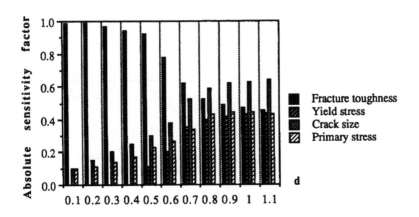

Fig. 4 Sensitivity Factors for Data Group 2 (Elastic-Plastic Fracture
Region)

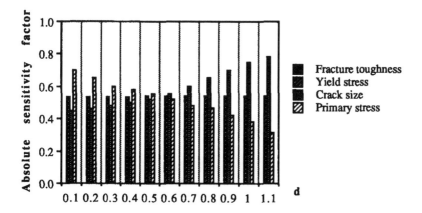

Fig. 5   Sensitivity Factors for Data Group 3 (Plastic Collapse Region)

Discussion

In the study, d has been changed by increasing the applied stress and θ has been changed by using different combinations of material properties (i.e. θ is large for high yield strength and low fracture toughness material). As can be seen from Figs. 3-5, the sensitivity factors for the four random variables involved in the analysis vary considerably depending on the region of the FAD in which the mean point 'A' lies and on the proximity of the mean point to the R6 curve. Not surprisingly, fracture toughness can be seen to be an important variable in the 'linear-elastic' region, and crack size becomes increasingly important for large values of d (i.e. as the mean value of the primary stress is increased and the probability of failure becomes larger). For those cases, it can be seen that the uncertainty in yield stress is totally unimportant.'

For the intermediate 'elastic-plastic' fracture region the fracture parameter is the dominant variable when d is small (i.e. at the low values of the primary stress), but when the mean point is close to the R6 curve, the uncertainty in crack size predominates, with the other three variables taking roughly the same values.

For failure in the plastic collapse region, the sensitivity to fracture toughness falls to zero (since there is no fracture toughness variable in the plastic collapse failure function). The factors are of roughly similar magnitude for the remaining variables, except for the increasing sensitivity to crack size at high values of applied stress.

The above results have been classified according to the parameter θ, based on the mean value of the input variables. It should be noted however, that the values of the variables at the 'design point' do not generally correspond to the same value of θ as at the mean point 'A'. When plotted on the R6 FAD, the 'design points' all lie on the R6 curve, taking up a range of positions depending on the position of the mean point 'A' and the types of distributions.

Because of the range of sensitivity factors obtained it is not possible to suggest fixed values for the partial safety factors which should be used in deterministic design procedures. In particular the proximity of the mean point in the R6 FAD to the R6 curve plays a significant role in determining the sensitivity factor and thus the required partial safety factor. It is clear, however, that the possibility of failure in each of the three distinct modes must be guarded against, even if deterministically only one mode is considered.

MATERIAL SENSITIVITY FACTORS FOR CRACKED COMPONENTS UNDER DYNAMIC LOADING

Failure Functions

There are basically two approaches for fatigue life prediction, the S-N approach and methods involving the use of fracture mechanics. From a reliability point of view the fracture mechanics approach has the advantage of including more variables explicitly in the failure function, and this allows their sensitivity factors to be explored. In this paper, only the fracture mechanics approach is considered.

Under dynamic loading, existing cracks in a component will tend to propagate. Final failure occurs when the crack has grown so large that the component cannot sustain the extreme loads applied. This may be assessed using the R6 failure criterion as for static loading cases. The failure function is

$$M = a_{cr} - a \tag{10}$$

where a_{cr} is the critical crack size based on the R6 method and a is the crack size after a period of crack growth.

The crack growth can be calculated from the Paris Law or from other relevant crack growth laws. However, the Paris Law remains simple and general, and is given by

$$\frac{da}{dN} = C\,(\Delta K)^m$$

$$= C\,(Y(a)\sqrt{\pi a}\,S)^m \tag{11}$$

where   N is the number of cycles of applied stress
            Y(a) is the geometry factor
            S is the stress range and
            C and m are the material dependent parameters.

Therefore

$$a = g_1(C,m,N,S,a_{in}) \tag{12}$$

From the R6 method, if only the fracture toughness, yield stress and primary stress are considered as variables,

$$a_{cr} = g_2(\sigma^P, \sigma_y, K_{Ic}) \tag{13}$$

where σ_y is the yield stress

The failure function may be therefore written as

$$M = a_{cr} - a = g_2(\sigma^p, \sigma_y, K_{Ic}) - g_1(C, m, N, S, a_{in}) \qquad (14)$$

Example

A crack is found at the weld toe of the tubular joint of an offshore structure. The Y factors are obtained from finite element analysis. The stress range and the corresponding number of cycles have been accurately modelled from the stress response spectra of the joint under wave action[9][10]. In this study 11 deterministic sea-states taken from recommendations by Wirsching have been used for initial input in the modelling of stress range and the number of cycles.

The scattering in the crack growth rate can be modelled by treating the material parameters C and m as random variables. Since these parameters are found to be strongly correlated it is generally sufficient to model only one as a random variable. It is found that C is a more sensitive parameter than m so C has been treated as a random variable while m is taken as a constant. According to a probabilistic study by Snijder *et al.*[11] for weld metal m=2.6 and C can taken as

$$C = C_0 \, 10^{F_c} \qquad (15)$$

where $C_0 = 0.8833 \; 10^{-12}$

F_c is normally distributed with zero mean and standard deviation equal to 0.102.

As assumed in the fracture reliability study the uncertainty in crack size is assumed to be that due to measurement error and is assumed to be normally distributed with COV = 0.1. The yield stress is assumed to be lognormally distributed with COV = 0.07. The fracture toughness has been taken as normally distributed with COV = 0.1. For the sake of simplicity, it is assumed that a random extreme load is applied to the structure after a period of fatigue crack growth. This load is assumed to be normally distributed with COV = 0.1 (the load process is a Gaussian process).

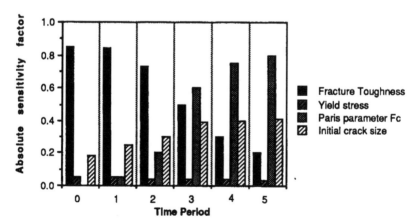

Fig. 6   Sensitivity Factors for Case of Fatigue/Fracture

Discussion

The results of the sensitivity analysis for the case of fatigue/fracture, based on the models given above, are shown in Fig. 6. It is clear that with increasing fatigue exposure time, there is a decrease in the sensitivity to fracture toughness and a corresponding increase in the sensitivity to the Paris Law parameter F_C. For the case considered there is also an increase in the sensitivity to intial crack size with exposure time.

With the safety margin defined as $M = a_{cr} - a$, failures due primarily to variations in a_{cr} may be thought of as fracture failures and failure due to increases in the crack size a as fatigue failures. With increasing exposure times (design lives) it can be seen that the fatigue fracture mode becomes increasingly dominant. Although the precise numerical results depend on the input values of the parameters in the models, Fig. 6 reflects the main trends of the sensitivity factors. More examples can be seen in [9].

CONCLUSIONS

The paper shows that reliability methods can be used for the quantitative assessment of fatigue and fracture problems in a simple and efficient way. In particular, first order reliability methods provide a direct measure of the sensitivity of the failure probability to the various input variables, knowledge of which can be used in various ways - in determining partial safety factors, in the design of quality control procedures for materials and in gaining an understanding of the dominant variables.

Using the CEGB R6 failure assessment method it has been shown that depending on the mean values of the input parameters the most likely failure mode will vary in its nature (linear-elastic, elastic-plastic or plastic collapse). The sensitivity of the failure probability to the input variables thus changes significantly over the various regions of the FAD. This means that a single set of partial safety factors is insufficient for use in design against fracture.

For components subjected to fatigue loading, it has been found that the material fracture toughness and yield stress become less important as the design life (exposure period) becomes longer, and the failure probability is then dominated by the uncertainties in the Paris Law parameters.

In conclusion, it can be seen that because of the difficulty in selecting partial safety factors for deterministic design methods, the direct use of reliability methods in design and assessment procedures has a number of distinct advantages.

REFERENCES:

1. Thoft-Christensen, P. and Baker, M.J., Structural reliability theory and its applications, Springer-Verlag, 1982.

2. Madsen, H.O., Krenk, S. and Lind, N.C., Methods of structural safety, Prentice-Hall, 1986.

3. Turner, R.C., The system reliability analysis of fixed offshore structures, Ph.D. Thesis, Department of Civil Engineering, Imperial College, London, 1990.

4. Hasofer, A.M. and Lind, N.C., Exact and invariant second moment code format, J. of Eng. Mechanics, ASCE, **100**, 1974.

5. Gates, R.S., The relationship between load factors and failure probabilities determined from a full elastic-plastic probabilistic fracture mechanics analysis. Int. J. Pres. Ves. & Piping, 1983, **13**, pp. 155-67.

6. R/H/R6-Rev. 3, Assessment of the integrity of structures containing defects, CEGB, Nov. 1985.

7. Newman.Jr., J.C., A review and assessment of the stress-intensity factors for surface cracks, in Part-through crack fatigue life prediction, ASTM STP 687, ed. Chang, J.B., 1979.

8. Gates, R.S., Statistical analysis of fracture toughness data, OED/STM/87/10125/N, Job No. 00-27, CEGB OED STB(midland area), Dec. 1987.

9. Zhao, W., Reliability analysis of fatigue and fracture under random loading. Ph.D. Thesis, Department of Civil Engineering, Imperial College, London, March 1989.

10. Zhao, W. and Baker, M.J., A new stress-range distribution model for fatigue analysis under wave loading, Proceedings, International conference on environmental forces on offshore structures and their prediction, London, 1990.

11. Snijder, H.H., Gijsbers, F.B.J., Gijkstra, O.D. and Ter Averst, F.J., Probabilistic fracture mechanics approach of fatigue and brittle fracture in tubular joints, in Steel in marine structures, ed. Noordhoek, C. and de Back, J. Elsevier Science Publishers B.V., Amsterdam, 1987.

For Product Safety Concerns and Information please contact our EU
representative GPSR@taylorandfrancis.com
Taylor & Francis Verlag GmbH, Kaufingerstraße 24, 80331 München, Germany